STATISTICS FOR EXPERIMENTALISTS

STATISTICS
FOR
EXPERIMENTALISTS

B. E. COOPER

Atlas Computer Laboratory,
Chilton, Didcot, Berkshire

PERGAMON PRESS

OXFORD · NEW YORK · TORONTO
SYDNEY · PARIS · BRAUNSCHWEIG

Pergamon Press Offices:

U.K.   Pergamon Press Ltd., Headington Hill Hall, Oxford, OX3 0BW, England

U.S.A.   Pergamon Press Inc., Maxwell House, Fairview Park, Elmsford, New York 10523, U.S.A.

CANADA   Pergamon of Canada Ltd., 207 Queen's Quay West, Toronto 1, Canada

AUSTRALIA   Pergamon Press (Aust.) Pty. Ltd., 19a Boundary Street, Rushcutters Bay, N.S.W. 2011, Australia

FRANCE   Pergamon Press SARL, 24 rue des Ecoles, 75240 Paris, Cedex 05, France

WEST GERMANY   Pergamon Press GmbH, 3300 Braunschweig, Postfach 2923, Burgplatz 1, West Germany

First edition 1969

Reprinted 1975

Library of Congress Catalog Card No. 68-18520

131 752

Printed in Great Britain by Biddles Ltd., Guildford, Surrey
ISBN 0 08 012600 6

2.6.88

CONTENTS

PREFACE

THE aim of this book is to provide the experimental scientist with a working knowledge of statistical methods and a searching approach to the analysis of data. The mathematical knowledge assumed is minimal, requiring only a little calculus and a limited capability in algebra. The book is arranged so that the reader with the required mathematics will be able to follow the mathematical derivations included in some parts of the book, but the reader without such mathematics will not be at a disadvantage in appreciating the application and the theoretical basis of the methods. The book should be well within the comprehension of most experimental scientists, both those who are working for their first or second degree and those who are currently working in the scientific field. The statistician wishing to make himself aware of the problems in applying his subject would also benefit from a study of some of the examples.

It is anticipated that the working knowledge derived by the reader, although practically oriented, will be soundly backed by an awareness of the theoretical assumptions made in its application. The subjects are arranged according to the type of experiment to which they refer rather than to the position they occupy in statistical theory. Thus an experimentalist wishing to analyse an experiment can follow the process of analysis as a continuous process without the necessity of referring to numerous different sections.

Many of the examples included in the book have been drawn from my experience as an applied statistician at the Atomic Energy Research Establishment at Harwell, and I am grateful to the United Kingdom Atomic Energy Authority for permission to include these examples. I would like to express my thanks to the many statisticians on both sides of the Atlantic who have influenced the production of this book, either by conversation, or by direct comment as reviewers. The manuscript was typed by the Mathematical Typing Section at Bell Telephone Laboratories, Inc., Holmdel, New Jersey, U.S.A. I am very grateful to them, and to Dr. M. E. Terry in particular, for this generous gesture.

CHAPTER 1

INTRODUCTION

THIS first chapter gives a wide coverage of many of the problems in the design of experiments and in the analysis of the results. It is intended to give the reader an overall appreciation of the subject. For this first chapter the reader should concern himself with basic concepts and leave the appreciation of the details until the various subjects are discussed again in later chapters.

1.1. EXPERIMENTAL RESULTS

It is well established that the repeat of an experiment does not exactly reproduce the original result. This difference between the results of apparently identical experiments causes concern in their interpretation. It is because of these differences that the statistical approach to the evaluation of experimental results has been developed. It is supposed that if it were possible to continue experimenting indefinitely under the same conditions the results would cluster around some fixed value. This would be the value we wish to obtain. It is clearly not possible, however, to continue experimenting indefinitely and we must base our interpretation on the finite number of results that we have obtained. The statistical problem is, then, on the basis of our finite set of results, what can we say about the infinite set? The finite set of results actually obtained is called a sample and the infinite set of results we would like to have obtained is called the population. The problem may be restated as follows: on the basis of the sample of results what can we say about the population?

1.1.1. Example 1

A certain object has been weighed five times and the following weights, in grams, were obtained:

$$2{\cdot}540 \quad 2{\cdot}538 \quad 2{\cdot}547 \quad 2{\cdot}544 \quad 2{\cdot}541$$

These results are a sample of five results and we are concerned to make statements about the infinite set of results we could have obtained had it been possible to continue weighing indefinitely. The infinite set of results, or population, would presumably cluster around the true weight of the object. This will be so provided there is no systematic bias introduced by the weighing device or by the operator performing the weighings. Assuming there is no bias in the results what can we say about the true weight? The true weight is a fixed value but the five weighings show variation from 2·538 to 2·547 g. It is clearly not possible to compute (or estimate) one value from the sample and to be certain that this is the true weight. There must always be some uncertainty associated with an estimate of the true value.

There are many possible estimates of the true weight that may be obtained from the sample, for example, the middle result (or median) 2·541; the average of the two extreme results 2·5425; the average of the centre three results 2·5417; or the arithmetic average of all five results 2·542 are all estimates of the true weight. The median is little affected by a "spurious" result since such a result will be one of the extremes, but the estimate is based on only one result. The average of the two extremes is based on two well-separated results but is greatly affected by a spurious result. The average of the centre three results is based on more values than the previous two estimates and not using the two extreme values is unaffected by a single spurious result. The arithmetic average is based on all the results available but can be affected to a certain extent by a spurious result. The choice between different estimation procedures is made on the basis of their statistical properties and a more detailed discussion of this problem is given in Chapter 4. For the purposes of this example the arithmetic mean will be used as the estimate of the true weight which is therefore estimated as 2·542 g.

1.1.2. Confidence Limits

A single estimate of the true value such as that computed above is insufficient on its own since it immediately raises the question "what values is it reasonable to believe that the true value could be?" The "reasonable values" are normally expressed as a range of values lying between two limits on either side of the estimate of the true value. Associated with these limits, which are called confidence limits, is the probability that the limits enclose the true value. This probability is called the confidence or the confidence probability. The success of the experiment depends on whether, or not, the interval defined by these limits is narrow enough for the estimated value to be useful.

Example confidence limits for the data given in Section 1.1.1 are given below:

There is a probability of 0·80 that the limits 2·537 to 2·547 enclose the true value.

There is a probability of 0·95 that the limits 2·532 to 2·552 enclose the true value.

There is a probability of 0·99 that the limits 2·526 to 2·558 enclose the true value.

The width of a confidence interval defined by confidence limits is determined by three factors:

1. The confidence probability (the more confident we wish to be that the limits enclose the true value, the wider the interval).
2. The variability in the results obtained (the greater the variability, the wider the interval).
3. The number of observations taken (the higher the number of observations, the narrower the interval).

If the interval is not narrow enough for the experimentalist's purpose and he is not prepared to accept a lower confidence, he may do one of two things:

1. Take more observations.
2. Reduce the variability in the results by improving the experimental technique.

It may occur that neither of these choices is a practical possibility, in which case the estimate is the best that may be obtained by this experimental procedure and it must be accepted with its limited precision.

It is important to consider what precision is required before the experiment is undertaken so that some assessment of the number of observations necessary to obtain the required precision can be made. Such assessment will require an estimate of the variability to be expected so that if this is not available from previous experience a pilot experiment should be performed to provide the required estimate of variability. With this information it is now possible to estimate the number of observations that will be required to obtain the given precision. If this number proves to be too high for the experiment to be practically, or economically, possible the experimentalist must be prepared to:

1. accept a lower precision, or
2. improve his experimental technique, or
3. accept the situation that it is not possible to obtain an estimate with the required precision.

It is worth noting that the variability in the first few results obtained by a new technique is frequently higher than in later results. The variability estimate obtained in a pilot experiment may well be an over estimate. It seems wise, therefore, to calculate the variability in the results at a number of stages in the experiment so that revised estimates of the number of observations required can be made. The experiment thus proceeds in stages and is terminated when the required precision is obtained.

The above procedure is an example of a "sequential" procedure in which the experiment continues until the required precision is obtained. There are more sophisticated sequential procedures available to the experimentalist and descriptions of such procedures developed for a variety of purposes are given in Wald (1947) and Wetherill (1966).

1.2. EXPERIMENTAL DESIGN

The following example experiment will form the basis of a discussion of a number of points needing careful consideration during the planning of an experiment. The experiment has as an objective the establishment, or otherwise, of the catalytic effect of the addition of a substance A on a certain chemical reaction. Interest is focused on the effect of the addition of substance A on the yield of substance B obtained from the reaction.

1.2.1. Replication

If we observe that the yield of B is 270 in the presence of A and 256 in the absence of A on two independent experiments we have no means of assessing how precisely we have estimated the yields under the two conditions. If we find, when we repeat the experiment a number of times, that the yield with A present varied from 268 to 272, and that the yield without A varied from 254 to 258, we will feel confident that the increase in yield of experiments with A over those without A will be maintained in the long run, that is in the population. If, however, we find that the with A results varied from 245 to 300 and the without A results varied from 230 to 280 we will have considerable doubt whether the presence of A increases the yield in the population.

A major requirement of an experiment is, therefore, that it should contain some means of assessing the precision of estimated quantities. A well-designed experiment will allow the variability in the results to be estimated so that an idea may be formed of the difference in yield that may be expected purely by chance. This will enable the experimentalist to judge whether the observed difference in yield between the two types of reaction represents a real difference due to the presence of A or whether the difference can be explained by chance variation normally observed in the yield.

1.2.2. Replicates and Repeats

In the experiment described above we may identify two sources of variability. Variability between results may be introduced during the reaction itself and also during the determination of the yield obtained. We have seen that the significance of the observed difference in yields obtained in the two types of reaction is judged against the variability normally present in results obtained under apparently identical conditions. It is important, therefore, that both sides of this comparison should contain the same sources of variability, Results obtained from different reactions contain both sources or variability, whereas two determinations of the yield obtained from the same reaction contain only variability from the determination stage and not from the reaction stage. If the experimental design had been to perform two reactions, one with A and one without A, and to make a number of determinations of the yield from each reaction the variability between determinations for one reaction would give a false impression of the variation normally present between yields obtained from different reactions. A significant result from such an experiment may arise because of the genuine effect of the presence of substance A or because of the greater variability between yield determinations from different reactions. Thus this design would not distinguish between these two possible interpretations. If, however, we perform a number of reactions with A and a number of reactions without A the variability in yield between reactions of the same type would now give a true impression of the natural variability. Duplicate measurements containing all sources of variability that exist in the comparison of interest, namely the difference in yield between reactions with and without A, will be called "replicates". Duplicate measurements with one or more source of variability less than exists in the comparison of interest will be called "repeats".

The distinction between repeats and replicates is important and will be further illustrated with a second example. The length of a straight line is measured by placing a rule by the line and reading off the length. If several results are obtained by reading the rule many times without moving and replacing the rule, these results will measure the variability experienced in reading the rule. If, however, the rule is picked up and replaced each time the results will measure the variability in reading the rule and in placing the rule by the line. To determine whether two straight lines are different in length the rule must be moved from one straight line to the other so that the estimate of variability required by this experiment must contain variability due to both the reading and the placing of the rule. Reading the rule many times without replacement would provide repeat measurements containing variability due to reading the rule, whereas reading the rule many times with replacement would provide replicate measurements containing variability due to reading and placing the rule. It is important for the experimentalist to ensure that the duplicate measurements are replicates rather than repeats.

It is interesting to note that neither set of results in the line measuring experiments measure variability experienced by using different rules to measure the lines.

1.2.3. Randomization

A chemical reaction may be influenced by a number of factors such as temperature, humidity and the carbon dioxide content of the air. If such factors exist, and are not controlled, their effect will be to increase the variability in the results and hence decrease the sensitivity of the experiment. If the effect of a rise in temperature, for example, is to increase the yield and if the temperature is steadily rising during the experiment we would expect a steady rise in yield in reactions of the same type. If we perform five experiments without *A* followed by five experiments with *A*, any effect the presence of *A* has would be added to the effect of temperature. In this case the effect of temperature would be to increase the yield in the last five results and apparently increase the effect of the presence of substance *A*. Thus we see that the presence of uncontrolled variables could be to increase, or decrease, the differences being studied. We may fail, therefore, to establish a real difference or we may attribute an observed difference to the presence of *A* when in fact no real difference exists. The experiment must be arranged (designed) to keep the effect of uncontrolled variables to a minimum.

One method of arrangement is to choose the order in which the experiments are to be performed at random. If we perform five experiments without *A* and five experiments with *A*, and we choose the order in which the ten experiments are to be performed at random, any difference observed between the two types of reaction may be attributed to the presence of *A* and not to an uncontrolled variable such as temperature.

1.2.4. Grouping Observations

A second method of reducing the effect of uncontrolled variables is to group the observations into a number of smaller self-contained experiments. The main principle in this respect is that the value of an uncontrolled variable is likely to change more between two points in time (or space) which are far apart than between two points which are close together. Applying this principle involves grouping the observations into a number of groups each normally consisting of one observation for each treatment studied. The order of the treatments within a group or BLOCK is decided at random. An experiment may consist of any number of blocks and such an experiment is called a randomized block experiment. A block in the above experiment would consist of the yield obtained from a reaction without *A* and the yield obtained from a reaction with *A*. Thus a reaction without *A* is paired, under as near identical conditions as possible, with a reaction with *A* even though there may be considerable differences in the conditions from block to block. The effect of uncontrolled variables in the randomized block design is to increase the variability between blocks rather than the difference between the two types of reaction. The estimate of general variability is not affected by block differences and may therefore be expected to be lower then the general variability estimate obtained from an experiment in which no blocking was made.

1.2.5. Concomitant Observations

A third method of reducing the effect of an uncontrolled variable is available to the experimentalist if the variable is known and if it can be observed every time the main variable is observed. Extra observations of uncontrolled variables are called concomitant observations and it is possible to analyse the main variable of interest (yield) allowing for the values of the concomitant variables (temperature). An essential condition for this method is that the concomitant variable is unaffected by the treatments applied. If the presence of A changes the temperature of the reaction then the temperature measurements will reflect the effect of A and the use of temperature as a concomitant observation will alter or even take out a real difference in yield. The measurement of temperature, however, just prior to the stage at which A is normally added to the reaction could be used as concomitant observations since the addition of A has had no chance to affect the temperature at this stage.

1.2.6. Balancing Other Variables

Suppose that in the above experiment the determination of the yield involved a titration with an end-point that was not well defined. It might be expected that two experimenters would differ in their determinations because each had taken a different end-point. If two experimenters are to take part in the experiment and if one performs the experiments with A and the other performs the experiments without A and a difference is observed, it is not possible, on this evidence alone, to say that the difference is due to the presence of A rather than to the experimenters. If, however, the experiment is arranged so that each experimenter performs the same number of experiments with A and without A, it is possible to isolate differences due to A, to isolate differences due to experimenters, and to measure the agreement between the experimenters about the magnitude of the effect of A. This third measure is usually called the interaction between the experimenters and the effect of A. Typical results for such an experiment may be:

	With A	Without A
Experimenter 1	304, 308	258, 266
Experimenter 2	320, 312	276, 264

The average result with A is 311 and without A is 266 and these suggest that the addition of A increases the yield. These averages both contain two observations made by each experimenter so that both are affected equally by experimenter differences. The comparison of the two treatment results is, then, unaffected by experimenter differences. The average result for experimenter 1 is 284 and for experimenter 2 is 293 and suggests that the experimenter 2 tends to produce higher observations than experimenter 1. These averages both contain two observations for each treatment so that both are affected equally by treatment differences.

The average values for each experimenter for each treatment are:

	With A	Without A
Experimenter 1	306	262
Experimenter 2	316	270

The difference between experimenter 2 and experimenter 1 with A is 10 and without A is 8, so that it is reasonable to believe that the two experimenters differ consistently for both treatments, or, in other words, that there is no interaction between the experimenters and the treatments. This type of experiment is called a factorial experiment. It consists of the two "factors" experimenters and treatments and each factor has two "levels". The analysis of results of factorial experiments will be given in detail in, Chapter 9.

1.2.7. Latin Square

The Latin square is a randomized block design with the further refinement that the order of the observations within a block is balanced. To achieve this balance the Latin square consists of as many blocks as treatments and each treatment appears once and once only in each position within a block. An example for four treatments A, B, C and D is given in Table 1.1.

TABLE 1.1. EXAMPLE LATIN SQUARE

		Block 1	2	3	4
Positions within a block	1	A	B	C	D
	2	C	D	A	B
	3	D	A	B	C
	4	B	C	D	A

Each treatment in this design appears in each column (block) once and each row (positions within a block) once so that the treatments are grouped in two directions. In many descriptions of the Latin square the groupings are called rows and columns but in practice these may be groupings according to any variable which must be balanced in the experiment. Examples of such variables are operators, machines, counters, days, rows and columns of a field, rows and columns of a tray of treatments processed together.

There are many other designs of a similar type to a Latin square which allow three or more groupings of the experimental units. One of these designs, the Graeco-Latin square in which three groupings are made, will be described in Chapter 10.

1.2.8. Incomplete Block Designs

The block size in a randomized block experiment is determined by the number of treatments to be compared. It may be that it is not possible to place all treatments in one block. A good comparison of the effect of four treatments on the wear of car tyres can be made by choosing tyres as blocks and allocating a treatment to each quarter of a tyre surface. It may be, however, that each treatment requires at least one-third of the surface of a tyre for an effective measure of wear to be made so that each tyre may contain at most three treatments. These form a large class of designs which allow treatments to be grouped in incomplete blocks and still permit balanced comparisons between the treatments.

Incomplete block designs of the Latin-square type, also available to the experimentalist, allow balancing in two directions. An example of this type of design may be obtained by deleting one row of a normal Latin-square design. This design is an example of one form of Youden square which is described in detail in Chapter 11.

The computation of the analysis of incomplete block designs is more involved than that of the complete designs. The use of electronic computers will reduce the labour of the computation and will, in the future, make these designs more attractive to experimentalists.

1.3. THE POWER OF AN EXPERIMENT

The statistical test applied to the experimental results obtained in the example experiment provides information from which we can judge the significance, or otherwise, of the increase in yield. If we conclude that the increase is not significant—that is, if A has no effect—we must satisfy ourselves that the experiment had a good chance of establishing an increase had there been one present to establish. The power function of the test is computed to provide the answer to this question. This is computed before the experiment is performed so that we may change the design if we judge the proposed design to be insufficiently "powerful".

The power function of the example experiment is the probability of detecting an increase in yield, plotted as a function of the increase. The experimental design must be chosen so that the probability of detecting the increase, which is just economically or scientifically important, is high. The choice between two or more possible designs should be made in favour of the experiment with the highest power.

1.4. Generality of the Conclusions

It is important to realize that the conclusions obtained from an experiment may not apply under conditions different from those prevailing at the time of the experiment. If the conditions are applied to other situations further assumptions are made. If in the above experiment an increase in yield was obtained when the temperature was controlled at one fixed value, this evidence is not sufficient on its own for us to be able to say that the increase would have been observed whatever the temperature had been. The increase may well vary with temperature, and further experiments would be necessary before we could say with confidence that the increase in yield is observed at all temperatures.

Questions

1. An experimentalist performs a number of determinations of a physical constant and from these he obtains a value for the physical constant (an estimate) and confidence limits associated with which there is a good chance (say 99 per cent) that the limits enclose the true value. All determinations are made by the same method and with the same apparatus. What population does his sample of results represent?

The experimentalist discovers that another experimentalist has also made several determinations of the same physical constant by a different method. On comparing the two sets of results it is found that the two estimates of the true value disagree and that the two sets of confidence limits do not overlap. Whilst it is possible for this to happen by chance what other explanations are available? What extra sources of variation are introduced in the comparison of the two sets of results that are not included in either set of confidence limits?

2. Six towns which are very similar in their geographical locations had the following rainfall, in inches, in the year 1965:

$$12{\cdot}43 \quad 11{\cdot}81 \quad 14{\cdot}08 \quad 10{\cdot}56 \quad 13{\cdot}17 \quad 12{\cdot}29$$

Another six towns, similar in geographical location amongst themselves but different location type from those above, had the following rainfall, in inches, in the year 1965:

$$15{\cdot}72 \quad 16{\cdot}80 \quad 14{\cdot}13 \quad 17{\cdot}05 \quad 13{\cdot}94 \quad 16{\cdot}47$$

(a) Plot these results. Do you believe the rainfall to differ in the two types of town?

(b) If the fourth result in the second group of towns had been 12·72 instead of 17·05 would your conclusion have changed?

3. The amounts of a substance Z dissolved in three solutions A, B and C are to be compared. Each solution is divided into three containers and a separate determination is to be made for each of the nine containers. Part of the determination process consists of passing hot air over the solutions and it is possible to take all nine containers through this part together. The nine containers are to be placed in a square tray in three rows and hot air is to be passed over the tray from one edge. Five possible arrangements of the containers are given below and the hot air is to be passed over the tray from left to right.

A	B	C	A	A	A	A	B	C	A	C	A	A	B	C
A	B	C	B	B	B	C	A	B	B	B	C	C	A	B
A	B	C	C	C	C	B	C	A	C	A	B	A	B	C

It is believed that it is possible for the air to cool to an important extent while it passes over the tray and thereby have a different effect on the three columns.

(a) Which of the above arrangements would you accept for this part of the determination and for what reasons?

(b) If at another stage in the determination hot air is passed over the tray from top to bottom which of the above arrangements would you now accept and for what reasons?

4. Twenty automobile tyres were used in an experiment to compare the effect of four treatments on the wear of tyres. They were used on the same automobile in four different periods during a total period of one year. The year was divided into four periods of 3 months each and the treatments T_1, T_2, T_3 and T_4 occupied the four possible wheel positions in the 3-month periods according to the following design:

		Wheel positions			
		1	2	3	4
3-month	1	T_1	T_2	T_3	T_4
periods	2	T_4	T_1	T_2	T_3
	3	T_3	T_4	T_1	T_2
	4	T_2	T_3	T_4	T_1

The wear of each tyre was measured at the end of each 3-month period.

(a) Identify the design as one described in this chapter and discuss the advantages it enjoys.

(b) If the automobile had had a fault, present for the whole year, which caused the tyre in position 4 to wear more than usual what effect would this have had on the comparison between the treatments?

(c) If, instead, the fault had developed after 9 months what effect would this have had on the comparison between the treatments?

5. Suppose there had been five treatments to compare in question 4 above and only one automobile available for the testing of tyre wear. Re-design the experiment to allocate the five treatments four at a time to the four positions such that each treatment appears in each position once only, and using only five tyre periods.

(a) Identify the new design as one described in this chapter and discuss its advantages.

(b) If the automobile had had a fault during the 15 months of the experiment which caused the tyre in position 4 to wear more than usual, what effect would this have had on the comparison between the tyres?

(c) If the tyres had been measured twice each at the ends of the five 3-month periods what extra information would have been obtained from the extra work? What sources of variability exist between two measurements of the same tyre after the same time period? What extra source of variability exists between two measurements on the same tyre made after different time periods?

(d) How should the experiment be performed to obtain replicate (rather than repeat) measurements which contain all sources of variability which exist between two measurements of wear on the same tyre but made after different time periods?

CHAPTER 2

PROBABILITY

THERE are normally considered to be two possible outcomes when tossing a coin, namely, the appearance of a head or of a tail. Prediction of the outcome of a particular toss of the coin is a matter of guesswork. However, if the coin is tossed a number of times it is observed that the proportion of occasions on which a head appears remains roughly the same as the number of tosses increases, so that this proportion may be used to predict the proportion of occasions on which a head will appear in future tossings of the coin.

When we say that the coin will probably show a head next time we are using our previous experience of the event to predict the outcome of a future event. In the past we may have observed a head more often than a tail so that we will say in future that a head is more likely to appear than a tail. The proportion of times a particular outcome of an event is observed can take any value from zero to unity. How likely this outcome is to occur in the future may be represented by this proportion. An outcome will be said to be unlikely if the proportion is small and most likely if the proportion is close to unity.

The value of the proportion obtained from observation of previous events will be based on a finite number of events and will vary as more events are observed. The observed experience is a sample of the outcomes of all possible events forming the corresponding population, and the observed proportion is an estimate of the proportion in the population of events. This population proportion is a fixed value and is referred to as the probability of the outcome. As more and more events are observed the precision of the estimate is increased, but the exact value can be obtained only from observing the infinite number of events in the population. The classical definition of probability, given below, is a suggested means of obtaining the population probability. Application of this definition produces probabilities that attempt to describe the outcomes of actual events. A few examples will help to explain this point.

An excellent discussion of probability can be found in David (1951) and is well worth the attention of the reader. Other sound introductions to probability can be found in Birnbaum (1962), Cramer (1951) and Feller (1957).

2.1. The Classical Definition of Probability

If an event can occur in N mutually exclusive and equally likely ways, and if n of these outcomes have a character A, then the probability of A is n/N.

2.1.1. Example 1

What is the probability of observing a head at the toss of a coin? The two outcomes ($N = 2$) of the toss may not occur together (mutually exclusive), and we assume them to be equally likely. The appearance of a head is one of these outcomes ($n = 1$), so that the probability of a head is $1/2$.

Application of the classical definition has produced a value for the population probability. If the assumption of equally likely is valid, then the proportion of heads in the population is a half and the proportion observed in a sample of tosses will be approximately a half. If we toss a coin forty times and observe a head on twenty occasions we will feel confident that the probability is a well-chosen value. If we observe only three heads we will feel that the value of a half is not well chosen and we will reject the equally likely assumption. If we observe a head sixteen times we will have some difficulty in deciding whether the probability is well chosen or not.

2.1.2. Example 2

What is the probability of observing a six at the throw of a die? Six different numbers may appear ($N = 6$), only one may appear at a time (mutually exclusive), we assume the six numbers are equally likely, one of the numbers is a six ($n = 1$) so that the probability of a six is $1/6$.

The practical interpretation of this probability is that if we throw a die a number of times we expect a six to appear on $1/6$th of the occasions.

2.1.3. Example 3

What is the probability of a card chosen at random from a pack of fifty-two playing cards being an ace? There are fifty-two cards ($N = 52$), we draw only one card (mutually exclusive), we assume that each card is equally likely to be selected, four of the cards are aces, so that the probability of an ace being selected is $4/52 = 1/13$.

The practical interpretation is that if we draw a card at random from a pack of fifty-two playing cards a large number of times we expect to draw an ace on $1/13$th of the occasions.

2.2. POPULATION AND SAMPLE

A sample is the collection of results we actually observe.

A population is the collection of results that we could have obtained had we collected observations, under the same conditions, indefinitely.

Statisticians often talk of a sample as being a selection from the population. Statistical methods are concerned to make statements about the population using the sample as the basis for such statements. Statistical tests have been designed to test statements about the population. Estimation procedures have been chosen to enable us to estimate population values and to assess the precision of such estimates. It is assumed in making inferences from the sample about the population that the sample is a *random* selection from the population. That is, each unit in the population has an equal chance of being selected for the sample. For example, the answers obtained from persons in Oxford Street chosen strictly according to a predetermined random process would be a random sample of answers from people who visit Oxford Street but would not necessarily be a random sample of answers from people in Great Britain. Some members of the population of Great Britain never visit Oxford Street, others rarely visit Oxford Street, whereas some members work in Oxford Street and are there every working day. Each member of the population of Great Britain does not, therefore, have an equal chance of being represented in this sample. Opinion expressed by this sample will not represent public opinion in Great Britain if different views are held by people in different parts of the country. Opinion expressed by this sample would, however, represent the opinion of the population of people who visit Oxford Street.

In the above sampling procedure we may reasonably assume that the population we are sampling from is not affected by the selection of an individual. In some situations this is not so. For example, if we select a card at random from a pack of fifty-two playing cards we leave fifty-one cards behind. If we now select another card without replacing the first we are selecting from a new population of fifty-one cards the make up of which depends on the card selected at the first selection. Many problems occur in which two individuals are selected from a *finite* population of individuals. In these problems the relevant probabilities are different for the selection of the first and the second individuals. It is not possible to select the same person twice. If, however, we select one individual and then replace the individual in the population before the second is selected, the relevant probabilities remain the same, and it is now possible to select the same individual twice.

Thus we may sample from an infinite population or we may sample from a finite population with replacement or without replacement. If we sample from an infinite population or if we sample with replacement from a finite population the relevant probabilities remain the same from selection to selection. If we sample from a finite population without replacement the relevant probabilities change from selection to selection. If, for example, we select an ace from a pack of fifty-two playing cards the probability that we select an ace from the remaining fifty-one cards is 3/51. If we had replaced the ace before selecting the second card the probability of selecting an ace would have remained equal to 1/13.

2.2.1. Example 4

The population in the die-throwing example (Example 2) consists of the outcomes of an infinite number of throws of which 1/6 show a six. The sample consists of a finite number of throws of which r show a six. We may derive a statistical test to test our belief that had we continued throwing the die indefinitely we would have observed a six on 1/6 of the throws. We may, on the other hand, have no previous beliefs about the probability of a six and wish to estimate both the probability and the precision of this estimate from actual throws. The evaluation of the precision of an estimate normally consists of limits which have a certain probability of enclosing the population value. Such limits must be narrow enough for us to be able to use the estimated probability. The maximum permissible width of such limits should, of course, be considered before observations are collected and the sample size chosen so that an acceptable precision can be obtained.

The concept of the random sample in this example is another way of stating the equally likely condition in the definition of probability. The population consists of an equal number of the numbers 1 to 6 and we assume that each unit in the population has an equal chance of being included in the sample—that is, each of the numbers 1 to 6 are equally likely. This assumption would be invalid, of course, if it were possible to influence the die, or the throwing of the die, in the favour of a six.

2.2.2. Example 5

What is the probability of scoring a total of seven at the throw of two dice? Each die can show any one of six numbers and may only show one number at a time so that two dice can show thirty-six combinations of the two numbers. Table 2.1 shows the total score of the two dice for each of the thirty-six combinations.

These thirty-six combinations are mutually exclusive and, we assume, equally likely. Six of these combinations result in a score of seven so that the probability of scoring seven is 6/36 = 1/6.

TABLE 2.1. TOTAL SCORES THROWING TWO DICE

		Die A score					
		1	2	3	4	5	6
Die B score	1	2	3	4	5	6	7
	2	3	4	5	6	7	8
	3	4	5	6	7	8	9
	4	5	6	7	8	9	10
	5	6	7	8	9	10	11
	6	7	8	9	10	11	12

2.3. PROBABILITY DISTRIBUTIONS

Table 2.1 above can be used to provide the probabilities of scoring 2, 3, ..., 12 at the throw of two dice. These probabilities are given in Table 2.2.

TABLE 2.2. PROBABILITY DISTRIBUTION OF TOTAL SCORES THROWING TWO DICE

Score	2	3	4	5	6	7	8	9	10	11	12
Probability	1/36	2/36	3/36	4/36	5/36	6/36	5/36	4/36	3/36	2/36	1/36

The scores 2, 3, ..., 12 are the only possible scores, so that these probabilities given above sum to unity and therefore form a probability distribution. A probability distribution is the description of a population. A sample corresponding to the above population would be the proportions of times each of the scores 2, 3, ..., 12 were obtained in a finite number of trials. The probability distribution has been obtained by application of the classical definition of probability and if the assumptions made in applying the definition are sound we expect the agreement between the proportions in the population and those in the sample to be good. A statistical measure which enables us to measure such agreement and a testing procedure to enable us to decide whether to accept the agreement as reasonable or not will be described in a later chapter.

2.4. PROBABILITY NOTATION

The probability of an outcome A is usually written $\Pr\{A\}$. The probability of an outcome A given that a previous outcome B has taken place is usually written $\Pr\{A|B\}$.

2.5. ADDITION OF PROBABILITIES

The probability that at least one of the outcomes A or B occurs is given by

$$\Pr\{A \text{ or } B \text{ or } (A \text{ and } B)\} = \Pr\{A\} + \Pr\{B\} - \Pr\{A \text{ and } B\}.$$

2.5.1. Example 5

What is the probability of at least one die showing a six?

Pr{die A and/or die B shows a six}

$\quad = $ Pr{die A shows a six} $+$ Pr{die B shows a six}

$\qquad - $ Pr{both dice show sixes}

$\quad = 1/6 + 1/6 - 1/36 = 11/36.$

A simple count of the scores in Table 2.1 that contain at least one six will confirm this probability.

2.6. Multiplication of Probabilities and Independence

The probability that a first event results in outcome A and a second event results in outcome B is given by

$$\text{Pr}\{A \text{ and } B\} = \text{Pr}\{A\} \times \text{Pr}\{B|A\}.$$

If the fact that A has occurred at the first event has no effect on the probability that B occurs at the second event then A and B are said to be independent and:

$$\text{Pr}\{A \text{ and } B\} = \text{Pr}\{A\} \times \text{Pr}\{B\}$$

or equivalently

$$\text{Pr}\{B|A\} = \text{Pr}\{B\}.$$

We normally assume that two throws of a die are independent events so that the probability of two sixes is therefore $1/6 \times 1/6 = 1/36$. An example of dependent events is given in Section 2.6.1.

2.6.1. Example 6

What is the probability of the second of two cards drawn at random without replacement from a pack of fifty-two playing cards being an ace? We consider two situations, in the first of which we know what the first card was and in the second situation this information is not available.

Situation 1

If the first card is an ace there are only three aces remaining in the fifty-one cards to choose from so that the probability of the second card being an ace is $3/51 = 1/17$.

If the first card is not an ace there are four aces remaining in the fifty-one cards to choose from so that the probability of the second card being an ace is $4/51$.

In this situation the selection of an ace as the second card depends on the result of the first selection, that is, the two events are dependent.

Situation 2

If the first card is not inspected we must combine the two possibilities above. We will denote by A_1 and A_2 the appearance of an ace as the first and second card respectively and by B_1 the appearance of a card other than an ace as the first card. Thus the probability of an ace as the second card is given by

$$\Pr\{A_2\} = \Pr\{A_1\} \times \Pr\{A_2|A_1\} + \Pr\{B_1\} \times \Pr\{A_2|B_1\}$$

$$\Pr\{A_2\} = 4/52 \times 3/51 + 48/52 \times 4/51$$

$$= 1/13.$$

Thus the probability of drawing an ace as the second card is the same in the case where the first card is not replaced and not inspected as it is in the case where the first card is replaced. If the first card is not inspected the two selections are therefore independent events.

2.7. BINOMIAL PROBABILITIES

Table 2.2 shows that the probability of scoring twelve when throwing two dice is 1/36. This probability may be obtained by applying the rule for the multiplication of probabilities. The probability that die A shows a six is 1/6 and the probability that die B shows a six is also 1/6. It is assumed that the score on die A is not affected by, nor influences, the score on die B (that is, the scores are independent) so that the probability that both dice show sixes is $1/6 \times 1/6 = 1/36$. The probability that a die shows a number other than a six is 5/6 so that, from a similar argument, it follows that the probability that neither die shows a 6 is $5/6 \times 5/6 = 25/36$.

The probability that just one die shows a six needs one further step in the calculation not previously necessary. The probability that die A shows a six and die B does not is $1/6 \times 5/6 = 5/36$. The probability that die B shows a six and die A does not is also 5/36 so that the probability that just one die shows a six is the sum 10/36, since both possibilities cannot happen simultaneously. The extra step necessary in this calculation is, then, to observe that we may choose the die that is to show the six in two ways so that the required probability is $1/6 \times 5/6 \times 2 = 10/36$.

Since we may only observe 0, 1, or 2 sixes when throwing two dice the probabilities of these three outcomes should sum to unity and form a *probability distribution*. This distribution, recorded in Table 2.3, is an example of a binomial distribution.

TABLE 2.3. DISTRIBUTION OF THE NUMBER OF SIXES WHEN THROWING TWO DICE

Number of sixes	0	1	2	Total
Probability	25/36	10/36	1/36	36/36 $=1$

2.7.1. Five-dice Example

We may observe 0, 1, 2, 3, 4, or 5 sixes when throwing five dice. The set of probabilities of observing 0, 1, ..., 5 sixes is a second example of a binomial distribution. The probability that no dice shows a six is $(5/6)^5 = 3125/7776$. The probability that all the dice show sixes is $(1/6)^5 = 1/7776$. The probability that one dice shows a six is $1/6 \times (5/6)^4 \times A_1$ where A_1 is the number of ways of choosing one die from five dice to show the six. This number of ways, A_1, is 5 so that the probability of observing one six is $1/6 \times (5/6)^4 \times 5 = 3125/7776$. (This just happens in this example to be equal to the probability of observing no sixes.)

The probabilities of observing 2, 3, 4 sixes are

$$(1/6)^2 \times (5/6)^3 \times A_2, \quad (1/6)^3 \times (5/6)^2 \times A_3, \quad (1/6)^4 \times (5/6) \times A_4,$$

where A_2, A_3, A_4 are the numbers of ways of choosing two, three, four dice from five dice to show the sixes. The probability of observing r sixes is $(1/6)^r \times (5/6)^{5-r} \times A_r$ where A_r, the number of ways of choosing r dice from five dice, is given by

$$A_r = \frac{5.4.3.2.1}{r(r-1)\ldots 2.1 \times (5-r)(4-r)\ldots 2.1} = \frac{5!}{r!(n-r)!}$$

[factorial $r = r! = r(r-1)(r-2)\ldots 3.2.1$; and $0! = 1$].

Thus we have:

$$A_2 = \frac{5.4.3.2.1}{2.1.3.2.1} = 10,$$

$$A_3 = \frac{5.4.3.2.1}{3.2.1.2.1} = 10,$$

$$A_4 = \frac{5.4.3.2.1}{4.3.2.1.1} = 5,$$

and hence obtain the binomial distribution in Table 2.4.

TABLE 2.4. DISTRIBUTION OF THE NUMBER OF SIXES WHEN THROWING FIVE DICE

No. of sixes	0	1	2	3	4	5	Total
Probability	$\frac{3125}{7776}$	$\frac{3125}{7776}$	$\frac{1250}{7776}$	$\frac{250}{7776}$	$\frac{25}{7776}$	$\frac{1}{7776}$	$\frac{7776}{7776}=1$

2.7.2. The General Case of the Binomial Distribution

If the probability of an outcome E of an event is p then the probability $P_{n,r}$ of observing outcome E r times in n independent events is given by

$$P_{n,r} = \frac{n!}{r!(n-r)!}\, p^r (1-p)^{n-r} \quad \text{for} \quad 0 \leqq r \leqq n.$$

A complete set of these probabilities $P_{n,r}$ for $r = 0, 1, \ldots, n$ forms the binomial distribution for n events and probability p and is usually denoted by $B(p, n)$.

It is assumed in the derivation of the binomial probability that the n events are independent. That is the outcome of any event has no effect on the outcome of any of the other events. It is important, therefore, that the independence of the events is established before the binomial distribution is used in a practical situation.

In the above example the event was the throwing of a die and the outcome of interest was the appearance of a six. It was assumed that the throws were independent, that is, the score on any die had no effect on any other die.

Tables of both the individual and the cumulative binomial probabilities are available for $n < 50$ in the Applied Mathematics Series No. 6 produced by the National Bureau of Standards, Washington, D.C. Limited tables are also included in Pearson and Hartley (1958).

2.7.3. Example $B(1/2, 5)$

Five events are considered and the probability of the outcome of interest is 1/2. This may be an attempt to describe the distribution of the number of heads observed when tossing five coins.

TABLE 2.5. BINOMIAL DISTRIBUTION $B(1/2, 5)$

r	0	1	2	3	4	5	Total
$P_{n,r}$	1/32	5/32	10/32	10/32	5/32	1/32	1

Notice that since $p = 1/2$ distribution is symmetric, that is $P_{n,r} = P_{n,n-r}$.

2.7.4. Example $B(0.4, 12)$

TABLE 2.6. BINOMIAL DISTRIBUTION $B(0.4, 12)$

r	0	1	2	3	4	5	6
$P_{n,r}$	0·0022	0·0174	0·0639	0·1419	0·2128	0·2270	0·1766
r	7	8	9	10	11	12	Total
$P_{n,r}$	0·1009	0·0420	0·0125	0·0025	0·0003	0·0000	1·0000

2.7.5. *Population-sample Agreement*

The above distributions are descriptions of populations. A corresponding sample could be obtained by observing the number of times outcome E (heads or a six) occurred in n events (tossing n coins or throwing n dice) on many occasions. It would then be possible to write down the proportion of times the outcome E occurred 0, 1, 2, ..., n times. These proportions would be expected to be in reasonable agreement with the population probabilities $P_{n,0}$, $P_{n,1}$, ..., $P_{n,n}$. The meaning of the word "reasonable" used in the previous sentence will be discussed in a later chapter.

2.7.6. *Models*

The binomial distribution in Section 2.7.3 is a description of a population and attempts to describe a practical situation. Five coins are tossed and it is desired to predict the proportion of occasions on which five heads will appear. To obtain such predictions the statistician sets up a model description of the practical situation. The model in this example is as follows. The tossing of a coin is an event and the outcome of interest is the appearance of a head. Five events, which are believed to be independent, are considered. The description of the practical situation obtained from this model is the distribution given in Table 2.5. One prediction obtained from the model is that the proportion of occasions on which five heads will appear is 1/32. The agreement between this proportion and that observed in a sample of a number of tosses of five coins will be a measure of the adequacy of the model.

2.8. Discrete and Continuous Distributions

The binomial distribution is an example of a *discrete* probability distribution. That is, the variable r may take only the discrete values 0, 1, 2, ..., n and may not take fractional values. It is the discrete nature of the values which is important, not the fact that the number of values r may take is finite. Discrete distributions for which r may take any of the infinite set of values 0, 1, 2, 3, ... exist and one such distribution (the Poisson distribution) is described in Chapter 13. Distributions which allow the variable to take all values within a finite or infinite range are called *continuous* distributions. An example of a continuous distribution (the normal distribution) will be described in the next chapter.

2.9. Mean and Variance of a Discrete Probability Distribution

There are two quantities of particular importance associated with every probability distibution, namely the mean and the variance. These quantities may be defined for a discrete probability distribution as follows.

The mean μ is given by

$$\mu = \sum r \, \Pr\{r\}$$

and the variance σ^2 by

$$\sigma^2 = \sum (r - \mu)^2 \, \Pr\{r\},$$

where the summations are taken over all values that r can take. The square root σ of the variance is known as the standard deviation.

The mean is a measure of the location of the distribution and provides a fixed point of reference in the distribution.

The variance is a measure of dispersion (or spread) about the mean. If most of the distribution extends over a large range the variance will be large, if the distribution is contained in a small range the variance will be small.

2.9.1. Mean and Variance of the Binomial Distribution

Mean

$$\mu = \sum_{r=0}^{n} r P_{n,r} = \sum_{r=0}^{n} r \frac{n!}{r!(n-r)!} p^r (1 - p)^{n-r}$$

$$= np \sum_{r=1}^{n} \frac{(n-1)!}{(r-1)!(n-r)!} p^{r-1} (1 - p)^{n-r}$$

$$= np.$$

The mean is the product of the number of events and the probability of the outcome of interest occurring at one event.

Variance

$$\sigma^2 = \sum_{r=0}^{n} (r - \mu)^2 P_{n,r}$$

$$= \sum_{r=0}^{n} r^2 P_{n,r} - 2\mu \sum_{r=0}^{n} r P_{n,r} + \mu^2 \sum_{r=0}^{n} P_{n,r}$$

$$= \sum_{r=0}^{n} r^2 P_{n,r} - (np)^2 \quad \text{since} \quad \sum_{r=0}^{n} P_{n,r} = 1 \quad \text{and} \quad \sum_{r=0}^{n} r P_{n,r} = np \, (=\mu)$$

$$= \sum_{r=0}^{n} [r(r-1) + r] P_{n,r} - (np)^2$$

$$= \sum_{r=2}^{n} r(r-1) \frac{n!}{r!(n-r)!} p^r (1 - p)^{n-r} + \sum_{r=1}^{n} r P_{n,r} - (np)^2$$

$$= n(n-1) p^2 \sum_{r=1}^{n} \frac{(n-2)!}{(r-2)!(n-r)!} p^{r-2} (1 - p)^{n-r} + np - (np)^2$$

$$= n(n-1) p^2 + np - (np)^2$$

$$\sigma^2 = np(1 - p).$$

The means and variances for the binomial distributions given above are given in Table 2.7.

2.10. Crude and Central Moments of Discrete Probability Distributions

These are quantities associated with probability distibutions which are used as measures of certain features of the distribution.

The sth crude (or absolute) moment μ'_s is defined:

$$\mu'_s = \sum r^s \Pr\{r\} \qquad s > 0$$

and the sth central moment μ_s is defined:

$$\mu_s = \sum (r - \mu)^s \Pr\{r\} \qquad s > 0,$$

where the summations are taken over all values that r may take.

The first crude moment μ'_1 is the mean and is normally written μ. The first central moment μ_1 is zero and the second central moment μ_2 is the variance σ^2.

The central moments contain information about the form of the distribution. If $\mu_s = 0$ for odd values of s the distribution is symmetrical. The even central moments, particularly μ_4, are used to describe the shape of the distribution, for example, the peakedness or the flatness. Moments are discussed in more detail in relation to continuous distributions in the next chapter.

2.11. Modal Value of a Discrete Probability Distribution

The modal value (or possibly values) of a probability distribution is the value (or values) of the variable r with the highest probability (the most likely value or values). The modal values (or values) r_n of the binomial distribution are given by the inequality:

$$(n + 1)p - 1 \leqq r_n \leqq (n + 1)p$$

which may be derived in two stages as follows.

Stage 1

The probability $P_{n,r}$ is greater than $P_{n,r-1}$ if

$$\frac{n!}{r!(n - r)!} p^r (1 - p)^{n-r} \geqq \frac{n!}{(r - 1)!(n - r + 1)!} p^{r-1}(1 - p)^{n-r+1}$$

$$\frac{p}{r} \geqq \frac{1 - p}{(n - r + 1)}$$

$$np - rp + p \geqq r - rp$$

$$p(n + 1) \geqq r.$$

Stage 2

The probability $P_{n,r}$ is greater than $P_{n,r+1}$ if

$$\frac{n!}{r!(n-r)!}p^r(1-p)^{n-r} \geqq \frac{n!}{(r+1)!(n-r-1)!}p^{r+1}(1-p)^{n-r-1}$$

$$\frac{(1-p)}{(n-r)} \geqq \frac{p}{(r+1)}$$

$$r + 1 - rp - p \geqq np - rp$$

$$r \geqq np + p - 1.$$

Thus the modal value is given by

$$(n+1)p - 1 \leqq r_n \leqq (n+1)p.$$

Note that it is possible for the binomial distribution to have two modal values.

The modal value for each of the binomial distributions given above is given in Table 2.7.

TABLE 2.7. MODAL VALUES OF CERTAIN BINOMIAL DISTRIBUTION

Distribution	Mean	Variance	Modal value
$B(1/6, 2)$	1/3	$5/18 = 0.28$	0
$B(1/6, 5)$	5/6	$25/36 = 0.69$	0 and 1
$B(1/2, 5)$	5/2	$5/4 = 1.25$	2 and 3
$B(0.4, 12)$	4.8	2.88	5

QUESTIONS

1. A large group of scientists consists of 15 per cent statisticians, 50 per cent physicists, and 35 per cent chemists. One scientist is picked at random. What is the probability that
  (a) a statistician is picked?
  (b) a physicist is picked?
  (c) a chemist is picked?
If two scientists are picked at random what is the probability that
  (d) two statisticians are picked?
  (e) two physicists are picked?
  (f) two chemists are picked?
  (g) two scientists from the same discipline are picked?
  (h) two scientists from different disciplines are picked?
If the group of scientists had consisted of three statisticians, ten physicists and seven chemists why would the answers to questions (d) to (h) have been changed and what would the new answers be?

2. A genetical model describes the inheritance of a characteristic such as red hair in the following way. Each person has two genes for the characteristic being considered. An individual gene may be one of two types which we will denote by A and a. Thus a person may have any of the three possible combinations AA, Aa or aa. A child receives one gene from each parent chosen at random from the two genes possessed by that parent. A person has red hair if both genes are a, that is if the person is aa. The model therefore predicts that if both parents are aa then all children will be aa.

(a) Given that both parents are Aa what are the probabilities that a child will be AA, Aa and aa, respectively?

(b) Given that both parents are AA what are the probabilities that a child will be AA, Aa and aa respectively?

(c) Given that one parent is AA and the other is Aa what are the probabilities that a child will be AA, Aa and aa respectively?

(d) Persons not possessing red hair may be either AA or Aa with probabilities 1/2, 1/2. Combine the answers to questions (a), (b) and (c) to obtain the probabilities that a child of non-red-haired parents will be AA, Aa or aa. Check that the three probabilities you obtain add to unity.

3. Plot the example binomial distributions recorded in Section 2.7 and mark the positions of the mean and mode.

4. Toss five coins as many times as stamina allows (preferably a multiple of 32) and observe the number of occasions on which 0, 1, 2, 3, 4 and 5 heads occur. Multiply each of the probabilities recorded in Section 2.7.3 by the number of times the five coins were tossed and compare these "expected" frequencies with those observed. Are you satisfied with the agreement between these two sets of frequencies? Save these results until Chapter 7 has been read.

5. The probability that an animal responds in a certain way to a stimulus is believed to be 1/3. If the stimulus is applied to each of eight animals:

(a) What is the probability that no animals will respond?

(b) What is the probability that one animal will respond?

(c) What is the probability that at least one animal will respond?

(d) What is the most likely number of animals to respond?

(Leave your answers as fractions of $3^8 = 6561$.)

CHAPTER 3

CONTINUOUS PROBABILITY DISTRIBUTIONS

THIS chapter describes the normal distribution and its use as the description of a population. Properties of variables possessing a normal distribution and the general results for continuous random variables are given. Distributions other than normal are briefly discussed and the mathematical properties of the chi-squared distribution are listed in preparation for its use in later chapters.

Mathematical expressions for distributions are given as well as some simple derivations. These are included for the reader with a little mathematics. The reader who does not follow these need have no fears that he has missed something which is vital to the understanding of later chapters. The reader should, however, appreciate the results of derivations since these will be used later.

3.1. THE NORMAL PROBABILITY DISTRIBUTION

The normal distribution is known by a variety of other names, the most common of which are the Gaussian distribution and the Error Function (some definitions of the Error Function omit the 2 and the $\frac{1}{2}$ which appear in the expression (3.1) given below). The normal distribution is the most important distribution in the application of statistical methods and is a basic assumption for many tests of significance and estimation procedures. The variable x_0 possessing a normal distribution may take any value in the range $-\infty$ to $+\infty$, and, therefore, x_0 is a continuous variable, and the normal distribution is a continuous distribution. The mathematical form of the normal distribution, as given in expression (3.1), defines the probability that the variable x_0 takes a value in an infinitesimally small interval x to $x + dx$ for all values of x. The variable x_0 is said to have a normal distribution if the probability that x_0 takes a value in the range x to $x + dx$ is given by expression (3.1) for all values of x.

$$\Pr\{x < x_0 < x + dx\} = \frac{1}{\sqrt{(2\pi)}\, m_2}\, e^{-\frac{1}{2}[(x-m_1)/m_2]^2}\, dx \qquad (3.1)$$

where, for the present discussion, m_1 and m_2 are constants. This probability

26

is usually denoted by $p(x)dx$ and it is represented by the shaded strip labelled (1) in Fig. 3.1. It follows from the law for the addition of probabilities (Section 2.5) that the probability that x_0 takes a value within the range x to $x + 2dx$ is the sum of the probability that x_0 takes a value within the range x to $x + dx$ and the probability that x_0 takes a value within the range $x + dx$ to $x + 2dx$ since these two intervals are mutually exclusive. It is clear that this argument may be extended to show that expression (3.1)

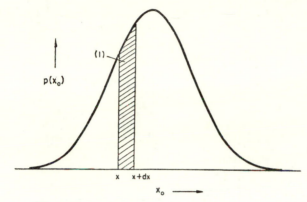

FIG. 3.1. The normal distribution.

may be summed (or, rather, integrated) to yield the probability that x_0 takes a value in a non-infinitesimal interval x_1 to x_2. This probability would be

$$\Pr\{x_1 < x_0 < x_2\} = \frac{1}{\sqrt{(2\pi)}\, m_2} \int_{x_1}^{x_2} e^{-\frac{1}{2}[(x-m_1)/m_2]^2}\, dx. \qquad (3.2)$$

3.1.1. A Normal Population

A convenient means of studying the distribution of values of a continuous variable x obtained in a sample is the histogram. The measurement scale is divided up into a number of adjacent and non-overlapping intervals and the numbers of values falling within each interval are counted. Columns are then drawn as in Fig. 3.2, each representing one interval, the bases of which cover the particular interval on the measurement scale and the volumes of which are proportional to the number of measurements counted in the interval. The width of the intervals are normally chosen to be the same so that the heights as well as the volumes of the columns are proportional to the number of observations observed in the intervals.

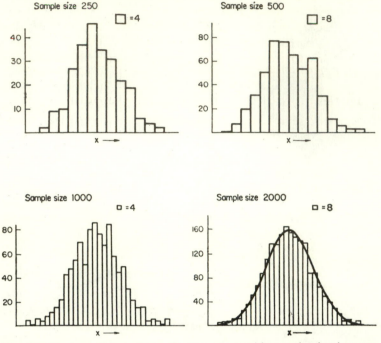

Fig. 3.2. The smoothing of sample histograms with sample size increase.

When the number of measurements are small the number of intervals must be small so that a reasonable number of observations fall in each interval. As the number of observations increases more intervals may be formed by subdivision and the outline of the histogram becomes smoother. The outline of the histogram will approach a smooth curve as the number of observations and the number of intervals increase. This curve will be the population distribution. The statement that a variable has a normal distribution is interpreted as saying that this limiting population distribution has the form given by expression (3.1). Figure 3.2 shows histograms for four different sample sizes and the last histogram has a normal distribution superimposed.

The normal distribution depends on two parameters m_1 and m_2 and the statement that a particular population has a normal distribution with known values of m_1 and m_2 is a complete description of that population. If it is known that the population of heights of men in Great Britain is normal with known values of m_1 and m_2 it is possible to make particular statements about the heights of men in Great Britain. It would be possible to compute the proportion of men whose heights fall within certain fixed ranges and the tailoring industry, for example, could base its production of ready-to-wear overcoats on these proportions.

If it were known that the lengths of pellets produced by a production process have a normal distribution with known values of m_1 and m_2 it would be possible to predict the proportion of pellets that will have lengths outside acceptance tolerances so that judgement may be made whether, or not, the present process is producing an unacceptable pellet too often. This prediction would, of course, make the additional assumption that the manufacturing conditions would not change so that the lengths of pellets would no longer have the normal distribution.

3.2. MEAN AND VARIANCE OF A CONTINUOUS PROBABILITY DISTRIBUTION

The mean and variance of a discrete probability distribution were defined in the previous chapter as follows:

$$\text{the mean} = \mu = \sum r \, \text{Pr}\{r\}$$

$$\text{and the variance} = \sigma^2 = \sum (r - \mu)^2 \, \text{Pr}\{r\},$$

where the summations are taken over all values that r can take. This definition also applies to continuous variables, but since the values that r can take vary continuously the summation signs are replaced by continuous sums, or integrals. The definitions for continuous distributions are therefore written as follows:

$$\text{the mean} = \mu = \int x \, p(x) \, dx, \tag{3.3a}$$

$$\text{the variance} = \sigma^2 = \int (x - \mu)^2 \, p(x) \, dx, \tag{3.3b}$$

where the integrals are taken over the range (or, conceivably, ranges) of values that x may take.

The mean is a measure of location of the continuous distribution and provides a fixed point of reference in the distribution.

The variance is a measure of dispersion (or spread) of the distribution about the mean. The square root of the variance is known as the standard deviation.

3.2.1. The Mean and Variance of the Normal Distribution

In this section we show that the parameters m_1 and m_2 of the normal distribution, as defined in expression (3.1), are the mean and standard deviation of the distribution. Applying the definitions (3.3) to the normal distribution we have the following derivations:

$$\text{The mean} = \mu = \frac{1}{\sqrt{(2\pi)}\,m_2} \int_{-\infty}^{\infty} x\, e^{-\frac{1}{2}[(x-m_1)/m_2]^2}\, dx$$

$$\mu = \frac{1}{\sqrt{(2\pi)}} \int_{\infty} (ym_2 + m_1)\, e^{-\frac{1}{2}y^2}\, dy \quad \text{where} \quad y = \frac{x - m_1}{m_2}$$

(that is $x = ym_2 + m_1$ and $dx = m_2 dy$)

$$\mu = \frac{m_2}{\sqrt{(2\pi)}} \int_{-\infty}^{\infty} y e^{-\frac{1}{2}y^2}\, dy + \frac{m_1}{\sqrt{(2\pi)}} \int_{-\infty}^{\infty} e^{-\frac{1}{2}y^2}\, dy$$

$$\mu = \frac{m_2}{\sqrt{(2\pi)}} [-e^{-\frac{1}{2}y^2}]_{-\infty}^{\infty} + m_1$$

$$\mu = m_1.$$

The mean of the normal distribution is the parameter m_1.

$$\text{The variance} = \sigma^2 = \frac{1}{\sqrt{(2\pi)}\,m_2} \int_{-\infty}^{\infty} (x - m_1)^2\, e^{-\frac{1}{2}[(x-m_1)^2/m_2]}\, dx$$

$$= \frac{1}{\sqrt{(2\pi)}} \int_{-\infty}^{\infty} y^2 m_2^2 e^{-\frac{1}{2}y^2}\, dy \quad \text{where} \quad y = \frac{x - m_1}{m_2}$$

$$= \frac{m_2^2}{\sqrt{(2\pi)}} [ye^{-\frac{1}{2}y^2}]_{-\infty}^{\infty} + \frac{m_2^2}{\sqrt{(2\pi)}} \int_{-\infty}^{\infty} e^{-\frac{1}{2}y^2}\, dy$$

$$\sigma^2 = m_2^2$$

(since the first expression is zero and the second is m_2^2 times the integral of a normal distribution over its entire range). The variance of the normal distribution is the square of the parameter m_2. The two parameters of the normal distribution are the mean and the standard deviation. The mean $\mu\,(= m_1)$ determines the position of the distribution and the standard deviation $\sigma\,(= m_2)$, or the variance $\sigma^2\,(= m_2^2)$ determines the spread of the distribution about the mean.

We may now rewrite expression (3.1) for the normal probability distribution in its more usual form:

$$\Pr\{x < x_0 < x + dx\} = p(x)\, dx = \frac{1}{\sqrt{(2\pi)}\,\sigma} e^{-\frac{1}{2}[(x-\mu)/\sigma]^2}\, dx. \quad (3.4)$$

The problem of obtaining values for μ and σ^2 from a sample of observations is an estimation problem and is discussed in the next chapter.

3.3. PREDICTION FROM A NORMAL DISTRIBUTION

The description of the population of lengths of pellets produced by a manufacturing process states that the distribution of pellet lengths is normal with mean 3·113 and variance 0·003962. From this description we may predict the proportion of pellets that will be made shorter than 3 cm and the proportion longer than 3·25 cm. The proportion of pellets less than 3 cm is represented by area (1) in Fig. 3.3 and the proportion of pellets longer than

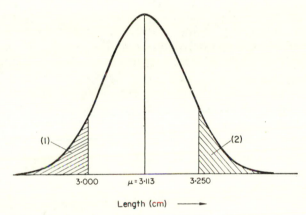

FIG. 3.3. The normal population distribution of lengths of pellets.

3·25 cm is represented by area (2). Tables of the normal probability distribution may be found in Pearson and Hartley (1958), Fisher and Yates (1953), Lindley and Miller (1953) and in Table 1 in the Tables Section (p. 315). These contain the proportion P of the distribution corresponding to values less than X for a normal distribution with zero mean and unit variance. That is for a standardized normal distribution. The proportions for other distributions may be found from these tables by calculating the value of X for a standardized distribution which corresponds to the required value of the variable of interest. The value X_0 of X corresponding to the value x_0 of the variable x may be calculated from the expression (3.5) below:

$$X_0 = \frac{x_0 - \mu}{\sigma}. \tag{3.5}$$

The value X_0 corresponding to 3·25 cm in the above example is

$$X_0 = \frac{3\cdot250 - 3\cdot113}{0\cdot063} = 2\cdot21$$

(note $\sqrt{0\cdot003962} = 0\cdot063$).

The proportion corresponding to values of X less than $X_0 = 2\cdot21$ given by the tables is 0·9864 so that the proportion corresponding to values greater than 2·21 is 0·0136. The proportion of pellets less than 3·25 cm is 0·9864 and the proportion greater than 3·25 cm is 0·0136.

The value X_0 corresponding to 3 cm is

$$X_0 = \frac{3\cdot000 - 3\cdot113}{0\cdot063} = -1\cdot80.$$

Tables of the normal probability distribution do not usually include negative values of X and proportions corresponding to negative values must be calculated by using the symmetry property of the normal distribution. The proportion corresponding to values less than $-1\cdot80$ is equal to the proportion corresponding to values greater than 1·80. The proportion of values less than 1·80 is tabulated as 0·9641 so that the proportion of values greater than 1·80 is 0·0351 and this is also the proportion of values less than $-1\cdot80$.

The proportion of pellets shorter than 3 cm is 0·0359 and the proportion of pellets longer than 3·25 cm is 0·0136 so that the proportion of pellets with lengths outside the range 3 cm to 3·25 cm is 0·0359 + 0·0136 = 0·0495.

3.4. PERCENTAGE POINTS AND SIGNIFICANCE LEVELS

If the probability of obtaining a value of the variable x greater than x_u is equal to α, the value x_u is said to be the upper α significance level of the distribution, or the upper 100α percentage point. If the probability of obtaining a value of the variable x less than x_l is equal to α, the value x_l is said to be the lower α significance level of the distribution, or the lower 100α percentage point. Percentage points are frequently tabulated instead of an extensive tabulation of the distribution itself. Commonly tabulated percentage points are the lower 0·1, 0·5, 1 and 5 percentage points and the upper 5, 1, 0·5 and 0·1 percentage points.

3.5. CENTRAL MOMENTS OF A CONTINUOUS DISTRIBUTION

Consider the term $(x - \mu)^s p(x)\, dx$. This is the sth power of the difference between the value x and the mean μ, times the probability, $p(x)\, dx$, that the variable takes the value x. The rth central moment of a continuous distribution is the sum (integral) of such terms for all values that x may take. The mathematical definition of the sth central moment is:

$$\mu_r = \int (x - \mu)^s \, p(x) \, dx. \tag{3.6}$$

The sth central moment for the normal distribution in particular is:

$$\mu_s = \frac{1}{\sqrt{(2\pi)}\,\sigma} \int_{-\infty}^{\infty} (x - \mu)^s \, e^{-\frac{1}{2}[(x-\mu)/\sigma]^2} \, dx.$$

The second central moment is, of course, the variance σ^2. The odd central moments are all zero since the distribution is symmetrical and the even central moments are all functions of σ^2. Use is made of the knowledge that the fourth central moment μ_4 is equal to $3\sigma^4$ and the ratio known as β^2 is an important measure of the shape of a distribution. This is defined as:

$$\beta_2 = \frac{\mu_4}{\mu_2^2}.$$

Its value in the case of the normal distribution is 3. A second important ratio is known as β_1 and it is defined as:

$$\beta_1 = \frac{\mu_3^2}{\mu_2^3}.$$

The ratio $\sqrt{\beta_1}$ is usually quoted instead of β_1 since this retains the sign of μ_3.

These two ratios measure the shape of the distribution. $\sqrt{\beta_1}$ measures the asymmetry, or skewness, of the distribution and β_2 measures the peakedness of the distribution.

The ratios β_1 and β_2 may be estimated (estimation is discussed in the next chapter) from a sample, and if the estimates obtained are approximately zero and three respectively the population distribution is usually believed to be normal, although, strictly, this does not necessarily follow.

3.6. NOTATION

The following notation is useful and will be used in future chapters:

$\sim \equiv$ is distributed as

$N(\mu, \sigma^2) \equiv$ a normal distribution with mean μ and variance σ^2,

$N(0, 1) \equiv$ a standardized normal distribution.

The statement $x \sim N(\mu, \sigma^2)$ is equivalent to the statement that the variable x has a normal distribution with mean μ and variance σ^2.

3.7. INDEPENDENCE OF TWO RANDOM VARIABLES

So far we have been concerned with the description of one variable at a time. For example, we may say

$$x \sim N(\mu, \sigma^2).$$

In the same way it is possible to describe the joint behaviour of two variables and we may say that the variables x and y have a particular joint distribution such as the bivariate normal distribution described in Chapter 12. A detailed discussion of bivariate distributions can be left until Chapter 12 but a definition of independence of two random variables is necessary for the present discussion. If we write the individual distributions of variables x and y as $p(x)$ and $p(y)$, respectively, and the joint distribution as $p(x, y)$ we may state the condition for independence of two random variables as:

$$p(x, y) = p(x)\,p(y). \tag{3.7}$$

That is, x and y are independent if the joint distribution of x and y can be separated into two parts, one depending on x alone, and the other on y alone. The part depending on x being the distribution of x and the part depending on y alone being the distribution of y. This is a natural extension of the definition of independence of two random events given in Section 2.6 stating that the probability of two events A and B is equal to the product of the probability of A and the probability of B, or, alternatively, that the probability of B does not depend on the occurrence of A. If x and y are independent standardized normal variables the joint bivariate normal distribution is given by:

$$p(x, y)\,dx\,dy = \frac{1}{2\pi} e^{-\frac{1}{2}(x^2+y^2)}\,dx\,dy$$

$$= \frac{1}{\sqrt{(2\pi)}} e^{-\frac{1}{2}x^2}\,dx\,\frac{1}{\sqrt{(2\pi)}} e^{-\frac{1}{2}y^2}\,dy$$

$$= p(x)\,dx\,p(y)\,dy.$$

3.8. PROPERTIES OF NORMAL VARIABLES

In this section we list and discuss properties of variables possessing a normal distribution that will be used in later chapters.

(a) If $x \sim N(\mu, \sigma^2)$ then $\dfrac{x - \mu}{\sigma} \sim N(0, 1)$.

This is the standardization process already described in Section 3.3.

(b) If $x \sim N(\mu, \sigma^2)$ then $ax + b \sim N(a\mu + b, a^2\sigma^2)$ where a and b are constants.

The addition of a constant b to a measurement x changes the mean of the distribution by the constant without affecting the variance. The multiplication of a measurement by a constant a multiplies the mean by a, and the variance by a^2 (the standard deviation is multiplied by a). The new variable also has a normal distribution.

(c) If $x \sim N(\mu, \sigma^2)$ the central moments are given by:

$$\mu_{2r+1} = 0 \quad \text{and} \quad \mu_{2r} = \frac{\sigma^{2r}}{2^r} \frac{(2r)!}{r!} \quad \text{for} \quad r \geqq 0$$

and the first two moment ratios are:

$$\beta_1 = 0 \quad \text{and} \quad \beta_2 = 3.$$

(d) If x_1 and x_2 are independent normal variables and if $x_1 \sim N(\mu_1, \sigma_1^2)$ and

$$x_2 \sim N(\mu_2, \sigma_2^2),$$

then

$$x_1 + x_2 \sim N(\mu_1 + \mu_2, \sigma_1^2 + \sigma_2^2)$$

and

$$x_1 - x_2 \sim N(\mu_1 - \mu_2, \sigma_1^2 + \sigma_2^2).$$

If two measurements each have normal distributions the new measurement formed by adding together the two measurements itself has a normal distribution. The mean and variance of this distribution are equal to the sum of the means of the two individual distributions and the sum of the individual variances respectively. If the new measurement is formed by subtracting the second from the first the mean of the new distribution is equal to the difference between the mean of the first and second distributions. The variance of the new distribution is the *sum* of the individual variances. The reader should be careful to distinguish between the distribution of the sum of two measurements and the sum of two distributions. The sum of two distributions is not usually normal and is frequently bi-modal (that is with two highest values). This property may be extended logically to more than two variables as in the next property to be discussed.

(e) If all measurements in a sample $\{x_1, x_2, \ldots, x_i, \ldots, x_n\}$ are drawn from the same normal population so that:

$$x_i = N(\mu, \sigma^2) \quad \text{for} \quad i = 1, 2, \ldots, n$$

then the sample mean

$$\bar{x} = \frac{1}{n} \sum_{i=1}^{n} x_i$$

has the following normal distribution:

$$\bar{x} \sim N(\mu, \sigma^2/n).$$

The mean of the distribution of the sample mean is the same as the mean of the distribution of the individual measurements but the variance of the sample mean is less than the variance of the individual measurements. This property shows the benefit of replicate measurements. The width of a distribution is measured in terms of the standard deviation rather than the variance so that to halve the standard deviation of the mean the number of measurements must be increased fourfold.

This section has introduced the concept of functions of measurements, and combinations of two or more measurements, possessing distributions themselves. The sample mean has a distribution itself and becomes only one observation with respect to its own population although the individual measurements are each observations with respect to their population.

Many of the properties listed above remain true for distributions other than the normal, and these are given below.

3.9. Properties of Continuous Random Variables

The properties listed below are properties (b), (d) and (e) discussed above but without the additional assumption of normality. The symbol $\mu(x)$ will be used to denote the mean of x, and $\sigma^2(x)$ will denote the variance of x.

(a) If $\mu(x) = M$ and $\sigma^2(x) = V$
then $\mu(ax + b) = aM + b$
and $\sigma^2(ax + b) = a^2 V$.

(b) If $\mu(x_1) = M_1$, $\mu(x_2) = M_2$
and $\sigma^2(x_1) = V_1$, $\sigma^2(x_2) = V_2$
and if x_1 and x_2 are independent
then $\mu(x_1 + x_2) = M_1 + M_2$, $\sigma^2(x_1 + x_2) = V_1 + V_2$
$\mu(x_1 - x_2) = M_1 - M_2$, $\sigma^2(x_1 - x_2) = V_1 + V_2$.

(c) If $\mu(x_1) = M$; $\sigma^2(x_1) = V$ for $i = 1, 2, ..., n$;

if $\bar{x} = \dfrac{1}{n} \sum_{i=1}^{0} x_i$, and if the x_i are mutually independent

then $\sigma^2(\bar{x}) = V/n$.

3.10. The Central Limit Theorem

This important theorem states that if n independent variables have finite variances then their sum will, when expressed in standard measure and under certain conditions, tend to be normally distributed as n tends to infinity. It is a necessary and sufficient condition for the validity of the theorem that the variances obey a condition which may be roughly expressed by saying that no one variance is large compared with their total. This definition is based on that given in Kendall and Buckland (1957). A more complete description may be found in Kendall and Stuart (1963) and many other statistical textbooks.

If we repeat measurements the distribution of their mean will be more nearly normal than the distribution of the individual measurements. The Central Limit Theorem together with the common occurrence of normal populations in practice gives the normal distribution its central position in statistical methods.

3.11. DISTRIBUTIONS OTHER THAN NORMAL

Although the normal distribution is the most common distribution occurring in practice, other distributions are of importance and are used as descriptions of populations. The population distribution may be obtained from a theoretical consideration of the practical situation. For example, the Poisson distribution, to be described in Chapter 13, is a description of a random process practically observed. Means of testing that the population distribution is normal are available and will be described in Chapter 5. It does occur, not infrequently, that a description of the population distribution is required after it has been established that the normal distribution is not an adequate description. There are two procedures available for this situation.

The first is to observe if it is reasonable to believe that the population distribution would be normal if certain values, very different from the bulk of the measurements, were excluded from the sample. Outliers due to factors not affecting the rest of the sample do occur in practice. The decision whether or not an extreme value is a genuine observation or an outlier is difficult, and not to be made lightly. Sound reasons for the rejection of an observation should apply. Conclusions drawn from samples from which observations have been excluded are only as sound as the reasons for the rejection. These reasons should, of course, be stated as well as the conclusions reached. Having rejected the outliers the normal distribution is available as the description of the bulk of the measurements.

The second possibility is to develop, either from theoretical considerations or by inference from the sample, an alternative mathematical form for the population distribution and to use this to describe the population. There are two systems of distributions available which provide a wide range of distribution forms and means of obtaining the values of parameters required to fit a distribution. These are the Pearson system described in Elderton (1935) and the Johnson system described in Johnson (1947).

3.12. THE CHI-SQUARED DISTRIBUTION

This distribution is used occasionally as a description of a population of measurements but is more frequently used in statistical tests of significance. Many uses of the chi-squared distribution will be described in Chapters 5 to 7. Only the mathematical form and properties will be given here.

The chi-squared distribution can be obtained from the normal distribution. If $x_1, x_2, \ldots, x_\nu$ are independent normal variables each with zero mean and unit variance (that is, standardized) then

$$\chi^2 = \sum_{i=1}^{\nu} x_i^2$$

has a chi-squared distribution with ν degrees of freedom.

The distribution depends on one parameter ν, the degrees of freedom, which may take only positive integer values. The distribution is defined for zero and positive values of the variable χ^2.

The distribution has the mathematical form:

$$\Pr\{\chi^2 \mid \nu\}\, d\chi^2 = \frac{e^{-\frac{1}{2}\chi^2}(\chi^2)^{\frac{1}{2}\nu-1}}{2^{\frac{1}{2}\nu}\Gamma(\frac{1}{2}\nu)}\, d\chi^2 \tag{3.8}$$

where the gamma function $\Gamma(x)$ satisfies the relations:

$$\Gamma(x) = (x-1)\,\Gamma(x-1)$$
$$\Gamma(1) = 1$$
$$\Gamma(\tfrac{1}{2}) = \pi^{\frac{1}{2}}.$$

So that $\Gamma(\tfrac{1}{2}\nu) = \begin{cases} (\tfrac{1}{2}\nu - 1)! & \text{if } \nu \text{ is even} \\[2mm] (\tfrac{1}{2}\nu - 1)(\tfrac{1}{2}\nu - 2) \cdots \tfrac{3}{2}\,\tfrac{1}{2}\pi^{\frac{1}{2}} & \text{if } \nu \text{ is odd.} \end{cases}$

3.12.1. Mean, Variance and Further Moments

The mean value of χ^2, applying expression (3.3a), is:

$$\mu(\chi^2) = \int_0^\infty \chi^2 \Pr\{\chi^2 \mid \nu\}\, d\chi^2$$

and it may be shown that $\mu(\chi^2) = \nu$.

The variance of χ^2, applying expression (3.3b), is:

$$\sigma^2(\chi^2) = \int_0^\infty (\chi^2 - \nu)^2 \Pr\{\chi^2 \mid \nu\}\, d\chi^2$$

and this may be shown to be 2ν. The other principal central moments and moment ratios are:

$$\mu_3 = 8\nu \qquad\qquad \beta_1 = \frac{(8\nu)^2}{(2\nu)^3} = \frac{8}{\nu}$$

$$\mu_4 = 48\nu + 12\nu^2 \qquad \beta_2 = \frac{48\nu + 12\nu^2}{(2\nu)^2} = \frac{12}{\nu} + 3.$$

Thus as $\nu \to \infty$, $\beta_1 \to 0$ and $\beta_2 \to 3$.

These are the values for a normal distribution and it may be further shown that the χ^2-distribution does in fact tend to normality as ν increases.

3.12.2. The Modal Value

The modal value of the distribution is given by $\nu - 2$ for values of ν of 2 and above. It is zero for $\nu = 1$.

3.12.3. Tables of the χ^2-distribution

The area under a χ^2-distribution corresponding to values of χ^2 from zero to the table argument is given in Table 7 of Pearson and Hartley (1958), and a useful range of percentage points in Table 4 in the Tables Section. Approximations for estimating percentage points outside the range of the tables are also given at the foot of Table 4.

3.12.4. The Shape of the χ^2-distribution

The shape of a χ^2-distribution depends on the degrees of freedom. The shapes for $\nu = 1$ and $\nu =: 2$ are not characteristic of the shapes for ν greater than 2. Figure 3.4 shows the shapes of χ^2-distributions for a number of values of the degrees of freedom.

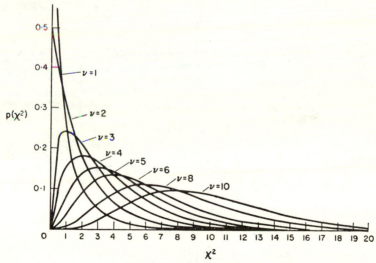

FIG. 3.4. χ^2-Distributions.

QUESTIONS

1. Cylindrical metallic pellets are produced in quantity for use in a reactor and the lengths of the pellets have a normal distribution with mean 0·342 cm and a standard deviation of 0·022 cm (variance 0·000484 cm^2).

(a) What is described by this normal distribution?

(b) What do we assume when we use this description to make predictions about pellets to be produced in the future?

(c) Pellets can only be used if their lengths are less than 0·400 cm. What proportion of pellets are predicted to satisfy this restriction?

(d) If a pellet can only be used if its length is between 0·300 cm and 0·400 cm what proportion of pellets is predicted to satisfy these restrictions?

Sixteen pellets are fitted end to end into a long cylindrical container and the length of a column of sixteen pellets must therefore satisfy conditions dictated by the manufacture of the container.

(e) What is the distribution of the lengths of the column of sixteen pellets, assuming that the ends of adjacent pellets are in intimate contact, and what are the mean, variance and standard deviation of this distribution?

(f) What proportion of columns of sixteen pellets are predicted to satisfy the requirement that their lengths lie between 5·460 cm and 5·480 cm?

2. The dry weights of 250 plants were determined and the following table shows the number of plants falling in each of twelve intervals. The dry weights are measured in grams and the mean weight is estimated as 16·10 and the variance is estimated as 7·1824 (standard deviation 2·680).

Interval	<11	11–12	12–13	13–14	14–15	15–16	16–17
Frequency	7	12	16	24	33	36	40

Interval		17–18	18–19	19–20	20–21	>21	Total
Frequency		30	21	19	9	3	250

(a) Draw the histogram representing this distribution and draw the normal distribution with mean 16·10 and standard deviation 2·680.

(b) Compute the proportions of this normal distribution falling in the twelve groups. Multiply these proportions by the sample size 250 and compare the expected frequencies with the observed frequencies recorded in the table above. These sets of frequencies will be compared formally in a question after Chapter 5.

3. The distribution of heights of men in a particular country is normal with mean 68 in. and standard deviation 3 in. A clothing firm plans to produce a large number of men's raincoats and the sizes will be determined by height. Sizes 1 to 12 will be made to correspond to heights 59–60·5, 60·5–62, 62–63·5, 63·5–65, 65–66·5, 66·5–68, 68–69·5, 69·5–71, 71–72·5, 72·5–74, 74–75·5 and 75·5–77 in. respectively.

(a) What proportion of coats should be produced in each of the twelve sizes?

(b) What proportion of the population is not catered for?

4. A variable x is normally distributed with mean 6 and standard deviation 2. Compute the lower and upper 2·5 per cent significance points of x.

5. A variable x has the rectangular distribution:

$$p(x)\, dx = 1\, dx \quad \text{for} \quad 0 \leq x \leq 1.$$

That is, all values of x in the interval 0 to 1 are equally likely.

(a) Calculate the mean and variance of x.

(b) Calculate the third and fourth central moments of x and from these the moments ratios β_1 and β_2.

6. The lengths of skulls of individuals in a particular country have a normal distribution with mean 9·5 in. and a standard deviation of 1·1 in. A skull of unknown origin has a length of 13 in. Is it reasonable to believe that this skull belonged to an individual in this country?

CHAPTER 4

ESTIMATION

WE HAVE seen in previous chapters that statistical methods are concerned with making inferences about the population using a sample of results as the basis for such inferences. The types of inference normally made may be divided into two sections, the first of which is a decision to accept, or reject, a hypothesis set up about some particular property of the population, and the second of which is the estimation from the sample of the value of a population parameter. To make this division clear we consider the following example. Suppose that a plant geneticist has observed that twenty-one plants out of a total of 100 plants have shown a particular characteristic. The population corresponding to his sample consists of an infinite number of plants of which a proportion p shows the characteristic of interest. If the geneticist suspects that the occurrence of the particular characteristic follows a genetical law which predicts that a quarter of this type of plant will show the characteristic, he will be interested in testing the hypothesis that $p = \frac{1}{4}$. If the geneticist has no belief in a particular value of p he will be interested to estimate the value of p from the sample of plants he has observed. Having obtained a single value (point estimate) for p the geneticist will be interested in the range of values (interval estimate) that p could reasonably take. Methods for testing hypotheses are described in the next chapters on significance tests. Some of the problems of estimation have been briefly discussed in the first chapter and are discussed in more detail here. This chapter is confined mainly to methods of obtaining point estimates, but interval estimates are discussed in Section 5.2.7 entitled "Confidence Limits" in the next chapter.

Before different methods of estimation and means of comparing them can be discussed it is necessary to consider properties that a good estimator should possess. It is also necessary to introduce a number of terms that will be referred to in the main discussion. This chapter therefore begins with introductory definitions, continues with a discussion of the desirable properties of estimators and finally introduces and compares different methods of estimation.

The mathematical level necessary for a detailed appreciation of estimation is, unfortunately, higher than that required in other chapters. It is, however, possible to appreciate the properties of good estimators and the bases of different methods without a high mathematical knowledge. For this reason the mathematics of most derivations is given in an appendix. It is believed, therefore, that the reader can gain an appreciation of the aims of estimation methods without a particularly high mathematical ability. The reader with a little mathematics should read this chapter once without the Appendix and then a second time with reference to the Appendix.

4.1. THE RANDOM SAMPLE

The fundamental requirement for a sound estimation procedure, and indeed for all statistical procedures, is the random sample. All estimation procedures are based on the assumption that the sample is a fair representation of the population. If a bias of unknown extent was present during the collection of the sample it is not possible to make reliable and useful inferences about the population. Suppose that during the collection of the sample a bias operated which gave preference to high values. If we estimate the population mean from this sample our estimate will be high by an amount which depends on the magnitude of the bias. Such an estimate is of little value unless the bias is known. Attention is therefore concentrated in this chapter on estimation of population parameters from random samples.

4.2. ESTIMATORS AND ESTIMATES

It is important to distinguish between the function of the observations (estimator) used to estimate the population parameter and the value obtained (estimate) in a particular application. An estimator is expressed as a function of the observations and is a random variable possessing a probability distribution. The estimate obtained in a particular application may be regarded as a sample of one drawn from the distribution of the estimator. We do not judge an estimator by the nearness to the true value of the population parameter it produces in a particular application, but rather by its performance in repeated applications. The distribution of an estimator is therefore an important property of the estimator.

4.2.1. Notation

The distinction between a population parameter and an estimator of the population parameter is an important one. The distinction is made by adding a circumflex to the symbol for the population parameter to denote the estimator. Thus, for example, we would write $\hat{\mu}$ to denote an estimator of the population parameter μ.

4.3. EXPECTED VALUES

The term "expected value" of an estimator (or of any random variable) is usually used instead of the "mean value", and the expected value of estimator t is written $\mathscr{E}(t)$. If $p(t)\,dt$ represents the probability distribution of t, the $\mathscr{E}(t)$ is given by:

$$\mathscr{E}(t) = \mu(t) = \int_{t\ \text{all}} t\,p(t)\,dt. \tag{4.1}$$

The variance σ_t^2 of t has been defined in (3.3) so far as:

$$\sigma_t^2 = \int_{t\ \text{all}} (t - \mu(t))^2\,p(t)\,dt. \tag{4.2}$$

We may regard this expression as the expected value of the function $(t - \mu(t))^2$ so that we may write the variance of t as:

$$\sigma_t^2 = \mathscr{E}(t - \mu(t))^2$$

or alternatively, and more usually, as:

$$\sigma_t^2 = \mathscr{E}(t - \mathscr{E}(t))^2. \tag{4.3}$$

4.3.1. Covariance and Correlation

Before properties of estimators can be discussed it is necessary to introduce the terms covariance and correlation. The covariance between two variables x_1 and x_2 is a measure of the joint variation between the two variables. The letter C will be used to refer to a covariance and the definition is:

$$\text{Covariance } (x_1, x_2) = C_{12} = \mathscr{E}\{(x_1 - \mathscr{E}(x_1))(x_2 - \mathscr{E}(x_2))\}$$
$$= \mathscr{E}\{(x_1 - \mu_1)(x_2 - \mu_2)\}$$
$$= \int_{\substack{\text{all all} \\ x_1\ x_2}} (x_1 - \mu_1)(x_2 - \mu_2)\,p(x_1, x_2)\,dx_1\,dx_2$$

where

$$\mu_1 = \mathscr{E}(x_1) \quad \text{and} \quad \mu_2 = \mathscr{E}(x_2).$$

The standardized version of the covariance between two variables is the correlation coefficients and is usually denoted by the Greek letter ϱ. The correlation coefficient ϱ_{12} betweeen variables x_1 and x_2 is defined as follows:

$$\varrho_{12} = \frac{C_{12}}{\sigma_1\sigma_2} = \frac{\mathscr{E}\{(x_1 - \mathscr{E}(x_1))(x_2 - \mathscr{E}(x_2))\}}{\sqrt{\{\mathscr{E}(x_1 - \mathscr{E}(x_1))^2\,\mathscr{E}(x_2 - \mathscr{E}(x_2))^2\}}},$$

where σ_1^2 and σ_2^2 are the variances of x_1 and x_2, respectively.

If x_1 and x_2 are independent random variables it may be shown, as in Section 4(a) of the Appendix, that the covariance and hence the correlation coefficient between them is zero. The converse, that if the covariance between x_1 and x_2 is zero then the variables are independent, is not necessarily true.

It should be noted that a covariance, as opposed to a variance, can take any value negative, zero or positive. A correlation coefficient may take any value in the range -1 to $+1$.

4.3.2. Useful Relations

Three useful relations between expected values (given already in a different form, and not so generally, in Section 3.9) are given below.

1. If (a) $\mathscr{E}(t) = \theta$,

   (b) the variance of t is σ_t^2,

   (c) a and b are constants,

then the expected value and variance of the variable $at + b$ are given by:

   (d) $\mathscr{E}(at + b) = a\mathscr{E}(t) + (b) = a\theta + b$,

   (e) the variance of $(at + b) = a^2\sigma_t^2$.

2. If (a) $\mathscr{E}(t_1) = \theta_1$,

   (b) $\mathscr{E}(t_2) = \theta_2$,

   (c) the variance of t_1 and t_2 are σ_1^2 and σ_2^2, respectively,

   (d) the covariance and correlation coefficient between t_1 and t_2 are C_{12} and ϱ_{12}, respectively,

then the expected values and variances of the variables $t_1 + t_2$ and $t_1 - t_2$ are given by:

   (e) $\mathscr{E}(t_1 + t_2) = \mathscr{E}(t_1) + \mathscr{E}(t_2) = \theta_1 + \theta_2$,

   (f) $\mathscr{E}(t_1 - t_2) = \mathscr{E}(t_1) - \mathscr{E}(t_2) = \theta_1 - \theta_2$,

   (g) $\sigma^2(t_1 + t_2) = \sigma_1^2 + \sigma_2^2 + 2C_{12} = \sigma_1^2 + \sigma_2^2 + 2\varrho_{12}\sigma_1\sigma_2$,

   (h) $\sigma^2(t_1 - t_2) = \sigma_1^2 + \sigma_2^2 - 2C_{12} = \sigma_1^2 + \sigma_2^2 - 2\varrho_{12}\sigma_1\sigma_2$.

Thus if t_1 and t_2 are independent the variance of $t_1 + t_2$ and the variance of $t_1 - t_2$ are both $\sigma_1^2 + \sigma_2^2$.

These relations may be extended to more than two variables. For example, with an obvious extension of notation the expected value and variance of $t_1 + t_2 + t_3$ are given by:

(i) $\mathcal{E}(t_1 + t_2 + t_3) = \theta_1 + \theta_2 + \theta_3$,

(j) $\sigma^2(t_1 + t_2 + t_3) = \sigma_1^2 + \sigma_2^2 + \sigma_3^2 + 2C_{12} + 2C_{13} + 2C_{23}$

$$= \sigma_1^2 + \sigma_2^2 + \sigma_3^2 + 2\varrho_{12}\sigma_1\sigma_2 + 2\varrho_{13}\sigma_1\sigma_3$$

$$+ 2\varrho_{23}\sigma_2\sigma_3.$$

3. It follows from a further extension of the above that if:

(a) $\mathcal{E}(t_i) = \theta$ for $i = 1, 2, ..., n$,

(b) $\sigma^2(t_i) = \sigma^2$ for $i = 1, 2, ..., n$,

(c) $\bar{t}$ is defined: $\bar{t} = \dfrac{1}{n} \sum\limits_{i=1}^{n} t_i$,

then   (d) $\mathcal{E}(\bar{t}) = \theta$.   (4.4)

If in addition the t_i are mutually independent then:

(e) $\sigma^2(\bar{t}) = \dfrac{\sigma^2}{n}$.

4.4. Unbiased Estimators

The most important class of estimators is the class of unbiased estimators. An estimator is unbiased if the mean of its distribution (its expected value) is equal to the true value of the parameter being estimated. If t is an estimator of the parameter θ, and if $p(t)\,dt$ is the distribution of the estimator t, then t is unbiased if:

$$\mathcal{E}(t) = \theta,$$

that is if

$$\mathcal{E}(t) = \int\limits_{\substack{\text{all} \\ t}} t\,p(t)\,dt = \theta.$$

The estimator most frequently used to estimate the population mean is the sample mean. We have already used the result (in Chapter 3) that the mean of a sample of size n is distributed with mean equal to the population mean and variance equal to the population variance divided by n. If the population mean is μ and the sample mean is $\bar{x}$ then

$$\mathcal{E}(\bar{x}) = \mu$$

and the sample mean is an unbiased estimator of the population mean.

In Chapter 1 problems of estimation were briefly discussed and several estimators of the population mean were considered. These included the sample median, the average of the two extreme values (the mid-range), the average of all values except the two extreme values, as well as the sample mean. Before discussing how we decide between a number of possible estimators we will first determine whether or not these estimators are biased.

It is clear that a wide class of estimators of the population mean μ, including those above, can be described as a weighted linear function of the observations. That is, estimator t is given by:

$$t = a_1x_1 + a_2x_2 + \cdots + a_nx_n = \sum_{i=1}^{n} a_ix_i, \qquad (4.5)$$

where the a_i $(i = 1, 2, ..., n)$ are constants.

If relation 2 given above in Section 4.3.2 is extended to more than two random variables we obtain the general rule that the expected value of a sum of n random variables is the sum of the expected values. So that:

$$\mathscr{E}(t) = \sum_{i=1}^{n} \mathscr{E}(a_ix_i).$$

If $\mathscr{E}(x_i) = \mu$ for $i = 1, 2, ..., n$ and using relation 1 in Section 4.3.2 we see that:

$$\mathscr{E}(t) = \sum_{i=1}^{n} a_i\mathscr{E}(x_i) = \mu \sum_{i=1}^{n} a_i.$$

From this result we see that t is an unbiased estimator of the population mean μ if

$$\sum_{i=1}^{n} a_i = 1.$$

This property is satisfied for each one of the estimators discussed above since:

1. For the sample mean $a_i = 1/n$ for $i = 1, ..., n$.

2. For the median value
   (a) $a_i = 1$ for the middle value if n is odd and zero otherwise.

   (b) $a_i = \frac{1}{2}$ for the two middle values if n is even and zero otherwise.

3. For the average of the two extremes (the midrange)

$$a_1 = \tfrac{1}{2}, \quad a_n = \tfrac{1}{2},$$
$$a_i = 0 \quad \text{for} \quad i = 2, ..., n - 1.$$

4. For the average of all observations except the two extreme values.

$$a_1 = 0, \quad a_n = 0$$
$$a_i = 1/(n - 2) \quad \text{for} \quad i = 2, ..., (n - 1).$$

Each one of these estimators is, therefore, an unbiased estimator of the population mean.

The estimators discussed above possess different distributions with the same mean μ and different variances. The selection of one estimator from a number of possible estimators will depend on the particular circumstances. The estimator frequently selected is that one with the smallest variance and this type of estimator is discussed below. In some situations it may be more serious to overestimate a parameter than to underestimate it so that we will choose that estimator which produces a seriously high estimate least often, and this may not be the estimator with the smallest variance. In some situations of this type it may be best to choose a biased estimator.

The variance of the estimator t given by:

$$t = \sum_{i=1}^{n} a_i x_i \quad \text{where} \quad \sum_{i=1}^{n} a_i = 1$$

is shown in Section 4(b) of the Appendix to be:

$$\sigma^2(t) = \sum_{i=1}^{n} a_i^2 \sigma_i^2 + 2 \sum_{i=1}^{n-1} \sum_{j=i+1}^{n} a_i a_j \, \varrho_{ij} \, \sigma_i \sigma_j, \tag{4.6}$$

where σ_i^2 is the variance of x_i and ϱ_{ij} is the correlation coefficient between x_i and x_j.

The terms specified by the double summation in expression (4.6) correspond to all possible ways of choosing different values for i and j when both are limited to the range 1 to n. If $n = 3$ there are three selections 1, 2; 1, 3 and 2, 3. If $n = 4$ there are six selections 1, 2; 1, 3; 1, 4; 2, 3; 2, 4; and 3, 4. Thus for $n = 4$ the variance of t is given by:

$$\sigma^2(t) = a_1^2 \sigma_1^2 + a_2^2 \sigma_2^2 + a_3^2 \sigma_3^2 + a_4^2 \sigma_4^2 + 2a_1 a_2 \varrho_{12} \sigma_1 \sigma_2 + 2a_1 a_3 \varrho_{13} \sigma_1 \sigma_3$$
$$+ 2a_1 a_4 \varrho_{14} \sigma_1 \sigma_4 + 2a_2 a_3 \varrho_{23} \sigma_2 \sigma_3 + 2a_2 a_4 \varrho_{24} \sigma_2 \sigma_4 + 2a_3 a_4 \varrho_{34} \sigma_3 \sigma_4.$$

If the x_i are mutually independent we have:

$$\sigma^2(t) = \sum_{i=1}^{n} a_i^2 \sigma_i^2. \tag{4.7}$$

From expression (4.7) we may obtain the variance of the sample mean since the x_i are independent. We substitute $a_i = 1/n$ and $\sigma_i^2 = \sigma^2$, and obtain:

$$\text{Var}(\bar{x}) = \sum_{i=1}^{n} \left(\frac{1}{n}\right)^2 \sigma^2 = \sigma^2/n. \tag{4.8}$$

The variances of the other three possible estimators of the population mean may not be obtained easily from expression (4.6) since they involve observations selected because of their position in an ordering of the n

observations. The selected observations are not, therefore, independent. The correlations between the observations in an ordered sequence are not easily obtained and, in fact, depend on the population distribution. Kendall and Stuart (1961), however, give on page 327, values recorded in Table 4.1 for the ratio of the standard deviations of the median and the mid-range to the standard deviation of the sample mean.

TABLE 4.1. STANDARD DEVIATION RATIOS

Sample size	Median to mean	Mid-range to mean
2	1·000	1·000
4	1·092	1·092
6	1·135	1·190
10	1·177	1·362
20	1·214	1·691
∞	1·253	∞

It will be seen from this table that, except for the case $n = 2$ when all three estimators are the same, the sample mean has the lowest standard deviation and the median has a lower standard deviation than the mid-range. It is worth noting that the ratio of the standard deviation of the median to that of the mean converges as n increases to the value 1·253 whereas the ratio of the standard deviation of the mid-range to that of the mean diverges. Thus, we may conclude that the sample mean is the best of the three possible estimates in the minimum variance sense. The question remains, however, whether or not there exists an estimator with smaller standard deviation than that of the sample mean.

4.5. MINIMUM VARIANCE LINEAR UNBIASED ESTIMATORS

Considering further the problem of finding an estimator for the population mean we proceed to find that estimator which is a linear function of the observations and which has minimum variance. The linear estimator

$$t = \sum_{i=1}^{n} a_i x_i$$

has variance given by expression (4.7) if the observations are independent, and is unbiased if

$$\sum_{i=1}^{n} a_i = 1.$$

The minimum variance linear unbiased estimator of the population mean may be found by minimizing the variance of t

$$\sigma^2(t) = \sum_{i=1}^{n} a_i^2 \sigma^2$$

subject to the condition $\sum_{i=1}^{n} a_i = 1$.

The method of minimization used to find the minimum of a function subject to conditions that must be satisfied is the method of Lagrangian Multipliers. The method applied to this problem is given in Section 4(c) of the Appendix and it is shown that the minimum variance estimator assuming the observations x_i to be independent is obtained when:

$$a_i = \frac{1}{n} \quad \text{for} \quad i = 1, 2, ..., n.$$

Thus we see that the sample mean is the minimum variance linear unbiased estimator of the population mean.

4.5.1. *Minimum Variance Quadratic Unbiased Estimator of the Population Variance*

By constructing a quadratic function of the data values as an estimator of the population variance, and by following the above procedure, it may be shown that the best quadratic unbiased estimator in the minimum variance sense is:

$$\hat{\sigma}^2 = \frac{1}{n-1} \sum_{i=1}^{n} (x_i - \bar{x})^2. \tag{4.9}$$

There is frequently some confusion between this estimator of the population variance and the quantity known as the *Sample Variance*. This is usually denoted by s^2 and is defined as:

$$s^2 = \frac{1}{n} \sum_{i=1}^{n} (x_i - \bar{x})^2.$$

The sample variance s^2 is used in some statistical procedures and the distinction between the two is important. The sample variance is a *biased* estimator of the population variance.

4.5.2. *Computation of a Sum of Squares*

The sum of squares

$$\sum_{i=1}^{n} (x_i - \bar{x})^2 \tag{4.10}$$

used in computing $\hat{\sigma}^2$ (and s^2) may also be written as:

$$\sum_{i=1}^{n} x_i^2 - \frac{1}{n}\left(\sum_{i=1}^{n} x_i\right)^2. \tag{4.11}$$

The second expression (4.11) is usually the more convenient to use on a desk machine since the two sums involved may be calculated simultaneously on most desk machines. Computation on an electronic computer with floating-point operations is, however, more accurately performed by calculating the sample mean first and adding up the squares of the differences $(x_i - \bar{x})$ as in the first expression (4.10). The reason for this is that the subtraction involved in expression (4.11) will lose more figures due to cancellation, than the subtractions involved in expression (4.10). On a desk machine the two terms in expression (4.11) are computed "double length" but only single length working is used on a computer. This point is illustrated in the following example.

We have observed the lengths of twelve pellets and wish to estimate the mean and variance of the population distribution of such pellets. The measurements, in centimetres, are:

3·023	3·142	3·173	3·098	3·001	3·072
3·204	3·149	3·192	3·103	3·112	3·087

From these values we obtain:

$$\sum_{i=1}^{12} x_i = 3\cdot023 + 3\cdot142 + \cdots + 3\cdot087 = 37\cdot356,$$

so that

$$\bar{x} = \frac{1}{12}\sum_{i=1}^{12} x_i = 37\cdot356/12 = 3\cdot113.$$

The population mean μ is estimated as $3\cdot113$.

The sum of squares calculated using expression (4.11) is

$$\sum_{i=1}^{12} x_i^2 - \frac{1}{12}\left(\sum_{i=1}^{12} x_i\right)^2 = 3\cdot023^2 + \cdots + 3\cdot087^2 - 37\cdot356^2/12$$

$$= 116\cdot332814 - 116\cdot289228 \tag{4.12}$$

$$= 0\cdot043586,$$

so that $\hat{\sigma}^2 = 0\cdot043586/11 = 0\cdot003962$ to four significant figures.

On a desk machine the two terms in the subtraction labelled (4.12) are computed in the product register to full (double-length) accuracy and no figures are lost. On an electronic computer working to say, seven significant figures, the two terms in subtraction (4.12) are rounded to $116\cdot3328$ and $116\cdot2892$, respectively, and the difference obtained is $0\cdot0436$ instead of

0·043586 and the variance estimate would be 0·003964 instead of 0·003962. In some cases the loss of accuracy at this stage in the computation is serious and it is possible, because of the different rounding errors made in the computation of the two terms, to produce a negative difference, and, hence, a negative variance estimate.

The sum of squares calculated using expression (4.10) is

$$\sum_{i=1}^{n} (x_i - \bar{x})^2 = (3 \cdot 023 - 3 \cdot 113)^2 + \cdots + (3 \cdot 087 - 3 \cdot 113)^2$$
$$= 0 \cdot 043586, \tag{4.13}$$

so that $\hat{\sigma}^2 = 0 \cdot 043586/11 = 0 \cdot 003962$ to four significant figures. In this method the subtractions are made at the stage labelled (4.13) and the loss of significant figures does not occur because there is room on the desk machine and in the computer for all the significant figures involved. In this method no numbers with more than seven significant figures are produced so that the same accurate answer is obtained on both the desk machine and the computer.

The conclusion is, therefore, to use expression (4.10) on an electronic computer and to use expression (4.11) on the desk machine although the use of expression (4.10) on a desk machine would produce an accurate value.

4.6. EFFICIENCY RATIO

The efficiency of an estimator is measured in terms of its variance. The lower the variance the more efficient the estimator is said to be. If the variances of two estimators t_1 and t_2 are $\sigma^2(t_1)$ and $\sigma^2(t_2)$ and if $\sigma^2(t_1)$ is less than $\sigma^2(t_2)$ for all sample sizes then t_1 is said to be more efficient than t_2. It is, of course, possible to have $\sigma^2(t_1)$ less than $\sigma^2(t_2)$ for some sample sizes and not for others. In this case t_1 is sometimes the more efficient estimator and t_2 is the more efficient on other occasions. We see from Table 4.1 in Section 4.4 that if the sample size is three or more the sample mean is a more efficient estimator of the population mean than both the sample median and the sample mid-range.

If an estimator whose variance is less than that of any other estimator exists it is called a most-efficient estimator and its variance forms a base against which the variances of other estimators are judged. If t_1 is the most efficient estimator and has variance $\sigma^2(t_1)$ then the efficiency $E(t_2)$ of estimator t_2 with variance $\sigma^2(t_2)$ is measured as:

$$E(t_2) = \frac{\sigma^2(t_1)}{\sigma^2(t_2)}. \tag{4.14}$$

For large sample sizes the variance of the median of a sample of size n randomly selected from a normal population is

$$\text{Variance (median} \mid n \text{ large)} = \frac{\pi\sigma^2}{2n},$$

where σ^2 is the population variance. The efficiency of the sample median as an estimator of the population mean is measured against the variance of the most efficient estimator, the sample mean. That is:

$$\text{Efficiency (median} \mid n \text{ large)} = \frac{2n}{\pi\sigma^2} \times \frac{\sigma^2}{n} = 0.637.$$

It is clear that where possible the most efficient estimator should be used. If the labour of computation involved in obtaining a most efficient estimate is considerable there may be a case for using an alternative estimator. In this situation the efficiency would play an important part in this decision.

4.7. CONSISTENT ESTIMATORS

An estimator is said to be consistent if its distribution clusters around the true value of the population parameter more and more closely as the sample size increases. If for any fixed small quantity ε, we can find a sample size n such that, for all samples of sizes greater than n, the probability that the estimator t differs from the true value θ by more than ε is as small as we choose, the estimator t is consistent. Representing the estimator t for sample size n by t_n we may say that t_n converges "in probability" to θ as n increases.

The consistency condition for an estimator seems, at first sight, to be possessed by all unbiased estimators. However, it is possible to construct unbiased estimators that are not consistent. If the variance of an unbiased estimator does not approach zero as n increases it is not consistent because we cannot choose a value of n which satisfies the condition given above. An unbiased estimator is also consistent if its variance approaches zero as n increases. Thus the sample mean and the sample median are consistent estimators, but the mid-range is not.

4.8. SUFFICIENT ESTIMATORS

A sufficient estimator of the population parameter θ is that estimator which contains all the information about the value of θ that it is possible to extract from the sample. Thus a sufficient estimator is the best possible estimator available. Unfortunately it is not always possible to find a sufficient estimator and an estimator not satisfying this condition must be accepted. The sample mean, however, may be shown to be a sufficient estimator of the population mean.

We may determine whether, or not, a given estimator t of the population parameter θ is sufficient by the following mathematical procedure. In Section 3.7 we considered the "bivariate" distribution of two variables x and y. In this section we consider the "multivariate" distribution of several variables and in particular the joint distribution of the n measurements $x_1, x_2, \ldots, x_n$ forming the sample. We write the probability that the n measurements jointly take the particular values observed, given the value of θ as:

$$p(x_1, x_2, \ldots, x_n | \theta) \, dx_1 \, dx_2 \ldots dx_n = p(t | \theta) \, dt \, p(x_2, x_3, \ldots, x_n) \, dx_2 dx_3 \ldots dx_n.$$

$$(4.15)$$

Since t is a function of $x_1, x_2, \ldots, x_n$ the rearrangement of the distribution to include t causes the deletion of one of the sample values. We have deleted x_1.

4.9. THE METHOD OF MAXIMUM LIKELIHOOD

The method of maximum likelihood is a method of obtaining estimators with certain desirable properties. It may be shown that maximum likelihood estimators are consistent, have minimum variances in the limit at least, tend to normality as n increases, and provide sufficient estimators where these exist. Maximum likelihood estimators are not always, however, unbiased.

The likelihood function (usually denoted by the letter L) is the joint distribution of the elements of the sample given the values of the population parameter θ. That is L is the probability of the observed results when the parameter takes the value θ. That is:

$$L = p(x_1, x_2, \ldots, x_n | \theta).$$

The maximum likelihood estimator t of θ is that function of $x_1, x_2, \ldots, x_n$ which maximizes the likelihood L. Mathematically we obtain the estimator t which maximizes L by differentiating L with respect to θ, equating the derivative to zero and solving the resulting equation. To confirm that the solution we obtain maximizes L (since the solution, or solutions, of the equation may minimize L instead of maximizing L) we accept the solution for which the second derivative (differentiate L twice) of L is negative. That is we solve

$$\frac{\partial L}{\partial \theta} = 0 \qquad\qquad (4.16)$$

and accept the solution for which

$$\frac{\partial^2 L}{\partial \theta^2} \text{ is negative.} \tag{4.17}$$

An alternative procedure which is often more convenient than that above is to obtain the estimator t which maximizes log L. That is we solve:

$$\frac{\partial \log L}{\partial \theta} = 0 \tag{4.18}$$

and accept the solution for which

$$\frac{\partial^2 \log L}{\partial \theta^2} \text{ is negative.} \tag{4.19}$$

4.9.1. Large Sample Variance of Maximum Likelihood Estimators

It may be shown for large samples that the variance of a maximum likelihood estimator may be obtained from either of the following expressions:

$$\sigma_t^2 \doteqdot \left[n \mathscr{E} \left(\frac{\partial \log p(x)}{\partial \theta} \right)^2 \right]^{-1} = \left[n \int \left(\frac{\partial \log p(x)}{\partial \theta} \right)^2 p(x)\, dx \right]^{-1}$$

$$\sigma_t^2 \doteqdot \left[-n \mathscr{E} \left(\frac{\partial^2 \log p(x)}{d\theta^2} \right) \right]^{-1} = \left[-n \int \frac{\partial^2 \log p(x)}{d\theta^2} p(x)\, dx \right]^{-1}, \tag{4.20}$$

where $p(x)$ denotes the population probability distribution, and the range of integration is all values of x.

4.9.2. Maximum Likelihood Estimator of the Population Mean

Let us again consider the problem of estimating the population mean from a sample of n observations. The likelihood function, if we assume the population distribution to be normal with mean μ and known variance σ^2, is:

$$L = \frac{1}{\sqrt{(2\pi)}\sigma} e^{-\frac{1}{2}[(x_1 - \mu)/\sigma]^2} \cdot \frac{1}{\sqrt{(2\pi)}\sigma} e^{-\frac{1}{2}[(x_2 - \mu)/\sigma]^2} \cdots \frac{1}{\sqrt{(2\pi)}\sigma} e^{-\frac{1}{2}[(x_n - \mu)/\sigma]^2}$$

$$= \left(\frac{1}{\sqrt{(2\pi)}\sigma} \right)^n e^{-\frac{1}{2} \sum_{i=1}^{n} [(x_i - \mu)/\sigma]^2}$$

$$\text{Log } L = -\frac{n}{2} \log (2\pi\sigma^2) - \frac{1}{2} \sum_{i=1}^{n} \left(\frac{x_i - \mu}{\sigma} \right)^2$$

$$\left(\frac{\partial \log L}{\partial \mu} \right) = 0 = \sum_{i=1}^{n} \left(\frac{x_i - \mu}{\sigma^2} \right).$$

That is

$$\hat{\mu} = \frac{1}{n} \sum_{i=1}^{n} x_i = \bar{x}.$$

We may confirm that this solution maximizes L by finding:

$$\frac{\partial^2 \log L}{\partial \mu^2} = -\frac{n}{\sigma^2} < 0.$$

Thus, we see that the sample mean is the maximum likelihood estimator of the population mean.

The large sample variance of this estimator may be computed as:

$$\text{Var}\,(\bar{x} \mid n \text{ large}) = \left[-n \int_{-\infty}^{\infty} \frac{\partial^2 \log p(x)}{\partial \mu^2} p(x)\,dx \right]^{-1}$$

$$= \left[n \int_{-\infty}^{\infty} \frac{1}{\sigma^2} p(x)\,dx \right]^{-1}$$

$$= \frac{\sigma^2}{n}.$$

This, as we have already seen, is true for all values of n.

4.9.3. Maximum Likelihood Estimator of the Population Variance

We again assume that the population distribution is normal and that the mean (known this time) is μ and the unknown variance to be estimated is σ^2. The likelihood function is the same as that given in Section 4.9.2 so that we may proceed to maximize:

$$\text{Log } L = -\frac{n}{2} \log\,(2\pi\sigma^2) - \frac{1}{2} \sum_{i=1}^{n} \frac{(x_i - \mu)^2}{\sigma^2},$$

with respect to σ^2 to obtain:

$$\frac{\partial \log L}{\partial \sigma^2} = 0 = -\frac{n}{2\sigma^2} + \frac{1}{2\sigma^4} \sum_{i=1}^{n} (x_i - \mu)^2.$$

That is

$$\hat{\sigma}^2 = \frac{1}{n} \sum_{i=1}^{n} (x_i - \mu)^2.$$

Note that this estimator differs from that given in Section 4.5.1 because in this example the population mean μ is assumed to be known. The case when μ is unknown is considered below.

4.10. THE JOINT ESTIMATION OF TWO PARAMETERS

So far we have considered the estimation of only one parameter. It frequently happens, of course, that the value of more than one parameter is unknown. The methods outlined above may be readily extended to the case of two or more parameters when the maximum likelihood estimators jointly maximize the likelihood function. For two parameters θ_1 and θ_2 we solve the equations

$$\frac{\partial L}{\partial \theta_1} = 0 \quad \text{and} \quad \frac{\partial L}{\partial \theta_2} = 0 \qquad (4.21)$$

or alternatively:

$$\frac{\partial \log L}{\partial \theta_1} = 0 \quad \text{and} \quad \frac{\partial \log L}{\partial \theta_2} = 0. \qquad (4.22)$$

We illustrate this extension by considering the joint estimation of the mean μ and variance σ^2 of a normal population. We have:

$$\text{Log } L = -\frac{n}{2} \log (2\pi\sigma^2) - \frac{1}{2} \sum_{i=1}^{n} \frac{(x_i - \mu)^2}{\sigma^2}$$

and we maximize this expression with respect to μ and σ^2 to obtain:

$$\frac{\partial \log L}{\partial \mu} = 0 = \sum_{i=1}^{n} \left(\frac{x_i - \mu}{\sigma^2} \right)$$

and

$$\frac{\partial \log L}{\partial \sigma^2} = 0 = -\frac{n}{2\sigma^2} + \frac{1}{2\sigma^4} \sum_{i=1}^{n} (x_i - \mu)^2.$$

The simultaneous solution of these equations provides the following estimates for μ and σ^2:

$$\hat{\mu} = \frac{1}{n} \sum_{i=1}^{n} x_i = \bar{x}$$

and

$$\hat{\sigma}^2 = \frac{1}{n} \sum_{i=1}^{n} (x_i - \bar{x})^2.$$

It should be noted that, in this case, the estimate of the population variance is biased, although the bias decreases in magnitude as the sample size increases.

The extension of this method to k parameters $\theta_1, \theta_2, ..., \theta_k$ would require the simultaneous solution of the k equations

$$\frac{\partial L}{\partial \theta_j} = 0 \quad \text{for} \quad j = 1, 2, ..., k, \qquad (4.23)$$

or alternatively:

$$\frac{\partial \log L}{\partial \theta_j} = 0 \quad \text{for} \quad j = 1, 2, ..., k. \qquad (4.24)$$

4.11. The Method of Least Squares

We will suppose that a sample of observations $\{y_1, y_2, ..., y_n\}$ has been collected and that the expected value of y_i, written as $\mathscr{E}(y_i \mid \theta_1, \theta_2, ..., \theta_p)$, depends on unknown parameters $\theta_1, \theta_2, ..., \theta_p$. The method of least squares provides estimators of the parameters by minimizing the sum of squares of the deviations of the y_i's from their expected values with respect to the p parameters. That is we minimize:

$$s = \sum_{i=1}^{n} (y_i - \mathscr{E}(y_i \mid \theta_1, \theta_2, ..., \theta_p))^2 \tag{4.25}$$

with respect to $\theta_1, \theta_2, ..., \theta_p$.

The connection between the method of least squares and the minimum variance method is clear from expression (4.25). In particular, if we are estimating the population mean μ from a sample $\{y_1, y_2, ..., y_n\}$, the sum of squares s may be written:

$$s = \sum_{i=1}^{n} (y_i - \mu)^2$$

since $\mathscr{E}(y_i) = \theta_1 = \mu$ for all values of i. We thus see that the two methods are equivalent in this situation.

The method of least squares is commonly applied to regression problems in which we seek to describe the variable y as a function of a variable x (or the r variables $v_1, v_2, ..., v_r$) and p unknown parameters $\theta_1, \theta_2, ..., \theta_p$. That is:

$$y_i = f(v_{1i}, v_{2i}, ..., v_{ri} \mid \theta_1, \theta_2, ..., \theta_p) + z_i.$$

Where the z_i are independent random errors with unknown, but constant, variance σ^2. We obtain estimates of $\theta_1, \theta_2, ..., \theta_p$ by minimizing the sum of squares of the error terms with respect to the parameters $\theta_1, \theta_2, ..., \theta_p$; that is:

$$s = \sum_{i=1}^{n} z_i^2 = \sum_{i=1}^{n} [y_i - f(v_{1i}, v_{2i}, ..., v_{ri} \mid \theta_1, \theta_2, ..., \theta_p)]^2. \tag{4.26}$$

Expressions (4.25) and (4.26) are equivalent since we may say that

$$\mathscr{E}(y_i \mid \theta_1, \theta_2, ..., \theta_p) = f(v_{1i}, v_{2i}, ..., v_{ri} \mid \theta_1, \theta_2, ..., \theta_p).$$

The method of least squares is considered in detail in Chapter 12 but we will illustrate the method here by considering the estimation of the parameters a and b in the function:

$$y_i = a + bx_i + z_i.$$

We estimate a and b by minimizing, with respect to θ_1 and θ_2, the sum of squares:

$$S = \sum_{i=1}^{n} (y_i - a - bx_i)^2.$$

Minimizing S we have:

$$\frac{\partial S}{\partial a} = 0 = -2 \sum_{i=1}^{n} (y_i - a - bx_i)$$

$$\frac{\partial S}{\partial b} = 0 = -2 \sum_{i=1}^{n} x_i(y_i - a - bx_i).$$

Thus to obtain estimators of a and b we solve the equations:

$$\sum_{i=1}^{n} y_i = \hat{a}n + \hat{b} \sum_{i=1}^{n} x_i$$

$$\sum_{i=1}^{n} x_i y_i = \hat{a} \sum_{i=1}^{n} x_i + \hat{b} \sum_{i=1}^{n} x_i^2.$$

The solution of these equations is readily shown to be:

$$\hat{b} = \frac{n \sum_{i=1}^{n} x_i y_i - \sum_{i=1}^{n} x_i \sum_{i=1}^{n} y_i}{n \sum_{i=1}^{n} x_i^2 - \left(\sum_{i=1}^{n} x_i\right)^2} = \frac{\sum_{i=1}^{n} (x_i - \bar{x})(y_i - \bar{y})}{\sum_{i=1}^{n} (x_i - \bar{x})^2}$$

$$\hat{a} = \frac{1}{n} \sum_{i=1}^{n} y_i - \frac{\hat{b}}{n} \sum_{i=1}^{n} x_i = \bar{y} - \hat{b}\bar{x},$$

where $\bar{x}$ and $\bar{y}$ are the mean x and mean y values respectively.

4.12. The Methods of Maximum Likelihood and Least Squares

It should be noted that no assumption of normality (or any other distributional form) was made in the application of the least-squares method. However, estimation procedures are often accompanied by tests of significance which do depend on the form of the population distribution. Thus, estimates obtained by the method of least squares do not involve assumptions concerning the population distribution but tests of significance involving the estimates usually do. If we assume that the error terms, e_i, in expression (4.26), are normally distributed with constant variance σ^2 then it may be shown that estimators obtained by minimizing (4.26) are the same as those obtained by the method of maximum likelihood starting from the assumption of normality for the population distribution.

Appendix

4. (a) To prove that if x_1 and x_2 are independent random variables the covariance, and hence the correlation between them is zero.

$$\text{Covariance } (x_1, x_2) = C_{12} = \int_{\substack{\text{all} \\ x_1}} \int_{\substack{\text{all} \\ x_2}} (x_1 - \mu_1)(x_2 - \mu_2) \, p(x_1, x_2) \, dx_1 \, dx_2.$$

If x_1 and x_2 are independent random variables we have from Section 3.7 that:

$$p(x_1, x_2) \, dx_1 \, dx_2 = p(x_1) \, dx_1 p(x_2) \, dx_2.$$

Substituting for $p(x_1, x_2) \, dx_1 \, dx_2$ in the expression for C_{12} above we have:

$$C_{12} = \int_{\substack{\text{all} \\ x_1}} \int_{\substack{\text{all} \\ x_2}} (x_1 - \mu_1)(x_2 - \mu_2) \, p(x_1) \, dx_1 p(x_2) \, dx_2$$

$$= \int_{\substack{\text{all} \\ x_1}} (x_1 - \mu_1) \, p(x_1) \, dx_1 \int_{\substack{\text{all} \\ x_2}} (x_2 - \mu_2) \, p(x_2) \, dx_2$$

$$= 0.$$

Both integrals above are zero since:

$$\int_{\substack{\text{all} \\ x_1}} (x_1 - \mu_1) \, p(x_1) \, dx_1 = \int_{\substack{\text{all} \\ x_2}} x_1 p(x_1) \, dx_1 - \mu_1 \int_{\substack{\text{all} \\ y_1}} p(x_1) \, dx_1$$

$$= \mu_1 - \mu_1$$

$$= 0.$$

4. (b) Derivation of the variance of the linear estimator of the population mean μ,

$$t = \sum_{i=1}^{n} a_i x_i \quad \text{where} \quad \sum_{i=1}^{n} a_i = 1$$

$$\sigma^2(t) = \mathscr{E}(t - \mathscr{E}(t))^2 = \mathscr{E}(t - \mu)^2$$

$$= \mathscr{E} \left(\sum_{i=1}^{n} a_i x_i - \mu \right)^2 = \mathscr{E} \left(\sum_{i=1}^{n} a_i x_i - \sum_{i=1}^{n} a_i \mu \right)^2$$

$$= \mathscr{E}(-\sum a_i(x_i - \mu))^{-2}$$

$$= \mathscr{E} \left(\sum_{i=1}^{n} a_i^2 (x_i - \mu)^2 \right) + 2\mathscr{E} \sum_{i=1}^{n-1} \sum_{j=i+1}^{n} a_i a_j [(x_i - \mu)(x_j - \mu)]$$

$$= \sum_{i=1}^{n} a_i^2 \, \mathscr{E} \, (x_i - \mu)^2 + 2 \sum_{i=1}^{n-1} \sum_{j=i+1}^{n} a_i a_j \, \mathscr{E} \, (x_i - \mu)(x_j - \mu)$$

$$\sigma^2(t) = \sum_{i=1}^{n} a_i^2 \sigma_i^2 + 2 \sum_{i=1}^{n-1} \sum_{j=i+1}^{n} a_i a_j \varrho_{ij} \sigma_i \sigma_j.$$

4. (c) Derivation of the minimum variance linear estimator of the population mean assuming independence between sample values.

The linear estimator t is given by

$$t = \sum_{i=1}^{n} a_i x_i \quad \text{where} \quad \sum_{i=1}^{n} a_i = 1.$$

The minimization is achieved by using the method of Lagrangian Multipliers and begins by constructing a function:

$$F = \sum_{i=1}^{n} a_i^2 \sigma^2 + 2\alpha \left(\sum_{i=1}^{n} a_i - 1 \right)$$

where the constant 2α is a Lagrangian Multiplier to be determined. The first term is the variance of t to be minimized (assuming independence between the sample values x_i), and the second term introduces the condition to be satisfied. That is:

$$\sum_{i=1}^{n} a_i - 1 = 0.$$

The function F has the same value as the variance of t since the second term is zero but the inclusion of this term in the minimization process ensures that the condition will be satisfied at the minimum. Minimizing F with respect to the a_i we have:

$$\frac{\partial F}{\partial a_i} = 2a_i \sigma^2 + 2\alpha \quad \text{for} \quad i = 1, 2, \ldots, n.$$

Equating the derivatives to zero gives:

$$a_i = \frac{\alpha}{\sigma^2} \quad \text{for} \quad i = 1, 2, \ldots, n.$$

We may most conveniently find the value of α by summing these expressions. That is:

$$\sum_{i=1}^{n} a_i = \sum_{i=1}^{n} \frac{\alpha}{\sigma^2} = \frac{n\alpha}{\sigma^2} = 1.$$

That is

$$\alpha = \frac{\sigma^2}{n}.$$

The minimum variance estimator for the population mean is obtained when:

$$a_i = \frac{\sigma^2}{n} \frac{1}{\sigma^2} = \frac{1}{n}$$

from which we see that the minimum variance linear unbiased estimator of the population mean is the sample mean.

QUESTIONS

1. Three observations x_1, x_2 and x_3 are made and from these we wish to estimate the population mean μ. The population variance is known to be σ^2. Three possible estimators t_1, t_2 and t_3 are defined as follows:

$$t_1 = \frac{1}{4} x_1 + \frac{1}{2} x_2 + \frac{1}{4} x_3$$

$$t_2 = \frac{1}{5} x_1 + \frac{2}{5} x_2 + \frac{2}{5} x_2$$

$$t_3 = \frac{1}{6} x_1 + \frac{2}{6} x_2 + \frac{3}{6} x_3$$

(a) Show that all three estimators are unbiased estimators of μ.
(b) Compute the variances of each estimator assuming the x_1, x_2 and x_3 are independent.
(c) Which of these three estimators would you prefer to use and why?
(d) Can you think of an even better estimator of μ? What is its variance?

2. Repeat question 1 for the situation in which the observations are correlated as follows:

$$\varrho_{12} = \tfrac{1}{2} \qquad \varrho_{13} = \tfrac{1}{3} \qquad \varrho_{23} = \tfrac{1}{4}.$$

3. Repeat question 1 for the situation in which the observations are correlated as follows:

$$\varrho_{12} = -\tfrac{1}{2} \qquad \varrho_{13} = -\tfrac{1}{3} \qquad \varrho_{23} = \tfrac{1}{4}.$$

4. The variable x is a binomial variable taking the value 1 for a success and 0 for a failure. The probability of a success at a single trial is p and we seek to find an estimator of p from a series of n trials. Find the minimum variance linear unbiased estimator of p by constructing the function:

$$t = \sum_{i=1}^{n} a_i x_i$$

and by the following steps.

(a) Show that $\mathscr{E}(x_i) = p$.
(b) Show that $\mathscr{E}(x_i^r) = p$.
(c) Show that $\mathscr{E}(t) = p$ if $\sum_{i=1}^{n} a_i = 1$.
(d) Assuming the x's are independent show that

$$\sigma^2(t) = p \sum_{i=1}^{n} a_i^2 + p^2 \sum_{i=1}^{n} \sum_{j=1}^{n} a_i a_j - p^2.$$

(e) Minimize $\sigma^2(t)$ subject to the condition

$$\sum_{i=1}^{n} a_i = 1$$

to show that $a_i = 1/n$.

(f) Hence show that the required estimator is

$$\hat{p} = \frac{1}{n} \sum_{i=1}^{n} x_i.$$

That is the proportion of successes observed in the sample.

5. (a) Estimate the population mean from the following sample of twelve results:

$$10 \cdot 023 \quad 10 \cdot 142 \quad 10 \cdot 173 \quad 10 \cdot 098 \quad 10 \cdot 001 \quad 10 \cdot 072$$
$$10 \cdot 204 \quad 10 \cdot 149 \quad 10 \cdot 192 \quad 10 \cdot 103 \quad 10 \cdot 112 \quad 10 \cdot 087$$

(b) Estimate the population variance by both methods described in Section 4.5.2.

(c) Estimate the population variance by both methods described in Section 4.5.2 keeping only six significant figure at ALL stages of the calculation.

6. (a) Denoting by r the number of successes in the n trials write down the likelihood of observing r successes.

(b) Differentiate the logarithm of this function to show that $\hat{p} = r/n$ is the maximum likelihood estimator of p.

7. A sample of size n measurements of two variables x and y has been taken and we wish to describe y by a straight line in x which passes through the origin. That is, we wish to estimate b in the model:

$$y_i = bx_i + e_i.$$

Show that the method of least squares gives the following estimate of b:

$$\hat{b} = \frac{\sum_{i=1}^{n} x_i y_i}{\sum_{i=1}^{n} x^2}.$$

CHAPTER 5

TESTS OF SIGNIFICANCE—I

THIS chapter describes the significance test method of testing hypotheses. Hypotheses to be tested may involve a property, or properties, of one or more populations. In this chapter, tests involving a single population will be described to answer the following questions:

1. Is the population mean equal to a particular value?
2. Is the population variance equal to a particular value?
3. Is the population distribution of a particular form?

The test of significance method follows the consequences of the assumption that the hypothesis being tested is true and provides means of judging whether, or not, it is reasonable to have observed the particular sample. For example, if we observe a sample mean of 36 and we wish to test the hypothesis that the population mean is 35 the significance tests method provides means of answering the question—if the population mean is 35 is it reasonable to expect to observe a sample mean of 36?

The method will be introduced by considering a simple coin-tossing example.

5.1. THE TEST OF SIGNIFICANCE METHOD

Let us suppose that we wish to test the hypothesis that the probability of "heads" at the toss of a coin is one-half, or less than one-half and that separate tosses are independent. One possible experiment would be to toss a coin a number of times and to observe how many times "heads" appears. If we toss a coin eight times there are nine possible results that may be obtained, that is, 0, 1, 2, ..., 8 heads may be observed. If the probability of "heads" is one-half the binomial distribution predicts that these results will occur with probabilities:

$$\frac{1}{256}, \quad \frac{8}{256}, \quad \frac{28}{256}, \quad \frac{56}{256}, \quad \frac{70}{256}, \quad \frac{56}{256}, \quad \frac{28}{256}, \quad \frac{8}{256}, \quad \frac{1}{256}.$$

If we observe four or less heads we will consider that this evidence is consistent with the hypothesis that the probability of "heads" is one-half or less. If we observe eight heads we will probably feel that the result is in bad agreement with the hypothesis since, if $p = \frac{1}{2}$ there is a probability of $1/256$ that we will observe eight heads (if $p < \frac{1}{2}$ this probability is less than $1/256$). With this small risk of being wrong we will reject the hypothesis. The rejection of a true hypothesis is called an error of the first kind, or a type 1 error. If we observe seven heads this result is in poor agreement with the hypothesis and will occur, if $p = \frac{1}{2}$, with probability $8/256$. There is one result which is in worse agreement with the hypothetical value of p, namely eight heads, and the probability of observing seven heads or more is $9/256$. If we are prepared to take this risk of being wrong we will reject the hypothesis if we observe seven or eight heads. Thus these two results are "critical" results which will cause us to reject the hypothesis, and if $9/256$ is the highest risk we are prepared to take of committing a type 1 error the results $0, 1, ..., 6$ heads will be acceptance results. Thus we have divided the possible results into two regions usually called the critical, or rejection region, and the acceptance region.

It is worth noting here that if we wish to test the hypothesis that $p = \frac{1}{2}$ then the critical region would consist of either:

0 or 8 heads with probability $2/256$, or
0, 1, 7, or 8 heads with probability $18/256$,
according to the risk we are prepared to take of committing a type 1 error.

5.1.1. Single-tailed and Double-tailed Tests

The first type of test described above in which the critical region consists of values on one side of the distribution is called a single-tailed test. The second type of test described above in which the critical region consists of values on both sides of the distribution is called a double-tailed test.

5.1.2. The Power of a Test

The single-tailed test described above has a critical region consisting of seven or eight heads. We will now consider the application of this test when the alternative hypothesis that $p > \frac{1}{2}$ is true. Let the actual value of p be p_0. If seven or eight heads are observed we will correctly reject the hypothesis that $p < \frac{1}{2}$ so that the probability that seven or eight heads occur will be the probability that we correctly reject the false hypothesis that $p \leqq \frac{1}{2}$. This probability P_R is:

$$P_R = p_0^8 + 8p_0^7(1 - p_0) = p_0^7(8 - 7p_0). \tag{5.1}$$

The probability that we observe $0, 1, ...,$ or 6 heads, namely

$$P_A = 1 - p_0^7(8 - 7p_0),$$

is the probability that we accept the hypothesis that $p \leq \frac{1}{2}$ when in fact it is false. The acceptance of a false hypothesis is called an error of the second kind, or a type 2 error.

The probability p_0 may take any value in the range $\frac{1}{2} < p_0 \leq 1$. For each value of p_0 the probabilities with which the nine results occur will be given by the binomial distribution $B(p_0, 8)$ and the probability that we reject the hypothesis $p \leq \frac{1}{2}$ is P_R. If we plot P_R against p_0 we obtain the Power Function for this test. This is plotted below in Fig. 5.1 and some values are given in Table 5.1.

TABLE 5.1. POWER FUNCTION VALUES

p_0		0·5	0·55	0·60	0·70	0·80	0·90	0·95
P_R	9/256 =	0·0352	0·0632	0·1064	0·2553	0·5033	0·8131	0·9428

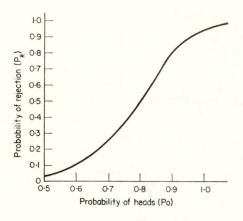

FIG. 5.1. Power function.

It is clear that if the hypothesis that $p \leq \frac{1}{2}$ is false we hope that the test will reject this hypothesis as often as possible. The power function is therefore a measure of how good the test is in detecting departures from the assumed hypothesis. If there are two tests designed to test the same hypothesis the better test is that one with the higher power function. It is, of course, possible for the two power functions to cross in which case the decision between the tests depends on the size of departure from the hypothesis value that it is required to detect.

The test of significance method will be further illustrated with examples of commonly occurring situations.

5.2. Is the Population Mean Equal to a Particular Value?

We have collected a sample of n measurements $\{x_1, x_2, ..., x_n\}$ and wish to test the hypothesis that the population mean is a particular value μ_p. There are two situations to be considered:

1. the population variance is known,
2. the population variance is unknown.

5.2.1. Is the Population Mean Equal to μ_p (Variance Known)?

A frequent situation in which this test is used is in the comparison of a new production method with a standard method that has been in use for some time. The standard method is well understood and the final product has been continually inspected. It seems reasonable to assume that the population characteristics of the product are known. The question to be answered is usually of the type: Is the new method better in some respect (for example, longer life of an electric light bulb, a greater yield in agricultural experiments, or a greater yield from a commercial chemical reaction) than the method in current use?

The best estimate of the population mean that can be obtained from the sample was shown in the previous chapter to be the sample mean $\bar{x}$. The problem may be stated: Is the sample mean "significantly" greater than the population mean μ_p or, alternatively, may we reasonably expect to observe the actual value of $\bar{x}$ or greater than this value when the population mean is in fact equal to μ_p? Let us consider how the significance test method is applied to this situation.

If we obtain a sample of size n which has been drawn from a population with mean μ_p and known variance σ^2 the sample mean, $\bar{x}_1$, will be a "reasonable" value; that is, a value that has occurred when the hypothesis is true. If a second sample is drawn from the same population the sample mean, $\bar{x}_2$, will be a second "reasonable" value. If a number of such samples are drawn, a sample of means $\bar{x}_1, \bar{x}_2, ..., \bar{x}_r$ will be obtained and these will be a set of "reasonable" values that have occurred when the hypothesis is true. This sample will begin to provide a picture of the values of $\bar{x}$ that can reasonably occur when the assumed hypothesis is true. In a practical application we may determine by comparison with this "reasonable" set of values whether, or not, the observed value of $\bar{x}$ is a "reasonable" value. If an infinite number of such samples could be drawn the sample means would form the population distribution of "reasonable" values of $\bar{x}$. With this distribution it would be possible to compute the proportion of reasonable values of $\bar{x}$ that are greater than the observed sample mean when the population mean is equal to μ_p. If this proportion is sufficiently small we will reject the assumed hypothesis that the population mean is equal to μ_p. This probability is the probability of committing a type 1 error.

The above method of obtaining the distribution of $\bar{x}$ is clearly not a practical possibility. The method has been described to give the reader some understanding of the nature of the distribution of a statistic ($\bar{x}$) and the meaning of the proportion on which the significance decision is based. This distribution can be obtained by mathematical methods if a particular functional form is assumed for the distribution of the individual measurements. We have seen in Chapter 3 that if the individual measurements are normally distributed, with mean μ_p and variance σ^2, the sample mean has a normal distribution with mean μ_p and variance σ^2/n. That is,

$$\bar{x} \sim N(\mu_p, \sigma^2/n)$$

or, alternatively,

$$X = \frac{\bar{x} - \mu_p}{\sqrt{(\sigma^2/n)}} \sim N(0, 1)$$

and the probability that a greater value of $\bar{x}$ than the observed value $\bar{x}_0$ will occur when the population mean is μ_p is given by

$$P = \int_{X_0}^{\infty} e^{-\frac{1}{2}u^2} \, du,$$

where

$$X_0 = \frac{\bar{x}_0 - \mu_p}{\sqrt{(\sigma^2/n)}} .$$

This probability is shown diagramatically in Fig. 5.2.

If this probability is small we will reject the hypothesis that the population mean is equal to μ_p. If α is the highest risk we are prepared to take of committing a type 1 error the α per cent critical region consists of the values of $\bar{x}$ contained in the highest α per cent of the distribution of $\bar{x}$.

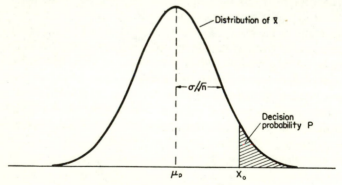

FIG. 5.2. Decision probability.

5.2.2. Example (Single-tailed)

A chemical plant has been producing a certain product B by a standard method for a number of years and a suggestion has been made that the yield may be improved by the addition of a certain quantity of a substance A. Ten experimental runs with substance A present have been performed and the following yields (in grams) were obtained:

$$235\cdot4 \quad 231\cdot3 \quad 222\cdot7 \quad 217\cdot9 \quad 214\cdot3$$

$$225\cdot4 \quad 234\cdot2 \quad 230\cdot1 \quad 221\cdot3 \quad 217\cdot4$$

The mean yield for the standard method is 218 g and the variance is $44\cdot52$. The question to be answered is: Does the addition of substance A to the reaction increase the yield of substance B?

The computation proceeds:

$$\bar{x} = 225\cdot0 \qquad n = 10$$

$$X = \frac{225\cdot0 - 218\cdot0}{\sqrt{(44\cdot52/10)}} = \frac{7\cdot0}{2\cdot11} = 3\cdot32$$

$$\Pr\{X \geqq 3\cdot32 \mid x_i \sim N(218, 44\cdot52)\} = 0\cdot0005.$$

If we reject the hypothesis that the population mean is 218 g we take a risk of $0\cdot0005$ of being wrong. If we are prepared to take this small risk of being wrong (remember—the smaller the risk we are prepared to take the less likely are we to detect a genuine difference) our decision will be to reject the hypothesis in favour of the alternative hypothesis that the yield of B is increased by the presence of A.

5.2.3. Example (Double-tailed)

The previous example asked the question: Is the new method better than the old method? Situations occur in which the question is asked: Is the new method different from the old method? In this double-tailed situation any difference is of interest, whether an improvement or not. If an α percentage critical region is to be used it will be made up of the lowest $\frac{1}{2}\alpha$ per cent and the highest $\frac{1}{2}\alpha$ per cent of the distribution.

A new method of producing pellets is to be compared with the method now in use. The new method is more economical but can only be accepted if it produces pellets of the same average length as the present method. A sample of pellets is produced by the new method and the following lengths (in cm) are obtained:

$$2\cdot50 \quad 2\cdot54 \quad 2\cdot67 \quad 2\cdot39 \quad 2\cdot40 \quad 2\cdot47 \quad 2\cdot46$$

The mean and variance of the lengths produced by the present method are assumed known to be

$$\mu = 2 \cdot 40$$

$$\sigma^2 = 0 \cdot 006294.$$

To test the hypothesis that the average length of pellets produced by the new method is also 2·40 we assume that the variance of lengths of pellets produced by the new method is the same as by the old method (this assumption could be tested by a test to be described in Section 5.3). We calculate

$$\bar{x} = 2 \cdot 49,$$

$$X = \frac{2 \cdot 49 - 2 \cdot 40}{\sqrt{(0 \cdot 006294/7)}} = 3 \cdot 00,$$

and

$$\Pr \{ X \geq 3 \cdot 00 \mid x_i \sim N(2 \cdot 40, \, 0 \cdot 006294) \} = 0 \cdot 0013.$$

Since a difference on either side of the hypothetical value of μ is of interest the critical region consists of both tails of the distribution, and the probability of a larger *difference* from μ is twice the probability given above, that is 0·0026.

5.2.4. Is the Population Mean Equal to μ_p (Variance Unknown)?

The test described above requires the population variance to be known but it may be modified to apply to the situation in which the variance is unknown. The modification consists of replacing the known variance σ^2 by an estimate $\hat{\sigma}_\nu^2$ with ν degrees of freedom. The estimate of σ^2 is normally obtained from the n sample measurements by the usual estimator

$$\hat{\sigma}_{n-1}^2 = \frac{1}{n-1} \sum_{i=1}^{n} (x_i - \bar{x})^2$$

although the test is still valid if the estimate is obtained from another source. The modified statistic is usually denoted by t and was first introduced by W. S. Gossett writing under the pen name of "Student". Thus:

$$t = \frac{\bar{x} - \mu_p}{\sqrt{(\sigma_\nu^2/n)}}.$$

The distribution of t (assuming the distribution of the sample measurements to be normal) depends on one parameter ν, the degrees of freedom associated with the estimate of variance, and has the mathematical form

$$p(t) \, dt = K \left(1 + \frac{t^2}{\nu} \right)^{-\frac{1}{2}(\nu+1)} dt,$$

where the constant K is

$$\frac{\Gamma(\tfrac{1}{2}(\nu + 1))}{\sqrt{(\nu\pi)}\,\Gamma(\tfrac{1}{2}\nu)}.$$

5.2.5. Properties of Student's t-distribution

1. The shape of the distribution is similar to that of the normal distribution.
2. The distribution is symmetric about $t = 0$.
3. As ν tends to infinity the t-distribution tends to the standardized normal distribution.

    (As ν increases the variance estimate becomes more precise and in the limit it is known and the statistic t is the same as the statistic X described above in Section 5.2.1. The approximation to the t-distribution by a normal distribution is good, for testing purposes, for $\nu \geq 30$.)
4. The mean value is zero.
5. The variance is $\nu/(\nu - 2)$ for $\nu \geq 3$.
6. Certain significance points of the t-distribution are given in Table 2 in the Tables Section. Tables of the t-distribution itself may be found in Pearson and Hartley (1958).

5.2.6. Example

It is possible that although the variance of the measurements obtained from the products in Examples 5.2.2 and 5.2.3 may be assumed known it is not reasonable to assume that this is the same as the variance obtained from the new method. (It is, of course, possible to test the hypothesis that the variance has remained the same and this would normally be done. Such a test is described in Section 5.3.) If this is the case the variance must be estimated and Student's t used instead of the statistic X. Testing the hypothesis that the mean yield obtained by the new method in Example 5.2.2 is 218 g under these new circumstances we have

$$\bar{x} = 225 \cdot 0,$$

$$\hat{\sigma}^2 = \frac{1}{9}\left[506750 \cdot 30 - 225 \cdot 0^2/10\right] = 55 \cdot 59,$$

$$t = \frac{225 \cdot 0 - 218 \cdot 0}{\sqrt{(55 \cdot 59/10)}} = \frac{7 \cdot 0}{2 \cdot 3577} = 2 \cdot 97,$$

and

$$\Pr\{t \geq 2\cdot97 \mid x_i \sim N(218, \sigma^2), \, \nu = 9\} = 0\cdot008.$$

If we are prepared to take the risk of 0·008 of being wrong we will reject the hypothesis that the mean yield is 218 g in favour of the alternative hypothesis that the addition of substance A increases the yield.

5.2.7. Confidence Limits

Having rejected the hypothesis that the population mean yield was 218 g in the above example we would normally ask what values is it reasonable to believe the population mean yield could be. An acceptable value would be, of course, the sample mean 225 g but there is clearly a range of values that are possible. The range of acceptable values, or confidence interval, may be obtained by repeated application of Student's t-test as described above. We apply the test in single-tail form for a range of values of the population mean using, for example, a 5 per cent critical region and obtain the results:

Population mean		$t =$		$p =$	
	= 218 g,	$t =$	2·97,	$p = 0\cdot008$	value rejected
	220 g,	$t =$	2·12,	$p = 0\cdot032$	value rejected
	222 g,	$t =$	1·27,	$p = 0\cdot118$	value accepted
	224 g,	$t =$	0·42,	$p = 0\cdot342$	value accepted
	225 g,	$t =$	0·00,	$p = 0\cdot500$	value accepted
	226 g,	$t =$	−0·42,	$p = 0\cdot342$	value accepted
	228 g,	$t =$	−1·27,	$p = 0\cdot118$	value accepted
	230 g,	$t =$	−2·12,	$p = 0\cdot032$	value rejected

The exact values of the population mean at which the decision changes may be found by repeating the test for values between 220 and 222 and between 228 and 230. These values will be found to be 220·68 and 229·32. The values of the population mean below 225 use the single-tailed test with critical region containing high values of t. The values above 225 use a critical region containing low values of t. Since a 0·05 probability is used as the critical region size for both limits there is a 0·90 probability that these limits enclose the true value. The probability associated with the confidence interval defined by confidence limits is called the confidence probability. It is an important part of the confidence limits approach to interval estimation that the confidence probability refers to the limits and not to the population mean. The statement "there is a 0·90 probability that the population mean lies within the confidence limits 220·68 and 229·32" is not a correct statement. If a second experiment were to be performed a second set of limits would be obtained, but the population mean would remain the same. The confidence probability may be interpreted practically to predict that if a large number of experiments were to be performed 90 per cent of the confidence limits obtained would enclose the population mean.

The above method of computation of confidence limits has been described to clarify their interpretation and to explain the general method. In this example the limits may be obtained directly by computing the values of the population mean which would be just significant at the chosen level of significance. The critical values of t using a 5 per cent critical region on both sides of the distribution and with nine degrees of freedom is $1 \cdot 833$. The $0 \cdot 90$ confidence limits may therefore be obtained by solving for μ in the equations:

$$\frac{255 - \mu}{\sqrt{(55 \cdot 59/10)}} = \pm 1 \cdot 833.$$

Limits are therefore given by

$$225 \pm 1 \cdot 833 \times \sqrt{(55 \cdot 59/10)}$$

and are found to be

$$220 \cdot 68 \quad \text{to} \quad 229 \cdot 32.$$

This direct method is only possible in situations where the distribution of the statistic (t in this case) does not change in shape as different values of the population parameter (mean μ in this case) are tried.

5.2.8. The Power of the Normal Test and the t-test

We have seen in Chapter 1 that if the conclusion from an experiment is to accept the null hypothesis of no difference, then this is only a meaningful result if we show that the experiment was able to detect a difference of a size that we would regard as important had such a difference existed. This information is provided by the power function of the test. The power function also plays an important part in choosing the sample size in the design stage of an experiment. We would choose that sample size which gives us a good chance of detecting a difference of important size if such a difference exists. In the case of the normal test described in Section 5.2.1 the power function depends on the significance level chosen in applying the test. In the case of the t-test described in Section 5.2.4 the power function also depends on the degrees of freedom associated with the variance estimate. In the experiment described in Examples 5.2.2 and 5.2.6 we tested the null hypothesis that the yield of substance B was the same with, or without, the addition of substance A. The alternative hypotheses of interest are that the yield of B is increased, by amounts represented by the symbol d, by the addition of substance A. The power function is a plot of the probability of rejecting the null hypothesis, when the alternative hypothesis that the yield is increased by d is true, against the value of d.

In the normal test case the calculation of these probabilities for various values of d is preceded by the calculation of that mean of a sample of ten observations that would be just significant at the 5 per cent level. That is, we solve for $\bar{x}$ in the equation

$$X = 1 \cdot 645 = \frac{\bar{x} - 218}{\sqrt{(44 \cdot 52/10)}}$$

and we obtain $\bar{x} = 221 \cdot 47$. Thus our testing procedure is to reject the null hypothesis if we observe a sample mean of $221 \cdot 47$, or greater. The distribution of X when the null hypothesis is true is $N(0,1)$, but when an alternative hypothesis is true the distribution of X is $N(d, 1)$. Thus the probability of a significant result is given by:

$$\Pr\left\{X' \geqq \frac{(221 \cdot 47 - 218 - d)}{4 \cdot 452} \mid X' \sim N(0,1)\right\}.$$

The first point on the power function is therefore seen to correspond to the null hypothesis. That is, when $d = 0$ this probability (or power) is equal to the significance level of the test. Probabilities computed in this manner for different values of d are recorded below and plotted in Fig. 5.3.

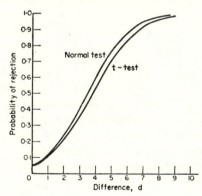

FIG. 5.3. Power function of the normal and t-tests (one-sample).

d	0·0	1·0	2·0	3·0	4·0	5·0	6·0	8·0
X	1·645	1·17	0·70	0·22	−0·25	−0·73	−1·20	−2·15
Probability	0·050	0·121	0·242	0·413	0·599	0·767	0·885	0·984

A similar procedure is carried out for the power function of the t-test. We first obtain that mean of a sample of ten observations that would be just significant at the 5 per cent level by solving for $\bar{x}$ in the equation:

$$t = 1\cdot833 = \frac{\bar{x} - 218}{\sqrt{(55\cdot59/10)}}.$$

From this we obtain $\bar{x} = 223\cdot2$. Our testing procedure is thus to reject the null hypothesis in favour of an alternative if we observe a sample mean equal to, or greater than, $223\cdot2$. The distribution of t under the null hypothesis is the t-distribution with $n - 1$ degrees of freedom as described in Section 5.2.4. Under the alternative hypothesis that the population mean is $218 + d$ the distribution of t is the non-central t-distribution depending on the two parameters v, the degrees of freedom, and δ, the non-centrality parameter. Figure 5.4 shows a sketch of a number of non-central t-distributions for different values of δ (a function of d to be discussed below). The proportions

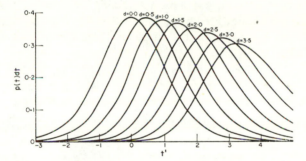

FIG. 5.4. Non-central t-distributions.

of these distributions to the right of the significance point $1\cdot833$ are there-quired probabilities of rejecting the null hypothesis when alternative hypotheses are true, and the power function consists of a plot of these probabilities against δ or d. Table 3 in the Tables Section gives non-central t probabilities required for power calculations for various values of δ and v and two values of t. The two values of t are different for each value of v and are chosen so that the probability of rejecting the null hypothesis when it is in fact true (that is the significance level of the test) is $0\cdot05$ and $0\cdot01$ respectively. A further complication arises in the computation of the power of the t-test which is not present in the normal test. In the normal case the population variance σ^2 is known whereas in the t-test it is not known. The power function of the t-test is, therefore, often plotted against δ, the non-centrality parameter, instead of against d, the difference between the population means under the alternative and null hypotheses. To interpret the power function, however, the value of σ^2 (or at least some estimate or reasonable guess) is required. A reasonable guess based on previous experience is usually adequate since power calculations are used as a basis for judging the adequacy of a proposed experiment rather than as an exact calculation. In fact one soon builds up a judgement based on the value of d measured in terms of the population standard deviation σ. For example, one may decide that an experiment is satisfactory if the probability of rejecting the null hypothesis when the population mean differs from the null hypothesis value by half the standard deviation is (say) 0·90. Such judgement can be built up consistently from experiment to experiment without consideration of the various population variances.

The relation between δ and d is

$$\delta = \frac{d\sqrt{n}}{\sigma}$$

and involves the unknown population standard deviation σ. The following power function is obtained from Table 3 at the end of the book and is given in terms of both δ and d using the value of σ used in the normal test. That is $\sigma = 6\cdot6723$ ($\sigma^2 = 44\cdot52$).

d	0·0	1·0	2·0	3·0	4·0	5·0	6·0	8·0
δ	0·00	0·47	0·95	1·42	1·90	2·37	2·84	3·79
Probability	0·050	0·114	0·221	0·371	0·542	0·706	0·835	0·967

This power function is plotted in Fig. 5.3.

5.3. IS THE POPULATION VARIANCE EQUAL TO A PARTICULAR VALUE?

The best estimate of the population variance obtained from a sample $\{x_1, x_2, x_3, ..., x_n\}$ has been shown in Chapter 4 to be given by:

$$\text{Estimate of } \sigma^2 = \hat{\sigma}^2 = \frac{1}{n-1} \sum_{i=1}^{n} (x_i - \bar{x})^2.$$

The hypothesis to be tested is that the population variance is equal to σ^2 or, alternatively, we seek to answer the question: Is the estimate $\hat{\sigma}^2$ significantly different from σ^2? The statistic most frequently used to test this hypothesis is

$$C = (n-1)\frac{\hat{\sigma}^2}{\sigma^2} = \frac{1}{\sigma^2} \sum_{i=1}^{n} (x_1 - \bar{x})^2.$$

We see that C is a constant times the ratio of the estimate to the hypothetical value. If the population distribution of the original measurements $\{x_i\}$ is normal the distribution of C may be shown to be a χ^2-distribution with $(n - 1)$ degrees of freedom. Most applications of this test are double-tailed, seeking to answer the question: Is the value of the estimate $\hat{\sigma}^2$ significantly lower or higher than the population variance σ^2? The α percentage critical region in a double-tailed test consists of the highest $\frac{1}{2}\alpha$ per cent of values and the lower $\frac{1}{2}\alpha$ per cent of values of χ^2. It should be noticed that the two parts of the critical region in this test are not symmetrically placed about the mean v. The 5 per cent critical region for $n = 10$ ($v = 9$) contains values less than 2·70 and values greater than 19·02. In single-tailed applications of this test the α percentage critical region consists of the extreme α per cent of values on one side of the distribution only. The examples discussed above, and further discussed below, provide an example of the double-tailed test.

5.3.1. Example

The problem discussed in Section 5.2.2 compared yields of a substance B produced by the new method with yields of B produced by a standard procedure. In Section 5.2.2 we rejected the hypothesis that the mean yield by the new method was 218 g—the mean yield obtained by the old method. It was assumed that the variance in the yield by the new method was equal to the known variance in the yield by the old method. (This assumption was relaxed in Section 5.2.4.) We may now test the validity of this assumption. We have

$$\hat{\sigma}^2 = 55\cdot59, \qquad v = 9$$

$$C = \frac{9 \times 55\cdot59}{44\cdot52} = 11\cdot24$$

$$\Pr\{C \geq 11\cdot24 \mid x_i \sim N(\mu, 44\cdot52)\} = \Pr\{\chi^2 \geq 11\cdot24 \mid v = 9\} = 0\cdot26.$$

If the population variance is equal to 44·52 there is a probability of 0·26 of observing a value of $\hat{\sigma}^2$ greater than 55·59. The value $\hat{\sigma}^2 = 55\cdot59$ is therefore contained in the $2 \times 0\cdot26$ per cent $= 0\cdot52$ per cent double-tailed critical region so that the assumption that the population variance is 44·52 is consistent with the observed results.

5.3.2. Confidence Limits for a Variance

It is possible, by repeated application of the above test, to obtain confidence limits for the population variance. Such limits are not often required but the possibility of their calculation is recorded here for completeness. These limits will not be symmetrical about the best estimate of the variance.

5.4. Is the Population Distribution of a Particular Form?

We have a sample of n observations $\{x_1, x_2, ..., x_n\}$ and we wish to test the hypothesis that the distribution of the population that this sample represents is of a particular form. The most frequent application is to test the hypothesis that the population distribution is normal and one of the examples given below tests this hypothesis. The method to be described is, however, a general test that may be used to test the adequacy of any particular distribution as the population description.

5.4.1. Is the Population Distribution Normal?

The method to be described requires a large number of measurements for the test to be affective. The procedure, stated briefly, consists of dividing the measurement scale into a number of groups, counting the numbers of observations that fall in each group, and comparing these observed frequencies with those calculated on the assumption that the distribution is normal. The procedure will be described in detail with reference to the following example. The lengths of 200 pellets were measured and form the sample. The length scale was divided into nine groups and the numbers of pellets with length within each of the groups were counted. The frequencies obtained are given in Table 5.2 and Fig. 5.5 shows a histogram of the sample distribution.

TABLE 5.2. PELLETS SAMPLE DISTRIBUTION

Length (cm)	Number of pellets
Less than 3·4	10
3·4 to 3·6	15
3·6 to 3·8	23
3·8 to 4·0	34
4·0 to 4·2	36
4·2 to 4·4	27
4·4 to 4·6	28
4·6 to 4·8	14
More than 4·8	13
Total	200

If the population of lengths of pellets is normal with known values of the mean and variance it is possible to compute, as described in Chapter 3, the proportions of lengths within each of the groups listed above. If these proportions are each multiplied by 200 we will obtain an ideal or "expected" set of frequencies which will be in good agreement with the observed set of frequencies if the population distribution is normal. The values of the

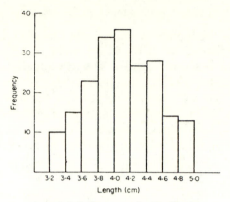

Fig. 5.5. Pellets sample distribution.

mean and variance to be used in the calculation of the expected frequencies may be estimated from the data in the usual manner, or they may be part of the hypothesis being tested. In the example problem these values were estimated from the data:

$$\hat{\mu} = \bar{x} = 4 \cdot 112$$

$$\hat{\sigma}^2 = 0 \cdot 1754$$

$$\hat{\sigma} = 0 \cdot 419.$$

The computation consists of calculating the values of X for a standardized normal distribution that correspond to the boundaries given (i.e. 3·4, 3·6, ..., 4·8), looking up the proportion of the normal curve to the left of these boundaries, taking the differences to obtain the proportions within each group, and multiplying these proportions by the sample size. The detailed calculations are given in Table 5.3.

TABLE 5.3. COMPUTATION OF EXPECTED FREQUENCIES

Boundary B	Standardized boundary	Cumulative proportion	Proportion	Expected frequencies	Observed frequencies
3·4	−1·699	0·0447	0·0447	8·94	10
3·6	−1·222	0·1090	0·0643	12·86	15
3·8	−0·745	0·2281	0·1191	23·82	23
4·0	−0·267	0·3947	0·1666	33·32	34
4·2	0·210	0·5832	0·1885	37·70	36
4·4	0·687	0·7540	0·1708	34·16	27
4·6	1·165	0·8780	0·1240	24·80	28
4·8	1·642	0·9497	0·0717	14·34	14
			0·0503	10·06	13
Totals			1·0000	200·00	200

5.4.2. A Goodness of Fit Test

If the population distribution is normal the observed and expected frequencies will agree reasonably. It is often possible to judge bad agreement and good agreement between these two sets of frequencies by eye, but for other cases some measure of agreement is needed. There is a number of possible measures that we may use, of course, but the usual measure is ϕ^2 as defined below. We will denote by O_i the ith observed frequency, and by E_i the ith expected frequency. Agreement between an observed set and an expected set of frequencies may be measured by calculating:

$$\phi^2 = \sum_{i=1}^{k} \frac{(O_i - E_i)^2}{E_i}$$

where k is the number of expected frequencies. If the agreement between the two sets is good the value of ϕ^2 is small, and if the agreement is bad the value of ϕ^2 is large. This test is a single-tailed test since interest lies in large values of ϕ^2 only. The α per cent critical region therefore consists of the α per cent of large values of ϕ^2. The problem remains to calculate the distribution of ϕ^2 and hence obtain the range of values of ϕ^2 forming the critical region.

It may be shown, without any assumptions involving the observed frequencies (other than that they are frequencies), that the distribution of ϕ^2 is approximately a χ^2-distribution. The degrees of freedom appropriate to a particular application depend on the number of groups k and the number of necessary relations r that exist between the two sets of frequencies. The degrees of freedom are $(k - r)$. In the example discussed above there are six degrees of freedom since there are nine groups and the following three necessary relations:

1. The total observed frequency equals the total expected frequency. (This relation is present in most applications of ϕ^2.)
2. The mean of the observed sample distribution equals the mean of the expected distribution $\bar{x} = \mu$.
3. The variance of the observed sample distribution equals the variance of the expected distribution $\hat{\sigma}^2 = \sigma^2$.

It is stressed that the distribution of ϕ^2 is approximately a χ^2-distribution and that the approximation may break down if low expected frequencies are present in some groups. There is a certain amount of disagreement in the literature on the subject of how low an expected value must be before it is regarded as "too low" but the most commonly quoted value is 5. If low expected frequencies occur the normal practice is to combine two or more adjacent groups to obtain higher values. This, of course, reduces the number of groups and hence the degrees of freedom.

The actual value of ϕ^2 in the above example is

$$\phi^2 = \frac{(8\cdot94 - 10)^2}{8\cdot94} + \frac{(12\cdot86 - 15)^2}{12\cdot86} + \cdots + \frac{(10\cdot06 - 13)^2}{10\cdot06} = 3\cdot38,$$

with

$$\nu = 6,$$

and

$$\Pr\{\chi^2 \geq 3\cdot38 \mid \nu = 6\} = 0\cdot75.$$

It is reasonable to believe, therefore, that the population distribution is normal. The mean and variance have been estimated as

$$\hat{\mu} = 4\cdot112 \quad \text{and} \quad \hat{\sigma}^2 = 0\cdot1754$$

so that the distribution $N(4\cdot112, 0\cdot1754)$ is taken to be a reasonable description of the lengths of pellets produced in the same way as those in the sample.

5.4.3. Is the Population Normal with a Particular Mean and Variance?

Testing the hypothesis of normality with particular values of the mean and variance is performed as described in Sections 5.4.1 and 5.4.2 above but with the following differences:

1. The particular mean and variance are used in place of the estimates
2. Relations 2 and 3 between the observed and expected frequencies do not exist so that the degrees of freedom are $k - 1$ instead of $k - 3$.

5.4.4. Estimation of Moments from Sample Data Presented as a Histogram

In some experimental situations instead of knowing the exact data values we have the data in the form of a histogram. If the mid-point of the ith histogram interval is x_i and there are n_i observations in the ith group we may estimate the mean and the central moments by the following expressions:

$$\hat{\mu}'_1 = \bar{x} = \frac{1}{N} \sum_{i=1}^{k} n_i x_i$$

$$\hat{\mu}_2 = \hat{\sigma}^2 = \frac{1}{N-1} \sum_{i=1}^{k} n_i(x_i - \bar{x})^2 = \frac{1}{N-1}\left[\sum_{i=1}^{k} n_i x_i^2 - \frac{1}{N}\left(\sum_{i=1}^{k} n_i x_i\right)^2 \right]$$

$$\hat{\mu}_r = \frac{1}{N-1} \sum_{i=1}^{k} n_i(x_i - \bar{x})^r$$

where there are k groups and where $N = \sum_{i=1}^{k} n_i$.

These expressions assume that each value in the ith group is equal to the mid-point of the group. Sheppard has pointed out that a correction to the moments given by these expressions is necessary since the mid-point of each interval is used instead of the centre of gravity. No correction is required for the odd moments. If the widths of the histogram intervals are all equal to h the corrections to the variance and the fourth central moments are:

$$\hat{\hat{\mu}}_2 = \hat{\mu}_2 - \frac{1}{12} h^2$$

$$\hat{\hat{\mu}} = \hat{\mu}_4 - \frac{1}{2} \hat{\mu}_2 h^2 + \frac{7}{240} h^4.$$

5.4.5. Probability Plotting

A second method available for investigating the normality of the population distribution consists of plotting the data on probability paper and judging whether, or not, the points lie approximately on a straight line. Such judgement should give more weight to the centre points than the extremes. An example of a probability plot is given in Fig. 5.6 and the data is that given in Table 5.3 and in rearranged form in Table 5.4.

TABLE 5.4. REARRANGED DATA FOR PROBABILITY PLOT

Boundary	3·4	3·6	3·8	4·0	4·2	4·4	4·6	4·8
Cumulative frequency	10	25	48	82	118	145	173	187
Cumulative proportion	0·050	0·125	0·240	0·410	0·590	0·725	0·865	0·935

The data to be plotted in the large sample situation is arranged in histogram form and we plot the proportion of observations less than each of the boundaries, that is, the cumulative proportions against the boundaries. Probability paper has a linear scale on which the boundaries are recorded and a non-linear scale of proportions (or percentages, or even sometimes 1000 times the proportion) on which we plot the cumulative proportions. The rearranged data in Table 5.4 shows the boundaries and the cumulative proportions.

For small samples it is more accurate to plot the individual observations against the ordinates $i/(n + 1)$ for $i = 1, 2, ..., n$. That is, if the sample size is ten we divide the probability scale into $n + 1$ equal intervals and plot the observations against the ten inner boundaries $1/11, 2/11, ..., 10/11$.

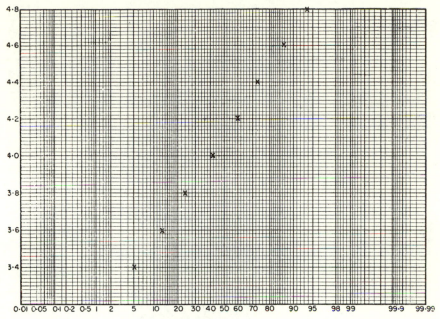

FIG. 5.6. Probability plot of pellet data.

This method provides a rough guide as to the adequacy of the assumption of normality for the population distribution. One criticism of this procedure is that no formal test of straightness exists at present. The judgement of straightness must of course take into account the sample size so that two identical plots for different sample sizes could produce different conclusions. Provided that these difficulties are recognized and no precise conclusions are drawn from a plot the method can be very useful as a guide to normality.

5.4.6. Is the Population Distribution Binomial?

Section 5.1 discussed the testing of the hypothesis that the probability of "heads" at the toss of a coin is one-half or less than one-half and that the separate tosses are independent. Section 5.1.1 introduced the double-tailed version of the test testing the hypothesis that the probability of "heads" is one-half. The test decision was based on the outcome of tossing eight coins once each. An alternative method of testing this hypothesis provides a second example of a population distribution test.

If eight coins are tossed a number of times and the frequencies with which 0, 1, ..., 8 heads occur are recorded, a sample distribution is obtained. The corresponding expected frequencies are obtained by multiplying the binomial probabilities given in Section 5.1 by the number of times the coins are tossed.

Eight coins have been tossed 256 times and the results obtained are given in Table 5.5.

TABLE 5.5. OBSERVED AND EXPECTED BINOMIAL DISTRIBUTIONS

Number of heads	0	1	2	3	4	5	6	7	8	Totals
Observed frequency	0	2	26	50	96	55	23	3	1	256
Expected frequency	1	8	28	56	70	56	28	8	1	256

The expected frequencies for 0 and 8 heads are low and will be combined with the expected frequencies for 1 and 7 heads, respectively, to give the amended distributions in Table 5.6.

TABLE 5.6. ADJUSTED DISTRIBUTIONS

Number of heads	0 and 1	2	3	4	5	6	7 and 8	Totals
Observed frequencies	2	26	50	96	55	23	4	256
Expected frequencies	9	28	56	70	56	28	9	256

It should be noted that this restriction applies to the expected frequencies and not to the observed frequencies. The observed frequencies may take any values including zero. The degrees of freedom in this application are six since there are seven groups, or frequencies, and one restriction on the totals. If the probability of "heads" had been estimated from the data instead of being part of the hypothesis, the degrees of freedom would be reduced, by one, to five. To measure the agreement we compute

$$\phi^2 = \frac{(2 - 9)^2}{9} + \frac{(26 - 28)^2}{28} + \cdots + \frac{(4 - 9)^2}{9}$$

$$= 19 \cdot 58$$

with

$$\nu = 6$$

and

$$\Pr\{\chi^2 \geqq 19 \cdot 58 \mid p = \tfrac{1}{2}, \quad \nu = 6\} = 0 \cdot 004.$$

If we are prepared to take this risk of being wrong we will reject the hypothesis that the probability of heads is one-half and that the trials are independent.

It is interesting to note in this example that there seems to be a preponderance of trials in which four heads were observed at the expense of extreme numbers of heads. The value of the probability of heads estimated from this data as described in Chapter 4 is $\hat{p} = 0 \cdot 5029$ and this value is close to

$p = \frac{1}{2}$. It would seem, therefore, that since the estimate is so close to the hypothetical value of $\frac{1}{2}$ that the discrepancy is caused by lack of independence between the trials. The variance of p is estimated as 0·02209 and the fact that this value is less (the significance of this difference could be tested by a suitably designed test) than the theoretical value of $p(1 - p)/n = 1/32 = 0·03125$ also suggests lack of independence between the trials.

QUESTIONS

1. A woman believes that she can tell the difference between hot tea made by putting milk into the cup first and adding tea to it and that made by adding milk to tea already in the cup. The woman has agreed to test her claim by tasting ten cups of tea and identifying the method of preparation. If her answers are merely guesses there is a probability of $\frac{1}{2}$ that she is correct in a guess.

(a) Write down the probabilities that she makes 0, 1, 2, ..., 10 correct replies.

(b) What is the null hypothesis you would test in order to establish, or deny, the truth of her claim?

(c) How would you divide the possible number of correct replies into a critical region and an acceptance region?

(d) If the woman makes nine correct replies would you accept her claim?

(e) What is the probability that you are wrong if you accept her claim when she makes nine correct replies?

(f) If her probability of a correct reply is 0·6 what is the probability that the number of correct replies will fall in your critical region?

(g) Draw the power function for your test.

2. (a) Estimate the population mean with 95 per cent confidence limits from the following sample of twelve results, assuming the population variance is known to be 0·0042:

$$10·023 \quad 10·142 \quad 10·173 \quad 10·098 \quad 10·001 \quad 10·072$$
$$10·204 \quad 10·149 \quad 10·192 \quad 10·103 \quad 10·112 \quad 10·087$$

(b) Repeat for the situation in which the population variance is not known.

(c) What assumptions have you made in answering this question?

3. A theoretical model of the breaking strengths of samples of a particular metal predicts that the breaking strength of a sample with a particular shape and size should be 523 lb. Ten samples of the shape and size are tested and the breaking strengths are:

$$438 \quad 491 \quad 528 \quad 543 \quad 490 \quad 465 \quad 506 \quad 517 \quad 454 \quad 532$$

(a) Are these values consistent with the prediction?

(b) If we know that the population variance is 810 how is your test altered and what is your decision now?

(c) Is the sample variance significantly greater than 810?

4. The lengths of the first leaf on each of 250 plants were measured, in inches, and grouped in the following table:

Length	3	4	5	6	7	8	9	10	Total
Frequency	3	19	34	42	60	49	28	13	2	250

(a) Draw the histogram and plot the data on probability paper. Are you satisfied that the population distribution is normal?

(b) Estimate the population mean and variance.

(c) Test the normality of the distribution formally by fitting a normal distribution to the sample histogram and test the adequacy of the description.

5. To test a method of producing random digits 1000 digits are produced and the numbers of times each of the digits 0–9 appear are counted. The counts are as follows:

Digit	0	1	2	3	4	5	6	7	8	9	Total
Frequency	104	91	82	114	131	95	86	101	92	104	1000

Is it reasonable to believe that each digit appears equally often in the population of all possible random digits produced by this method?

6. Compare the observed and expected frequencies computed in question 2 after Chapter 3 and decide whether, or not, it is reasonable to believe that the population distribution is normal.

CHAPTER 6

TESTS OF SIGNIFICANCE—II

THIS chapter presents further tests of significance concerned with comparing properties of different populations.

The following questions will be considered:

1. Are the means of two populations equal?
2. Are the variances of two populations equal?
3. Are the variances of more than two populations equal?

It is assumed in applying these tests as described here that the population distributions are normal. Section 7.6 at the end of the next chapter discusses the importance of the normality assumption in these tests as well as those described in Chapter 5.

6.1. ARE THE MEANS OF TWO POPULATIONS EQUAL?

We will denote the population means by μ_1 and μ_2 and the population variances by σ_1^2 and σ_2^2. The data consists of two samples of observations $\{x_1, x_2, ..., x_n\}$ and $\{y_1, y_2, ..., y_m\}$ and the hypothesis to be tested is that $\mu_1 = \mu_2$. Three different situations will be considered.

1. The variances of σ_1^2 and σ_2^2 are known.
2. The variances σ_1^2 and σ_2^2 are unknown but may be assumed to be equal.
3. The variances σ_1^2 and σ_2^2 are unknown and may not be assumed to be equal.

6.1.1. Are the Means of Two Populations Equal (Variances Known)?

The best estimates of μ_1 and μ_2 that can be obtained from this data are $\bar{x}$ and $\bar{y}$, respectively, the two sample means. If the null hypothesis that $\mu_1 = \mu_2$ is true $\bar{x} - \bar{y}$ has a distribution with zero mean and variance $\sigma_1^2/n + \sigma_2^2/m$ since the variance of $\bar{x}$ is σ_1^2/n, the variance of $\bar{y}$ is σ_2^2/m and the variance of the difference $\bar{x} - \bar{y}$ is the sum of these variances.

The statistic chosen to measure agreement between μ_1 and μ_2 is, therefore:

$$X = \frac{\bar{x} - \bar{y}}{(\sigma_1^2/n + \sigma_2^2/m)^{\frac{1}{2}}}.$$

The distribution of X depends on the distributions of the variables x and y. If we assume these to be normal the distribution of X is a standardized normal distribution $N(0, 1)$. The derivation written formally is:

(a) $x_i \sim N(\mu_1, \sigma_1^2)$ for $i = 1, 2, ..., n$

   $y_j \sim N(\mu_2, \sigma_2^2)$ for $j = 1, 2, ..., m$

(b) $\bar{x} \sim N(\mu_1, \sigma_1^2/n)$ where $\bar{x} = \dfrac{1}{n} \sum_{i=1}^{n} x_i$

   $\bar{y} \sim N(\mu_2, \sigma_2^2/m)$ where $\bar{y} = \dfrac{1}{m} \sum_{j=1}^{m} y_j$.

(c)   thus, $(\bar{x} - \bar{y}) \sim N(\mu_1 - \mu_2, \; \sigma_1^2/n + \sigma_2^2/m)$

(d)   and if $\mu_1 = \mu_2$

$$X = \frac{\bar{x} - \bar{y}}{(\sigma_1^2/n + \sigma_2^2/m)^{\frac{1}{2}}} \sim N(0, 1). \tag{6.1}$$

6.1.2. Example

An experiment has been performed to investigate the possible catalytic effect of the presence of a substance A in a chemical reaction. The final product of the reaction is a substance B and the presence of A will be regarded as useful if it increases the yield of B. The yields of B from eight separate reactions with A present and the yields of B from ten reactions with A absent have been determined. The order in which the eighteen determinations were made was chosen at random and the results are given in Table 6.1 and plotted in Fig. 6.1.

TABLE 6.1. EXAMPLE TWO-SAMPLE DATA, I

With A	304·1, 318·8, 292·7, 284·9, 309·3, 321·6, 318·1, 296·1
Without A	304·3, 298·2, 282·9, 273·1, 281·9, 301·4, 298·2, 288·1, 278·9, 290·0

The hypothesis to be tested is that the presence of A increases the yield of B. We will assume that the population variances are both equal to 150 and that the population distributions are normal.

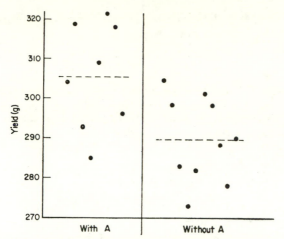

FIG. 6.1. Plot of example two-sample data, I.

The two sample means are

$$\bar{x} = 305.7 \qquad \bar{y} = 289.7.$$

Hence

$$X = \frac{305.7 - 289.7}{(150/8 + 150/10)^{\frac{1}{2}}} = \frac{16.0}{5.81}$$

$$= 2.75$$

and

$$\Pr\{X \geq 2.75 \mid X \sim N(0, 1)\} = 0.003.$$

We will almost certainly regard this probability as small and reject the hypothesis that the means of the two populations are the same in favour of the alternative hypothesis that the presence of A increases the yield of B.

6.1.3. Are the Means of Two Populations Equal (Variances Unknown but Equal)?

The above test assumed that the population variances were known. If the population variances are not known, but may be assumed equal, a version of Student's t-test may be used to test for equality of the population means. The variance assumption may be tested by the test described in Section 6.2. If this assumption is not valid the t-test should not be used. Tests are available for the unequal variance situation and one is described in Section 6.1.5.

The statistic used to measure agreement between sample means is defined as follows:

$$t = \frac{\bar{x} - \bar{y}}{(\hat{\sigma}^2/n + \hat{\sigma}^2/m)^{\frac{1}{2}}}, \tag{6.2}$$

where $\hat{\sigma}^2$ is an estimate of the common variance of the two populations. This is usually estimated from the two samples as:

$$\hat{\sigma}^2 = \frac{\sum\limits_{i=1}^{n} (x_i - \bar{x})^2 + \sum\limits_{j=1}^{m} (y_j - \bar{y})^2}{n + m - 2}. \tag{6.3}$$

If the two populations have normal distributions with common variance then t has a Student's t-distribution with $n + m - 2$ degrees of freedom.

6.1.4. Example

Let us now repeat the analysis of the data given above in Section 6.1.2 and relax the assumption that the population variances are known. We will retain the assumption that the population distributions are normal, and further assume that the population variances are equal. The computation proceeds as follows:

$$\sum_{i=1}^{8} (x_i - \bar{x})^2 = 1287{\cdot}50; \qquad \sum_{j=1}^{10} (y_j - \bar{y})^2 = 996{\cdot}48$$

$$\hat{\sigma}^2 = \frac{1}{16}(1287{\cdot}50 + 996{\cdot}48) = 142{\cdot}74$$

$$t = \frac{305{\cdot}7 - 289{\cdot}7}{\left[142{\cdot}74\left(\dfrac{1}{8} + \dfrac{1}{10}\right)\right]^{\frac{1}{2}}} = \frac{16{\cdot}0}{5{\cdot}67}$$

$$= 2{\cdot}82$$

$$\Pr\{t \geq 2{\cdot}82 \mid \nu = 16\} = 0{\cdot}006.$$

From this small probability we will almost certainly conclude that the addition of substance A increases the yield of B.

6.1.5. Are the Means of Two Populations Equal (Variances Unknown and Unequal)?

Student's t-test as applied in Section 6.1.3 makes the assumption that the two population variances are equal and may give misleading results if this assumption is not valid. Two tests have been proposed for this situation where the population variances are believed to be different. Details of one of these tests, the Fisher–Behren's test, may be found in Fisher and Yates (1953) and the test due to Welch (1947) is described below.

The variance of $\bar{x} - \bar{y}$ is $\sigma_1^2/n + \sigma_2^2/m$, where σ_1^2 and σ_2^2 are the variances of the two populations, and the ratio Z, defined below, no longer has a t-distribution.

The statistic Z, a natural extension of t, is defined by

$$Z = \frac{\bar{x} - \bar{y}}{(\hat{\sigma}_1^2/n + \hat{\sigma}_2^2/m)^{\frac{1}{2}}},$$

where $\hat{\sigma}_1^2$ and $\hat{\sigma}_2^2$ are the usual estimates of σ_1^2 and σ_2^2.

The distribution of Z assuming $\mu_1 = \mu_2$ and that both population distributions are normal depends on two numbers of degrees of freedom ν_1 and ν_2 and a ratio C. The degrees of freedom are those associated with the variance estimates $\hat{\sigma}_1^2$ and $\hat{\sigma}_2^2$ (that is, $\nu_1 = n - 1$ and $\nu_2 = m - 1$, usually) and the ratio C is defined as

$$C = \frac{\hat{\sigma}_1^2/n}{\hat{\sigma}_1^2/n + \hat{\sigma}_2^2/m}.$$

Five and one per cent points of the distribution of Z are given in Table 11 in Pearson and Hartley (1958) and Tables 5 in the Tables Section. In these tables the letter v is used instead of the letter Z used above.

6.1.6. Example

We will apply this test to the data given in Section 6.1.2 and relax the condition that $\sigma_1^2 = \sigma_2^2$ (although this assumption would appear to be justified on this data).

$$\hat{\sigma}_1^2 = 183\cdot93; \quad n = 8; \quad \nu_1 = 7$$

$$\hat{\sigma}_2^2 = 110\cdot72; \quad m = 10; \quad \nu_2 = 9$$

$$C = \frac{183\cdot93/8}{183\cdot93/8 + 110\cdot72/10} = 0\cdot674$$

$$Z = \frac{305\cdot7 - 289\cdot7}{\{183\cdot93/8 + 110\cdot72/10\}^{\frac{1}{2}}} = \frac{16\cdot0}{5\cdot84}$$

$$Z = 2\cdot74$$

$$\Pr\{Z \geqq 2\cdot74 \mid C = 0\cdot674, \quad \nu_1 = 7, \quad \nu_2 = 9\} < 0\cdot05.$$

If we are prepared to take a risk of at most 0·05 of being wrong we will reject the hypothesis that the two population means are the same.

6.1.7. The Correlated t-test

This form of the t-test is used in situations where there are equal numbers of observations in the two samples and where there is correspondence between the ith member of each sample. In Section 1.2.4 the principle of grouping observations was discussed and it was suggested that the observations of the chemical reactions should be made in pairs, or blocks, each consisting of one reaction with A and one reaction without A. Let us suppose that eight such blocks were performed and the yields obtained were the first eight results in each sample in the data given above in Section 6.1.2. That is, the yields are as given in Table 6.2.

TABLE 6.2. EXAMPLE TWO-SAMPLE DATA, II

Pair	1	2	3	4	5	6	7	8
With A	304·1	318·8	292·7	284·9	309·3	321·6	318·1	296·1
Without A	304·3	298·2	282·9	273·1	281·9	301·4	298·2	288·1

In the original design in Section 6.1.2 the order in which the eighteen yields were collected was chosen at random so that no connections existed between observations from different samples. In the present design the yields are determined in pairs so that the ith members of both samples are collected as near together in time as possible. By grouping the reactions into eight pairs we ensure that the comparisons between the two types of reaction are made under as near identical conditions as possible. However, if we perform the t-test as described in Section 6.1.3 on this data we will gain nothing from the grouping since the estimate of variance $\hat{\sigma}^2$ we obtain by this method will include variability due to the changing conditions. To gain from such a design we require some means of estimating the variability that is independent (or as nearly so as possible) of the changing conditions. Such an estimate is obtained in the correlated t-test by assuming that the ith observation from each sample is made up of an effect due to the particular type of reaction, an effect due to the conditions prevailing at the time the pair of observations were taken, and a random error. That is if x_i and y_i are the ith observations of the first and second samples

$$x_i = \mu_1 + C_i + Z_i$$
$$y_i = \mu_2 + C_i + Z_i'.$$

If we perform the t-test as described in Section 6.1.3 we will include the condition terms C_i in the estimate of error. If we take the differences:

$$d_i = x_i - y_i = \mu_1 - \mu_2 + Z_i - Z_i'$$

we obtain a single sample of observations d_i which do not include the condition terms C_i.

By this means the two sample test has become a one sample test testing the hypothesis that the values d_i have been randomly selected from a population of differences with zero mean and unknown variance σ_d^2. This form of the t-test has already been described in Section 5.2.4. Thus,

$$t = \frac{\bar{d} - 0}{(\hat{\sigma}_d^2/n)^{\frac{1}{2}}} \tag{6.4}$$

has a t-distribution with $n - 1$ degrees of freedom, where n is the size of both samples, where

$$\bar{d} = \text{the mean difference} = \frac{1}{n} \sum_{i=1}^{n} d_i = \bar{x} - \bar{y}$$

and

$$\hat{\sigma}_d^2 = \frac{1}{n - 1} \sum_{i=1}^{n} (d_i - \bar{d})^2.$$

The degrees of freedom have been reduced from $2n - 2$ (that is $n + m - 2$) to $n - 1$ by the pairing of observations, but the estimate of variability $\hat{\sigma}_d^2$ does not include the conditions terms C_i. The variance estimate of the mean difference, $\hat{\sigma}_d^2/n$, may be expected to be smaller than the corresponding variance estimate, $\hat{\sigma}^2(1/n + 1/n)$, in the uncorrelated t-test described in Section 6.1.3.

6.1.8. Example

Let us reconsider the example data and suppose that the first eight results in each sample had been obtained in pairs as described above.

TABLE 6.3. EXAMPLE TWO-SAMPLE CORRELATED DATA

Pair	1	2	3	4	5	6	7	8	i
1st Sample	304·1	318·8	292·7	284·9	309·3	321·6	318·1	296·1	x_i
2nd Sample	304·3	298·2	282·9	273·1	281·9	301·4	298·2	288·1	y_i
Differences	−0·2	20·6	9·8	11·8	27·4	20·2	19·9	8·0	d_i

$$\bar{d} = 14{\cdot}69 \qquad \hat{\sigma}_d^2 = 78{\cdot}95$$

$$t = \frac{14{\cdot}69}{(78{\cdot}95/8)^{\frac{1}{2}}} = \frac{14{\cdot}69}{3{\cdot}141}$$

$$= 4{\cdot}68$$

$$\Pr\{t \geq 4{\cdot}68 \mid \nu = 7\} = 0{\cdot}0011.$$

Thus, we would draw the same conclusions from this test as we have drawn from the other tests applied to this data.

The value of $\hat{\sigma}^2$ obtained in the application of the uncorrelated t-test (Section 6.1.3) to this data was 142·74 (this contains two more observations than in the present example). The value obtained in this section was $\hat{\sigma}_d^2 = 78\cdot95$ but this is an estimate of the variance of a difference between two observations and is, therefore, an estimate of twice the variance of a single observation. Thus the grouping has reduced the variance of an individual observation from 142·74 to 39·48. The value of t has been increased from 2·82 to 4·68 and in spite of a reduction in degrees of freedom the significance probability has been lowered from 0·006 to 0·0011. The grouping in this example has had the desired effect of reducing the variability estimate against which the significance of the difference between the sample means is judged. The reduction in variance has more than compensated for the loss of degrees of freedom. Other examples of the effect of grouping are given in later chapters.

6.1.9. Power of the Two-sample Normal Test and the t-test

The concept of a power function has been introduced in Section 5.1.2. The power function is a plot of the probability of rejecting the null hypothesis when the population means differ by an amount d assuming a single-tailed test) against the difference d. The power function enables us to determine if the experiment has a good chance of detecting a difference of important size should such a difference exist. If the experiment is performed and does not reject the null hypothesis we will feel confident in accepting such a conclusion only if we have established that the experiment was powerful enough to detect such a difference had one existed. Saying no is not meaningful when there was very little chance of saying yes. The computation is similar to that described in Section 5.2.8. In the case of the normal test described in Section 6.1.1 the power function depends on the significance level chosen in applying the test, whilst in the case of the t-test described in Section 6.1.3 it depends on the significance level and the degrees of freedom associated with the estimate of variance.

Computing the difference between sample means (assuming a one-sided test) that would be just significant at the chosen level of significance for the normal test we solve for $(\bar{x} - \bar{y})$ in the equation:

$$X = 1\cdot645 = \frac{\bar{x} - \bar{y}}{\sqrt{(\sigma_1^2/n + \sigma_2^2/m)}}.$$

For the example data in Section 6.1.2 we have:

$$1\cdot645 = \frac{\bar{x} - \bar{y}}{\sqrt{(150/8 + 150/10)}}.$$

That is, $(\bar{x} - \bar{y}) = 9 \cdot 56$. Thus our testing procedure is to reject the null hypothesis if we observe $(\bar{x} - \bar{y})$ to be equal to, or greater than, $9 \cdot 56$. The distribution of X under the null hypothesis and alternative hypothesis are $N(0, 1)$ and $N(d, 1)$, respectively. Thus the probability of a significant result is given by:

$$\Pr\{\bar{x} - \bar{y} \geqq 9 \cdot 56 \mid X \sim N(d, 1)\} =$$

$$\Pr\left\{X' \geqq 1 \cdot 645 - \frac{d}{\sigma}\sqrt{\left(\frac{nm}{n+m}\right)}\,\middle|\, X' \sim N(0, 1)\right\}.$$

Again it is clear that when $d = 0$ this probability is equal to the significance level of the test. Probabilities computed in this manner for different values of d are recorded below and plotted in Fig. 6.2.

d	0·0	2·0	4·0	6·0	8·0	10·0	12·0	16·0
X	1·645	1·30	0·96	0·61	0·27	−0·08	−0·42	1·11
Probability	0·050	0·097	0·169	0·271	0·394	0·532	0·663	0·867

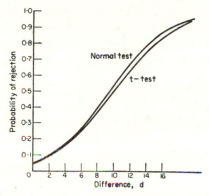

FIG. 6.2. Power functions of the normal and t-tests (two-sample).

For the t-test we compute the critical value of $(\bar{x} - \bar{y})$ that would be just significant at the chosen level of significance (5 per cent) by solving for $(\bar{x} - \bar{y})$ in the equation:

$$t = \frac{\bar{x} - \bar{y}}{\sqrt{(\hat{\sigma}^2/n + \hat{\sigma}^2/m)}} = 1 \cdot 746.$$

From this we obtain $(\bar{x} - \bar{y}) = 9 \cdot 90$ for the example data of Section 6.1.2. Our testing procedure is thus to reject the null hypothesis in favour of the alternative if we observe a value of $(\bar{x} - \bar{y})$ equal to, or greater than, $9 \cdot 90$. The distribution of t under the null hypothesis and alternative hypothesis is the t-distribution with $m + n - 2$ degrees of freedom and the non-central t-distribution with $n + m - 2$ degrees of freedom and non-centrality parameter δ given by:

$$\delta = \frac{d}{\hat\sigma} \sqrt{\left(\frac{nm}{n + m}\right)}.$$

This relation, as in Section 5.2.8, involves the unknown population variance σ^2 and we must plot the power function in units of σ instead of absolute units. To translate this into absolute terms some estimate or informed guess is necessary for σ. This could be based on previous experience or a trial set of observations. This question of the interpretation of the scale of the power function is discussed in Section 5.2.8. A sketch of certain non-central t-distributions was given in Fig. 5.4 and the proportions of these distributions to the right of the significance point ($1\cdot746$ in this case) are the required probabilities of rejecting the null hypothesis when alternative hypotheses are true. The power function consists of a plot of these probabilities against δ or d, and Table 3 at the end of the book gives non-central t probabilities required for power calculations for various values of ν and δ and for two values of t chosen so that the probability of rejecting the null hypothesis when in fact it is true (that is the significance level of the test) is $0\cdot05$ and $0\cdot01$ respectively. The following power function is obtained from Table 3 and is given in terms of δ and d using the estimate of σ obtained from the two samples ($\hat\sigma^2 = 142\cdot74$)

δ	0·0	0·34	0·69	1·03	1·38	1·72	2·07	2·75
d	0·0	2·0	4·0	6·0	8·0	10·0	12·0	16·0
Probability	0·050	0·094	0·162	0·256	0·372	0·501	0·630	0·839

This power function is plotted in Fig. 6.2.

6.2. Are the Variances of Two Populations Equal?

The data consist of two samples $\{x_1, x_2, ..., x_n\}$ and $\{y_1, y_2, ..., y_m\}$ and we wish to test the null hypothesis that the two population variances σ_1^2 and σ_2^2 are equal.

If the null hypothesis is true the two estimates $\hat\sigma_1^2$ and $\hat\sigma_2^2$, estimated in the usual way, should be in close agreement. The ratio of these variances $\hat\sigma_1^2/\hat\sigma_2^2$ (usually denoted by F) will be distributed about the value unity. Departure of this ratio from unity, in either direction, is a measure of agreement between $\hat\sigma_2^2$ and $\hat\sigma_2^2$. The distribution of the ratio F depends on the distributions of the two populations but if we assume these to be normal F has the following distribution:

$$P(F) \, dF = kF^{\frac{1}{2}\nu_1 - 1} (\nu_2 + \nu_1 F)^{-\frac{1}{2}(\nu_1 + \nu_2)} \, dF \quad 0 < F < \infty \qquad (6.5)$$

where the constant k is given by:

$$k = \frac{\Gamma(\frac{1}{2}(\nu_2 + \nu_1))}{\Gamma(\frac{1}{2}\nu_1)\,\Gamma(\frac{1}{2}\nu_2)} \, \nu_1^{\frac{1}{2}\nu_1} \, \nu_2^{\frac{1}{2}\nu_2}.$$

The F-distribution depends on two parameters ν_1 and ν_2, the degrees of freedom associated with the estimates $\hat{\sigma}_1^2$ and $\hat{\sigma}_2^2$. The degrees of freedom associated with the variance on the top of the ratio is ν_1 and the degrees of freedom associated with the variance on the bottom of the ratio is ν_2. Some upper percentage points of the F-distribution may be found in Pearson and Hartley (1958), Fisher and Yates (1953), Lindley and Miller (1953) and Table 6 in the Tables Section. These tables may be used in this two-tailed application of the F-test if the bigger variance estimate is taken as the top of the ratio. The probability level obtained from the tables is a single-tailed probability and the probability required for a double-tailed test is twice the tabulated level. Care must be taken in the application of this test to ensure that ν_1 is the degrees of freedom associated with the variance on the top of the ratio.

Most applications of the F-test described in later chapters are single-tailed. The null hypothesis is that $\hat{\sigma}_1^2$ and $\hat{\sigma}_2^2$ are both estimates of the same variance σ^2, and the alternative hypothesis is that $\hat{\sigma}_1^2$ is an estimate of $(\sigma^2 + f)$ where f is a positive quantity. We are only interested in establishing whether, or not, $\hat{\sigma}_1^2$ is significantly greater than $\hat{\sigma}_2^2$. In these applications the tabulated level is the actual level required.

6.2.1. Properties of the F-distribution

(a) The mean is $\nu_2/(\nu_2 - 2)$ for $\nu_2 > 2$.

(b) The variance is $\dfrac{2\nu_2^2(\nu_2 + \nu_1 - 2)}{(\nu_2 - 2)^2\nu_1(\nu_2 - 4)}$ for $\nu_2 > 4$.

(c) The distribution tends to normality as $\nu_1, \nu_2 \to \infty$.

(d) If $x = \nu_1 F$, and ν_2 is infinite, then $x \sim \chi^2$ with ν_1 degrees of freedom.

(e) If $x = \sqrt{F}$, and $\nu_1 = 1$, then $x \sim$ "$2t$" with ν_2 degrees of freedom. Here "$2t$" denotes that the range of t is $0 < t < \infty$ and the usual probability is to be doubled.

(f) $\displaystyle\int_0^f p(F \mid \nu_1, \nu_2)\, dF = \int_{1/f}^{\infty} p(F \mid \nu_2, \nu_1)\, dF$ and hence,

$$\int_0^1 p(F \mid \nu_1 = \nu_2)\, dF = 0.5.$$

The fourth property relating the χ^2-distribution to the F-distribution may be simply demonstrated as follows. A variance estimate with infinite degrees of freedom is no longer an estimate but a known value. The variance ratio is thus

$$F = \frac{\hat{\sigma}_1^2}{\sigma_2^2}.$$

The statistic F may be defined in more generality than indicated above as the ratio of two χ^2 variables each standardized by division by their degrees of freedom. That is,

$$F = \frac{\chi_1^2/\nu_1}{\chi_2^2/\nu_2}.$$

We have seen in Section 5.3 that $\nu\hat{\sigma}^2/\sigma^2 \sim \chi^2$ with ν degrees of freedom so that the earlier definition of F is seen to be consistent with the more general definition just given.

Thus,

$$\nu_1 F = \frac{\nu_1 \hat{\sigma}_1^2}{\sigma_2^2} = C,$$

where C is as defined in Section 5.3 and has a χ^2-distribution with ν_1 degrees of freedom.

The fifth property relating the t-distribution to the F-distribution may be demonstrated by showing that the expression for t omitting $\hat{\sigma}^2$ is an estimate of σ^2, the common population variance, with one degree of freedom. We show this for the form of t given in Section 6.1.3 and leave the reader to consider the other forms of Student's t.

If
$$x_i \sim N(\mu, \sigma^2) \quad \text{for} \quad i = 1, 2, \dots, n,$$
$$y_j \sim N(\mu, \sigma^2) \quad \text{for} \quad j = 1, 2, \dots, m,$$

then
$$\bar{x} - \bar{y} \sim N(0, \sigma^2(1/n + 1/m)).$$

Thus
$$\frac{\bar{x} - \bar{y}}{(1/n + 1/m)^{\frac{1}{2}}} \sim N(0, \sigma^2)$$

and hence
$$t_1^2 = \frac{(\bar{x} - \bar{y})^2}{\sigma^2(1/n + 1/m)} \sim \chi^2$$

with 1 degree of freedom. We have seen already from Section 5.3 that

$$\frac{\nu\hat{\sigma}^2}{\sigma^2} \sim \chi^2$$

with ν degrees of freedom so that

$$t^2 = \frac{t_1^2 \sigma^2}{\hat{\sigma}^2} = \frac{\chi_1^2/1}{\chi_\nu^2/\nu} \sim F$$

with 1 and v degrees of freedom. When t is squared the original sign is lost and the range becomes $0 \leq t^2(=F) < \infty$ so that

$$\Pr\{F \geq A^2\} = \Pr\{t \geq A\} + \Pr\{t \leq -A\} = 2\Pr\{t \geq A\}.$$

This relationship between the t- and F-distributions can, of course, be demonstrated easily by mathematical derivation of the t-distribution from the F-distribution. The method used above, however, has the merit of underlining the practical aspects of the relationship.

6.2.2. Example

Let us consider the data given in Section 6.1.2 and test the hypothesis that the variance of the yield with substance A is equal to the variance of the yield without substance A.

$$\hat{\sigma}_1^2 = 183 \cdot 93 \quad \text{with} \quad v_1 = 7$$

and

$$\hat{\sigma}_2^2 = 110 \cdot 72 \quad \text{with} \quad v_2 = 9$$

$$F = \frac{183 \cdot 93}{110 \cdot 72} = 1 \cdot 66$$

$$\Pr\{F \geq 1 \cdot 66 \mid v_1 = 7, \quad v_2 = 9\} = 0 \cdot 25.$$

The double-tailed probability is twice this value, that is, $0 \cdot 5$, and our decision will be to accept the null hypothesis that the two population variances are the same.

6.3. ARE THE VARIANCES OF MORE THAN TWO POPULATIONS EQUAL?

The data consists of k samples of observations representing k populations and we wish to test the hypothesis that these populations have equal variances. We will denote the ith observation in the jth sample by x_{ij} and the number of observations in the jth sample by n_j. The data is represented in Table 6.4.

TABLE 6.4. NOTATION FOR MANY-SAMPLE DATA

Sample 1	$x_{11}, x_{21}, x_{31}, \ldots, x_{i1}, \ldots, x_{n_1 1}$
Sample 2	$x_{12}, x_{22}, x_{32}, \ldots, x_{i2}, \ldots, x_{n_2 2}$
Sample j	$x_{1j}, x_{2j}, x_{3j}, \ldots, x_{ij}, \ldots, x_{n_j j}$
Sample k	$x_{1k}, x_{2k}, x_{3k}, \ldots, x_{ik}, \ldots, x_{n_k k}$

We may estimate from each sample the variance of the corresponding population; that is:

$$\hat{\sigma}_j^2 = \frac{1}{n_j - 1} \sum_{i=1}^{n_j} (x_{ij} - \bar{x}_{\cdot j})^2,$$

where

$$\bar{x}_{\cdot j} = \frac{1}{n_j} \sum_{i=1}^{n_j} x_{ij} = \text{the mean of the } j\text{th sample.}$$

Thus, we may obtain estimates of the variances of the k populations which, if the null hypothesis is true, are all estimates of the same variance. Two tests available for testing the null hypothesis are described below. Both tests assume that the k populations have normal distributions.

6.3.1. Bartlett's Test

The statistic proposed by Bartlett (1937) is defined as:

$$M = \sum_{j=1}^{k} v_j \log_e \hat{\sigma}^2 - \sum_{j=1}^{k} v_j \log_e \hat{\sigma}_j^2 \tag{6.6}$$

or

$$M = \left[\sum_{j=1}^{k} v_j \log_{10} \hat{\sigma}^2 - \sum_{j=1}^{k} v_j \log_{10} \hat{\sigma}_j^2 \right] \log_e 10 \tag{6.7}$$

where

$$\hat{\sigma}^2 = \frac{\sum_{j=1}^{k} v_j \hat{\sigma}_j^2}{\sum_{j=1}^{k} v_j}$$

and $\log_e 10 = 2 \cdot 302585$ to six decimal places and v_j is the degrees of freedom associated with $\hat{\sigma}_j^2$ (in this application $v_j = n_j - 1$).

Expression (6.7) is the more convenient for computation but it is easier to appreciate what the statistic is measuring from the rearranged form:

$$M = \sum_{j=1}^{k} v_j \log_e \left(\frac{\hat{\sigma}^2}{\hat{\sigma}_j^2} \right). \tag{6.8}$$

From this expression we see that M is a simple function of variance ratios. It may be shown that for two populations the significance tests based on M and F are equivalent and we may now see that M is a reasonable extension of F.

Large values of M occur when the agreement between the variance estimates is poor so that the α per cent critical region will consist of the highest α per cent of values of M. If the null hypothesis is true, and if all the populations have normal distributions, it may be shown that the distribution of M is approximately a χ^2-distribution with $k - 1$ degrees of freedom. Hartley (1940) has investigated means of improving the approximation and suggested the modified statistic M' defined as:

$$M' = M\left(1 + \frac{d}{3(k-1)}\right)^{-1}$$

where

$$d = \sum_{j=1}^{k} \frac{1}{v_j} - \frac{1}{\sum_{j=1}^{k} v_j}.$$

The distribution of M' is closer to a χ^2-distribution with $k - 1$ degrees of freedom than the distribution of M. Hartley also suggested further improvements using terms of the form

$$\sum_{j=1}^{k} \frac{1}{v_j^3} \quad \text{and} \quad \left(\sum_{j=1}^{k} v_j\right)^{-3}$$

and details may be found in Pearson and Hartley (1958).

We will confine ourselves in applications, however, to the use of the statistic M'.

6.3.2. Example

Eight different experimentalists have performed six replicate determinations of the yield of substance B from a chemical reaction and the data is summarized in Table 6.5.

TABLE 6.5. EXAMPLE MANY-SAMPLE DATA

Experi-mentalist	Mean determination	Variance estimate	Degrees of freedom	Log₁₀ (variance estimate)
j	$\bar{x}_j$	$\hat{\sigma}_j^2$	v_j	$\mathrm{Log}_{10}\ \hat{\sigma}_j^2$
1	232·4	80·48	5	1·9057
2	286·1	189·23	5	2·2770
3	221·3	190·38	5	2·2796
4	242·1	60·23	5	1·7798
5	227·8	156·39	5	2·1942
6	304·1	204·70	5	2·3111
7	254·9	120·88	5	2·0824
8	271·3	151·93	5	2·1816

Computing the first term we have:

$$\hat{\sigma}^2 = \frac{1}{40} (5 \times 80 \cdot 48 + 5 \times 189 \cdot 23 + \cdots + 5 \times 151 \cdot 93)$$

$$= 144 \cdot 28.$$

Thus

$$\log_{10} \hat{\sigma}^2 = 2 \cdot 1592$$

and

$$\sum_{j=1}^{k} \nu_j \log_{10} \hat{\sigma}^2 = 40 \times 2 \cdot 1592 = 86 \cdot 3680.$$

Computing the second term we have:

$$\sum_{j=1}^{k} \nu_j \log_{10} \hat{\sigma}_j^2 = 5 \times 1 \cdot 9057 + 5 \times 2 \cdot 2770 + \cdots + 5 \times 2 \cdot 1816$$

$$= 85 \cdot 0570.$$

Taking the difference we have:

$$M = 2 \cdot 302585 \times (86 \cdot 3680 - 85 \cdot 0570)$$

$$\underline{M = 3 \cdot 019.}$$

Computing the modified statistic M' we have:

$$\sum_{j=1}^{k} \frac{1}{\nu_j} = \frac{8}{5} \quad \text{and} \quad \left[\sum_{j=1}^{k} \nu_j \right]^{-1} = \frac{1}{40}.$$

Thus, $M' = M \left[1 + \left(\frac{8}{5} - \frac{1}{40} \right) \Big/ 21 \right]^{-1} = 40M/43 = 2 \cdot 81$

$$\Pr\{M \geqq 2 \cdot 81 \mid k = 8\} \simeq \Pr\{\chi_\nu^2 \geqq 2 \cdot 81 \mid \nu = 7\} = 0 \cdot 905.$$

We conclude from this high probability that it is reasonable to believe that the k populations have equal variances. The common variance is estimated as 144·28 with 40 degrees of freedom.

6.3.3. A Second Variance Test

The second test for testing the equality of variances of normal populations uses the ratio of the largest variance estimate to the smallest estimate. The upper 5 per cent and 1 per cent points of the distribution of this ratio are given in Table 31 of Pearson and Hartley and in Table 8 in the Tables Section. These tables assume that the variance estimates all have the same degrees of freedom so that only two parameters, k the number of samples, and ν the common degrees of freedom, are involved. The tables may be used approximately, however, when different degrees of freedom occur, by entering the table with an average value for the degrees of freedom v.

6.3.4. Example

The ratio of the largest estimate to the smallest estimate for the data summarized above is:

$$R = \frac{\hat{\sigma}^2_{max}}{\hat{\sigma}^2_{min}} = \frac{204 \cdot 70}{60 \cdot 23} = 3 \cdot 34$$

$$\Pr\{R \geq 3 \cdot 34 \mid k = 8, \quad v = 5\} >> 0 \cdot 05.$$

The conclusion from this test is therefore the same as that obtained from Bartlett's test.

QUESTIONS

1. The following data given by O. H. Latter (1902 *Biometrika*, vol. 1) record the lengths (in mm) of cuckoo's eggs found in nests made by reed-warblers or wrens.

Host bird   Lengths of eggs in mm
Reed-warblers 23·2, 22·0, 22·2, 21·2, 21·6, 21·9, 22·0, 22·9, 22·8
Wrens   19·8, 22·1, 21·5, 20·9, 22·0, 21·0, 22·3, 21·0, 20·3, 20·9, 22·0, 20·0,
                     20·8, 21·2, 21·0

(a) Is there any evidence that the variability in lengths differs in the two types of nest?
(b) Is there any evidence that the average length differs in the two types of nest?
(c) Plot the two sets of results as separate sets but on the same page of probability paper. Does this plot reflect the conclusions you have come to in questions (a) and (b)?
(d) Do you feel that the assumption of normality made in (a) and (b) above is reasonable?

2. Add the following data of the lengths of cuckoo's eggs found in nests of hedgesparrows to that given in question 1.

Host bird   Lengths of eggs in mm
Hedge-sparrows 22·0, 23·9, 20·9, 23·8, 25·0, 24·0, 21·7, 23·8, 22·8, 23·1, 23·1, 23·5,
                      23·0, 23·0

(a) Repeat question (a) for the three types of nest.
(b) Add this third set of data to the probability plot drawn in question 1 (c). Do you consider it reasonable to believe that the population represented by this set of data is normally distributed?
(c) The question of equality of population means for the three sets of results will be considered as a question to Chapter 7; can you anticipate the results from the probability plot?

3. Consider the three samples of results given in questions 1 and 2 above. One possible method of analysis for mean differences using our present knowledge would be to perform three *t*-tests comparing each possible pair of samples. If we use a 5 per cent significance level for each of these tests there is a 5 per cent chance of a significance result when, in fact, the null hypothesis is true.

(a) If the null hypothesis is true, what is the probability that we obtain at least one significant result from the three tests, assuming the tests to be independent?

(b) If we perform a series of twenty such tests (each test on different pairs of samples) how many significant results would you expect if the null hypothesis is true (again assuming that the tests are independent)?

4. An experiment is to be performed to compare the response (in terms of weight) of rats to irradiated and unirradiated food and it is believed that the comparison will reveal no differences. Two groups of rats will be chosen and one group will be given irradiated food and the other unirradiated food. After a number of weeks the rats in both groups will be weighed and compared using the test described in Section 6.1.1 since the variances of the two populations of rats are known (from previous experience) to be equal to 100. The test will use a 5 per cent significance level, and will be single-tailed.

(a) What is the probability that the experiment will detect a difference of 5 in the population means if both groups are chosen to contain ten rats?

(b) Repeat question 4 (a) for sample sizes 20, 30, 40, 50, 60, 80 and 100.

(c) What sample sizes should be used (both equal) in the experiment if we wish to have a 90 per cent chance of detecting a difference of 10 between the population means?

(d) Plot the probabilities obtained in questions 4 (a) and 4 (b) against the sample size.

(e) Is this plot the power function of the test?

5. Eight lengths of a cotton material have been chosen one from each of eight rolls. Each length was cut into two halves, one half was dyed with a newly developed red dye, and the other was dyed with a red dye that had been in use for some time. Each of the sixteen pieces of material was then washed and the amount of dye washed out recorded for each piece. The results were:

Roll	1	2	3	4	5	6	7	8
New dye	12·5	14·3	16·8	14·9	17·4	11·4	15·6	15·2
Old dye	13·2	13·7	15·4	13·5	16·8	10·2	14·8	16·0

Observation of the fastness of the old dye has shown that the amount of dye washing out has a normal distribution with mean equal to 13·5 and standard deviation equal to 1·9.

(a) Are the observations of the old dye consistent with past experience?

(b) Are the observations of the new dye consistent with the past experience of the old dye?

(c) Test the null hypothesis that there is no difference between the two dyes, making allowance for possible differences between rolls and ignoring previous experience.

(d) What normality assumption is made by the test used in (c)?

CHAPTER 7

TESTS OF SIGNIFICANCE—III

THIS chapter continues the description of tests of significance and in particular considers the following questions:

1. Are the distributions of several populations identical?
2. Are the means of several populations equal?

The description of the general form of the procedure for testing the first hypothesis is followed by consideration of special cases which are often described separately. The testing procedure for the second hypothesis is introduced as a further test of significance. It is, in fact, an example of the technique of analysis of variance and it will be described again from this approach in Chapter 8.

7.1. ARE THE DISTRIBUTIONS OF SEVERAL POPULATIONS IDENTICAL?

The testing procedures described so far in the preceding chapters test hypotheses concerning particular parameters of the population such as mean and variance, and they can be applied to small or large samples alike. Testing procedures for hypotheses concerning population distributions require much larger numbers of observations and cannot be applied satisfactorily to small samples. The data available consist of k samples of observations representing k populations and the null hypothesis to be tested is that the k populations have the same distribution. The measurement scale of the observations is divided into a number of groups by a chosen sequence of boundary values, and numbers of observations falling in each group in each sample are counted. The form of the data and the notation to be used are given in Table 7.1.

The number of observations in the jth sample is the sum of the frequencies $n_{1j}, n_{2j}, \ldots, n_{hj}$ and these individual frequencies must be of a reasonable size if an adequate picture of the distribution is to be obtained. If four groups are employed, and no frequency is less than ten, the sample size is at least forty. Thus, with this modest number of groups, and these modest frequencies, the sample size is quite high. Certainly a sample size of forty would be considered large in testing mean or variance differences.

TABLE 7.1. FREQUENCY DATA FOR TESTING POPULATION DISTRIBUTION DIFFERENCES

Group	1	2		i		h	Totals
Sample 1	n_{11}	n_{21}	...	n_{i1}	...	n_{h1}	$N_{\cdot 1}$
Sample 2	n_{12}	n_{22}	...	n_{i2}	...	n_{h2}	$N_{\cdot 2}$
...	...	...	...	...	...	...	...
Sample j	n_{1j}	n_{2j}	...	n_{ij}	...	n_{hj}	$N_{\cdot j}$
...	...	...	...	...	...	...	...
Sample k	n_{1k}	n_{2k}	...	n_{ik}	...	n_{hk}	$N_{\cdot k}$
Totals	$N_1 \cdot$	$N_2 \cdot$	...	$N_i \cdot$	...	$N_h \cdot$	$N_{\cdot \cdot}$

If the k populations have the same distribution the best estimates of the common distribution is obtained from the total frequencies $N_1 \cdot, N_2 \cdot, ..., N_h \cdot \cdot$. The proportion of observations falling in the ith group in the common population distribution is estimated by the ratio $N_i \cdot / N_{\cdot \cdot}$ and the expected number of observations in the ith group in the jth sample is, therefore, $N_i \cdot N_{\cdot j} / N_{\cdot \cdot \cdot}$ If the null hypothesis is true the expected frequencies $N_i \cdot N_{\cdot j} / N_{\cdot \cdot}$ may be expected to be in reasonable agreement with the observed frequencies n_{ij}. The agreement between the expected frequencies and the observed frequencies may be measured by the statistic ϕ^2 described in Section 5.4.2 and here defined as:

$$\phi^2 = \sum_{j=1}^{k} \sum_{i=1}^{h} \frac{(n_{ij} - N_i \cdot N_{\cdot j} / N_{\cdot \cdot})^2}{N_i \cdot N_{\cdot j} / N_{\cdot \cdot}}. \tag{7.1}$$

The statistic ϕ^2 in this application has approximately a χ^2-distribution with $(k-1)(h-1)$ degrees of freedom. The approximation *may* be invalid if expected frequencies of less than five are used.

7.1.1. Example

The lengths of a number of pellets from each of three different production processes were measured and the three population distributions of lengths of pellets were to be compared. There were 200, 300 and 250 pellets from the three production processes respectively, and the lengths were divided into five groups. The data recorded is the number of pellets in each group in each sample, and is given in Table 7.2.

TABLE 7.2. EXAMPLE 5 × 3 FREQUENCY TABLE

Group	1	2	3	4	5	Totals
Process A	21	40	78	46	15	200
B	40	61	90	74	35	300
C	28	52	106	46	18	250
Totals	89	153	274	166	68	750

The expected frequencies corresponding to this set of observed frequencies are calculated as the product of the appropriate row and column total divided by the grand total. The expected frequencies for groups 1 and 2 for Process A are, for example, $200 \times 89/750 = 23.73$ and $200 \times 153/750 = 40.80$, respectively. The complete computation is now given in Table 7.3.

TABLE 7.3. EXAMPLE COMPUTATION FOR A 5 × 3 FREQUENCY TABLE

	Group	1	2	3	4	5	Totals
A	Observed	21	40	78	46	15	200
	Expected	23.73	40.80	73.07	44.27	18.13	200.00
	Difference	−2.73	−0.80	4.93	1.73	−3.13	0.00
	Contr. to ϕ^2	0.31	0.02	0.33	0.07	0.54	1.27
	Observed	40	61	90	74	35	300
	Expected	35.60	61.20	109.60	66.40	27.20	300.00
	Difference	4.40	−0.20	−19.60	7.60	7.80	0.00
	Contr. to ϕ^2	0.54	0.00	3.51	0.87	2.24	7.16
	Observed	28	52	106	46	18	250
	Expected	29.67	51.00	91.33	55.33	22.67	250.00
	Difference	−1.67	1.0	14.67	−9.33	−4.67	0.00
	Contr. to ϕ^2	0.09	0.01	2.36	1.57	0.96	4.99
	Totals	89	153	274	166	68	750

$$\phi^2 = \frac{(21 - 23.73)^2}{23.73} + \cdots + \frac{(18 - 22.67)^2}{22.67}$$

$$= 18.42 \quad \text{and} \quad \nu = 2 \times 4 = 8$$

$$\Pr\{\chi^2 \geq 13.42 \mid \nu = 8\} = 0.098.$$

If we reject the null hypothesis that the population distributions are identical we take a risk of 0.098 of being wrong so that the decision to accept, or reject, depends on whether, or not, we regard this risk as being too great.

7.1.2. Special Case—Two Samples

If there are only two samples involved in the test the computation of ϕ^2 may be shortened by use of the following formula:

$$\phi^2 = \phi_{.1}^2 \left(1 + \frac{N_{.1}}{N_{.2}} \right), \tag{7.2}$$

where $\phi_{.1}^2$ is the value of ϕ^2 for the first sample only, and where $N_{.1}$ and $N_{.2}$ are the two sample sizes.

7.1.3. Special Case—Two Groups

The computation of ϕ^2 may be similarly shortened if only two groups are involved in the test. The following formula for ϕ^2 may be used:

$$\phi^2 = \phi_1^2 \cdot \left(1 + \frac{N_1.}{N_2.}\right), \tag{7.3}$$

where $\phi_1^2.$ is the value of ϕ^2 for the first group only, and where $N_1.$ and $N_2.$ are the total frequencies in the two groups.

7.1.4. Special Case—Two Samples and Two Groups

This type of data is usually referred to as the 2×2 table and although it may be analysed by the methods described above, it is usually considered separately.

7.2. THE 2×2 TABLE

Data of this form is normally referred to by different notation from that used above and this is given in Table 7.4.

TABLE 7.4. THE 2×2 FREQUENCY TABLE

Group	1	2	Totals
Sample 1	a	c	m
Sample 2	b	d	n
Totals	r	s	N

For given values of the marginal totals m, n, r and s it is only necessary to know the value of one of the quantities a, b, c or d, since the values of the other three may then be calculated. Thus, one of these may be chosen as the statistic to be used in testing the null hypothesis. The choice of statistic from among these four possibilities is arbitrary but the conventional choice is b. It may be shown that the probability of a particular value of b, if the null hypothesis is true, is

$$p(b) = \frac{m!n!r!s!}{a!b!c!d!N!}. \tag{7.4}$$

A complete set of these probabilities for $b = 0, 1, ..., t$ (where t is the smaller of r and n) forms a "hypergeometric" distribution. Tables of the upper 5, 2·5, 1, and 0·5 per cent point of the distribution of b are given in Table 38 of Pearson and Hartley and the notation used in the table is A

for m above; B for n above; a and b used as above. It is worth recording that for small values of N it is possible to calculate the distribution of b, and, in fact, only the part of the distribution corresponding to the critical region need be computed. The test may, of course, be single-tailed or double-tailed since with only two groups and one degree of freedom we may specify a direction for the alternative hypothesis. That is, for example, we may consider the alternative hypothesis to be that b is lower than expectation, or greater than expectation, or different from expectation.

This test involving only two groups can be applied to smaller sample sizes than was suggested above but two groups do not give a good representation of a distribution. With two groups the test becomes mainly a test for location differences between the populations rather than a test for distributional differences.

7.2.1. Example

The following data consists of the number of pellets of length less than 300 cm and the number longer than 300 cm in two samples of pellets produced by different processes. It is interesting to note that although in most experiments the actual lengths would be available, permitting the use of a better test, this type of data could arise in situations where the products are divided into groups according to set gauges. In this type of process the actual measurements such as length would not be available and we would have only the number of items within each group. The data is given in Table 7.5.

TABLE 7.5. EXAMPLE 2 × 2 FREQUENCY TABLE

	Lengths less than 300 cm	Lengths greater than 300 cm	Totals
Process 1	10	5	15
Process 2	4	11	15
Totals	14	16	30

Table 38 of Pearson and Hartley shows that for $A = m = 15$, $B = n = 15$, $a = 10$, the value of $b = 4$ is significant at the 5 per cent significance level so that if we accept the risk of at most 5 per cent of being wrong we reject the null hypothesis that the proportion of pellets less than 300 cm is the same in both processes.

7.2.2. Approximation for 2 × 2 Tables

Table 38 of Pearson and Hartley includes values of A and B not greater than 15. For values of A and B greater than 15 the following approximation may be used.

It may be shown that the mean and variance of b are

$$\mu(b) = \frac{nr}{N}$$

$$\sigma^2(b) = \frac{mnrs}{N^2(N-1)} \tag{7.5}$$

and that the hypergeometric distribution of b tends to normality with these moments as N increases.

That is:

$$b \approx N(\mu(b), \; \sigma^2(b)),$$

and hence:

$$X = \frac{b - \mu(b)}{\sqrt{\sigma^2(b)}} \approx N(0, 1).$$

7.2.3. Continuity Correction

The above approximation involves the description of a discrete distribution by a continuous distribution. Approximations of this type are always improved by the use of a continuty correction. Figure 7.1 shows the discrete upper tail of the distribution of b together with the continuous normal distribution.

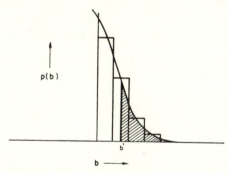

$p(b)$

b'

$b \longrightarrow$

Fig. 7.1. The hypergeometric distribution and the normal tail approximation.

The probability that b takes a value equal to or greater than b' is the sum of the areas of the three rightmost columns shown in Fig. 7.1. The normal approximation, without correction, to this area is the shaded portion and it will be seen that an improvement to this approximation could be made if the integration of the normal curve was taken from $b' - \frac{1}{2}$ instead of from b'. That is, from the boundary of the column instead of the mid-point. If an approximation is required to a lower-tail area instead of to an upper-tail area the continuity correction consists of adding a half to the statistic value. In the case of the distribution of b the discrete values of b differ by unity and the continuity correction is plus or minus a half according to which tail area is required. It is important to note, however, if the discrete values of the statistic differ by h that the continuity correction is $\pm\frac{1}{2}h$.

7.2.4. Example

The value of b observed in the data given in Section 7.2.1 is four. Applying the normal approximation with continuity correction to obtain an approximation to the probability of observing a value of b equal to or less than four we have:

$$\mu(b) = \frac{15 \times 14}{30} = 7$$

$$\sigma^2(b) = \frac{15 \times 15 \times 14 \times 16}{30 \times 30 \times 29} = 1\cdot9310$$

$$X' = \frac{4 - 7 + \frac{1}{2}}{\sqrt{(1\cdot9310)}} = -1\cdot80$$

$$\Pr\{X' \leqq -1\cdot80\} = 0\cdot036.$$

This value is in good agreement with the exact value of $0\cdot033$ given in Table 38 of Pearson and Hartley and the decision is the same as in Section 7.2.1.

7.3. A Test for Independence

The tests described so far in this chapter have been concerned to test the hypothesis that the distributions of two or more populations are identical. These tests, however, are also available as tests of independence between two characters. As an example of this application we will consider the hypothesis that the proportion of people in four blood groups are the same in three different countries or, alternatively, that the probability that a person falls into a particular blood group is independent of his nationality. If a number of individuals from each of the three countries are classified into the four blood groups a table of the form of Table 7.6 is obtained.

The null hypothesis states that the probability p_i that an individual falls in blood group i is the same in each country. If this hypothesis is true the expected number of individuals in blood group i in country j is $P_i N_{.j}$, where $N_{.j}$ is the number of individuals observed in the jth country. If, as is usually the case, the values of the p_i's are unknown, their values are estimated

TABLE 7.6. EXAMPLE TEST OF INDEPENDENCE

Blood group	1	2	3	4	Totals
Country A	43	75	58	24	200
B	71	105	87	33	296
C	39	42	50	21	152
Totals	153	222	195	78	648

from the total proportions observed in each blood group. That is,

$$\hat{p}_i = N_{i.}/N_{..},$$

where $N_{i.}$ is the total number of individuals in the ith blood group, and $N_{..}$ is the total number of individuals. The expected number of individuals in blood group i in country j is, therefore, $N_{i.}N_{.j}/N_{..}$ and this value is the expected frequency found in the previous applications. Thus, we see that there are no computational differences between the calculation of ϕ^2 in the two applications. Following the procedure described in Section 7.1.1 the value of ϕ^2 is 10·95 and the significance probability is 0·090.

7.4. TESTING PARTICULAR GROUP PROPORTIONS

Situations arise in both applications of the $h \times k$ table in which it is desired to test that a particular set of known proportions describe all samples (or countries). In the first application this corresponds to testing the hypothesis that each population has a particular distribution. In the second application this corresponds to testing the hypothesis that the blood group proportions, predicted (say) by genetical models with known constants, are the same in all countries. If the proportions p_i are known the expected number of individuals in the ith group in the jth country is $p_i N_{.j}$ and

$$\phi^2 = \sum_{j=1}^{k} \sum_{i=1}^{h} \frac{(p_i N_{ij} - N_{.j})^2}{p_i N_{.j}}. \tag{7.6}$$

The restrictions existing previously between the observed and expected frequencies that the sum of the expected frequencies in each group equals the sum of the observed frequencies no longer exist in this application. The degrees of freedom of ϕ^2 are therefore $k(h - 1) + 1$.

7.5. ARE THE MEANS OF SEVERAL POPULATIONS EQUAL?

Procedures have been described to test the hypothesis that the means of two populations are equal and three situations were considered:

1. Population variances known.
2. Population variances unknown but assumed equal.
3. Population variances unknown and unequal.

The extension of these tests to more than two populations introduces difficulties which are not encountered in the case of two populations. In fact no entirely satisfactory testing procedure exists for the third situation. The description given here is concerned mainly with the second situation although the first situation is briefly discussed.

The notation to be used in this description will be the same as in Section 6.3. It is redefined here for convenience:

$x_{ij} \equiv$ the ith observation in the jth sample.
$k \equiv$ the number of samples.
$n_j \equiv$ the number of observations in the jth sample.
$\bar{x}._j \equiv$ the mean of the jth sample,

$$\left(\text{that is, } \bar{x}._j = \frac{1}{n_j} \sum_{i=1}^{n_j} x_{ij}\right).$$

$\hat{\sigma}_j^2 \equiv$ the estimate of the common population variance σ^2 obtained from the jth sample,

$$\left(\text{that is, } \hat{\sigma}_j^2 = \frac{1}{n_j - 1} \sum_{i=1}^{n_j} (x_{ij} - \bar{x}._j)^2\right).$$

Each sample provides an estimate of the common population variance which may be combined as in Bartlett's test to provide one estimate.

$$\hat{\sigma}_2^2 = \sum_{j=1}^{k} \nu_j \hat{\sigma}_j^2 \bigg/ \sum_{j=1}^{k} \nu_j \quad \text{(where } \nu_j = n_j - 1)$$

or alternatively

$$\hat{\sigma}_2^2 = \frac{\displaystyle\sum_{j=1}^{k} \sum_{i=1}^{n_j} (x_{ij} - \bar{x}._j)^2}{\displaystyle\sum_{j=1}^{k} (n_j - 1)}.$$

It will be seen from this expression that $\hat{\sigma}_2^2$ is estimated as a function of differences between observations and their sample mean so that the differences between the population means do not affect this estimate of σ^2. This estimate has

$$\sum_{j=1}^{k} (n_j - 1)$$

degrees of freedom.

If the population means are all equal the sample means $\bar{x}_{.j}$ vary about the common population mean in a predictable manner and it is possible to obtain another estimate $\hat{\sigma}_1^2$ of σ^2 from the variability shown by these means. This estimate is increased by differences between the population means, so that if it is significantly larger than the estimate $\hat{\sigma}_2^2$ we will conclude that the population means are different. This estimate $\hat{\sigma}_1^2$ may be obtained by the following argument. The variance of $\bar{x}_{.j}$ is σ^2/n_j, so that

$$\frac{1}{k} \sum_{j=1}^{k} n_j(\bar{x}_{.j} - \mu)^2$$

(where μ is the population mean) provides an estimate of σ^2. The population mean, however, is usually unknown and σ^2 is, therefore, estimated by the following expression:

$$\hat{\sigma}_1^2 = \frac{1}{k-1} \sum_{j=1}^{k} n_j(\bar{x}_{.j} - \bar{x}_{..})^2 = \frac{1}{k-1} \left[\sum_{j=1}^{k} n_j \bar{x}_{.j}^2 - \bar{x}_{..}^2 \sum_{j=1}^{k} n_j \right]$$

$\left(\text{where } \bar{x}_{..} \text{ is the grand mean. That is: } \bar{x} = \sum_{j=1}^{k} n_j \bar{x}_{.j} \middle/ \sum_{j=1}^{k} n_j \right).$

This estimate of σ^2 has $k-1$ degrees of freedom.

The two estimates of σ^2 may be shown to be independent estimates of the common population variance, so that if we make the further assumption that the population distributions are normal we may compare the two estimates of σ^2 by an F-test. That is, $F = \hat{\sigma}_1^2/\hat{\sigma}_2^2$ with $\nu_1 = k-1$ and

$$\nu_2 = \sum_{j=1}^{k} (n_j - 1).$$

If the value of F is significantly greater than unity we will conclude that the population means are different. It should be noted that this application of the F-test is single-tailed since interest is confined to significantly large values of F.

This test will be considered again as an example of the Analysis of Variance approach to experimental data and a more detailed description of the computational procedure will be given there.

7.5.1. Example

The data summarized in Section 6.3.2 consisted of six determinations of the yield of substance B from a certain chemical reaction made by each of eight experimentalists. Bartlett's test applied to this data in Section 6.3.2 showed that it is reasonable to believe that the eight populations have a common variance. To test the hypothesis that the eight populations have the same mean we apply the F-test described above. The assumption of common population variance necessary in this test is seen to be reasonable. The additional assumption of normality, also necessary, is assumed to be reasonable but is considered again in the next section.

The combined estimate of variance was calculated in Section 6.3.2 as

$$\hat{\sigma}_2^2 = 144\cdot28 \quad \text{with} \quad \nu_2 = 40.$$

The estimate of variance measuring difference between population means may be calculated (on a desk machine—see Section 4.5.2) as follows:

$$\bar{x}.. = 255\cdot00$$

$$\hat{\sigma}_1^2 = \frac{1}{7}[5 \times 232\cdot4^2 + \cdots + 5 \times 271\cdot3^2 - 40 \times 255\cdot00^2]$$

$$= 4426\cdot01.$$

The ratio of these two estimates is

$$F = \frac{4426\cdot01}{144\cdot28} = 30\cdot68,$$

and

$$\Pr\{F \geq 30\cdot68 \mid \nu_1 = 7, \quad \nu_2 = 40\} < 0\cdot001.$$

We will usually conclude from this small probability that the population means are not all the same.

7.6. Robustness of Tests Assuming Normal Populations

Nearly all of the significance tests described above have made the assumption that the populations involved have normal distributions. Strictly, therefore, these tests should only be applied when the assumption of normality is justified. It is a necessary part of the study of these tests to seek an answer to the important question: "What would be the effect of applying these tests to situations with non-normal populations?" The effect of the assumption of a particular non-normal population is to change the distribution of the statistic (t, F, etc.). There have been two approaches to this question. The first approach is purely numerical consisting of computing the value of the statistic for each of a number of samples (or sets of several samples as required by the particular test) drawn randomly from the particular non-normal populations and building up a sample distribution of the statistic. This distribution is then compared with that obtained in the normal situation. The second approach is to mathematically calculate the moments of the statistic and study the conditions under which these moments are close to those of the statistic when the normality assumption applies. Both types of studies agree on the general conclusions that tests for population means are insensitive to departures from normality whereas tests on population variances may be very sensitive to departures from normality. Tests on population means are therefore described as robust under non-normal assumptions whereas tests on population variances may not be so described. As an example of the differences in distributions of a variance statistic under normal and non-normal conditions the results of the following sampling experiment are given. Eight samples of size four were randomly drawn from a standardized normal population. Bartlett's modified statistic M' as described in Section 6.3.1 for testing for population variance differences was computed on these results. The selection of eight samples was repeated and a second value of M' obtained. This process was continued until 10,000 values of M' had been obtained. The distribution of M' values obtained by this process is shown in the upper portion of Fig. 7.2. To obtain the distribution of M' under a non-normal assumption the whole of the process described above was repeated with the observations selected from the exponential population

$$p(x) = e^{-x}\, dx, \qquad 0 \leqq x < \infty$$

instead of the normal. The sample distribution thus obtained is shown in the lower portion of Fig. 7.2.

It will be seen from these two distributions that the distribution of Bartlett's M' is very different in the two situations. The sample 5 per cent significance point in the normal situation is 14·04 and in the exponential situation it is 27·34. Thus application of Bartlett's M' in non-normal situations may provide very misleading results. There is evidence that significant results can be obtained by Bartlett's M' which are due to non-normality rather than to differences between population variances. There is some justification in the belief that a non-significant result obtained from Bartlett's test is evidence in favour of the assumption of normality.

7.7. TRANSFORMATIONS AND DISTRIBUTION-FREE TESTS

We have seen in the previous section that tests on population means are robust so that we may, with reasonable safety, apply tests on means in all situations where the population distributions are not violently non-normal. The same procedure may not be recommended for variance tests. A second approach to the question of non-normality is to find a transformation of the data such that the population distribution in the transformed scale is normal. To do this, however, either requires knowledge of the form of the

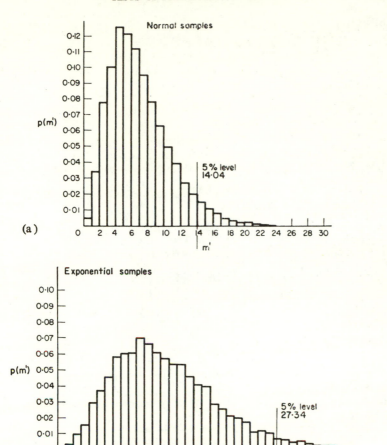

Fig. 7.2. A comparison of sample distributions of Bartlett's statistic.

population distribution, or requires a number of transformations to be tried until a successful transformation is found. If the form of the population distribution is known it is possible to redevelop tests such as the t-test and the F-test assuming the particular non-normal distributional form for the population distributions. This redevelopment would change the distribution of the statistic (t or F) from that obtained from the normal assumption. Whilst it is easy to state that this could be done, in practice the derivation of the distribution of the statistic often involves considerable mathematics and numerical integration.

A further approach currently receiving much attention is the development of tests which make no assumption of a particular form for the population distribution. Such tests are available in all situations. A number of such "distribution-free" tests have been developed particularly for comparing properties of two populations. A few tests have been developed for other situations and many problems remain to be solved. The general approach to distribution-free tests is discussed in Chapter 14 and many tests are described.

QUESTIONS

1. (a) Perform the test for the equality of population means referred to in question 2 (c) after Chapter 6.

(b) Does the result of this test confirm the conclusion drawn from the plot referred to in question 2 (c) after Chapter 6?

2. Four groups of individuals are observed and classified within each group according to a genetical characteristic for which the three types Dominant, Hybrid and Recessive can be identified. The following frequencies were obtained:

Group	1	2	3	4	Totals
Dominant	54	82	40	86	262
Hybrid	124	134	101	178	537
Recessive	68	60	55	74	257
Totals	246	276	196	338	1056

(a) Is there any evidence for believing that the proportions of Dominants, Hybrids and Recessives differ from group to group?

(b) Genetical theory predicts that the proportions of Dominants, Hybrids and Recessives in the population are 0·25, 0·50 and 0·25, respectively. Compute the table of expected frequencies corresponding to the above table of observed frequencies and test the hypothesis that the population proportions are as given by theory. (What are the degrees of freedom in this case?)

(c) How would the test procedure differ if the Dominants and Hybrids had been indistinguishable?

(d) An alternative genetical theory to that in question 2 (b) predicts that the proportions of the three types are $(1 - p)^2$, $2p(1 - p)$ and p^2, respectively, where the parameter p is unknown. Write down the logarithm of the likelihood for the three totals and hence show that the estimator of p obtained is $\hat{p} = (h + 2r)/2n$ where h, r and n are the numbers of Hybrids, Recessives and the grand total respectively.

(c) What changes are necessary to the analysis because of the use of the estimate of p instead of a given value? What are the degrees of freedom now? (Do not perform the analysis since the estimate of p is very close to 0·5 and the revised test is clearly not worth while.)

3. The following table shows the numbers of smokers and non-smokers contracting lung cancer. Assume that these fictitious results were obtained by randomly choosing 100 smokers and 100 non-smokers from the population and observing whether, or not, they had lung cancer.

	Smokers	Non-smokers	Totals
With cancer	24	10	34
Without cancer	76	90	166
Totals	100	100	200

Would you conclude on this evidence that a smoker was more likely to contract lung cancer than a non-smoker? (Use the Normal approximation to the hypergeometric distribution to test significance.)

CHAPTER 8

ANALYSIS OF VARIANCE—I.
HIERARCHICAL DESIGNS

A PROCEDURE to test the hypothesis that the means of two or more populations are the same has been described in the last chapter. This hypothesis will be reconsidered in this chapter and will be used as an introductory example of the technique of analysis of variance. Further, more complicated hypotheses corresponding to experimental designs of the hierarchical (or nested) type will be described and illustrated.

8.1. ARE THE MEANS OF TWO OR MORE POPULATIONS EQUAL?

The test described in the last chapter assumed that the populations have normal distributions with the same but unknown variances. The notation used in the last chapter, and to be used in this chapter, is described again:

$x_{ij} \equiv$ the ith observation in the jth sample.
$k \equiv$ the number of samples.
$n_j \equiv$ the number of observations in the jth sample.
$N \equiv$ the total number of observations.
$\bar{x}._j \equiv$ the mean of the jth sample

$$\left(\text{that is } \bar{x}._j = \frac{1}{n_j} \sum_{i=1}^{n_j} x_{ij} \right)$$

$\hat{\sigma}_j^2 \equiv$ the estimate of the common population variance from the jth sample

$$\left(\text{that is } \hat{\sigma}_j^2 = \frac{1}{n_j - 1} \sum_{i=1}^{n_j} (x_{ij} - \bar{x}._j)^2 \right).$$

The F-test described in the last chapter consisted of comparing two independent estimates of the common population variance σ^2, namely:

$$\hat{\sigma}_1^2 = \frac{1}{k - 1} \sum_{j=1}^{k} n_j (\bar{x}._j - \bar{x}..)^2$$

$$\hat{\sigma}_2^2 = \frac{\sum\limits_{j=1}^{k} \sum\limits_{i=1}^{n_j} (x_{ij} - \bar{x}._j)^2}{\sum\limits_{j=1}^{k} (n_j - 1)}.$$

The estimate $\hat{\sigma}_1^2$ estimates σ^2 if the null hypothesis of no difference between population means is true and is increased if the means are different. The estimate $\hat{\sigma}_2^2$ estimates σ^2 whether the null hypothesis is true or not. If the populations are normally distributed and the null hypothesis is true the ratio $\hat{\sigma}_1^2/\hat{\sigma}_2^2$ has an F-distribution. The occurrence of a significantly high value of F will be followed by the conclusion that the population means are different.

This form of data is the simplest form of data analysed by the technique of analysis of variance and illustrates the analysis of variance approach of computing a number of independent estimates of σ^2 which have interesting properties. Estimate $\hat{\sigma}_1^2$ measures the agreement between the sample means and is an estimate of σ^2 if, and only if, the null hypothesis is true. It over-estimates σ^2 if the null hypothesis is not true. Estimate $\hat{\sigma}_2^2$ estimates σ^2 irrespective of whether the hypothesis is true or not and is independent of $\hat{\sigma}_1^2$. In this example two estimates of σ^2 are obtained but in general a number of independent estimates of σ^2 are obtained. One estimate ($\hat{\sigma}_2^2$ in this example) normally provides an estimate of σ^2 no matter which experimental effects are present. Each of the other estimates allows the presence of a particular experimental effect to be tested.

8.1.1. The Analysis of Variance Approach (Fixed-effect Model)

Three types of analysis of variance models may be distinguished, namely the fixed-effect model, the random-effect model and the mixed model. The description that follows describes an example of a fixed-effect model and the differences between the three types of model are clarified later.

The ith member of the jth sample may be represented by the following equation, or model:

$$x_{ij} = A + E_j + z_{ij}, \tag{8.1}$$

where A is a general level common to all observations,

E_j is a "fixed" or constant departure from A due to the jth population and common to all members of the jth population,

z_{ij} is a random error having a normal distribution with zero mean and variance σ^2 (σ^2 is often called the residual variance).

The null hypothesis that the population means are equal is equivalent to the hypothesis that the E_j's are all zero and hence that the population means are all equal to A.

From the model (8.1) it is possible to derive similar expressions for the mean of the jth sample and for the grand mean:

$$\bar{x}._j = \frac{1}{n_j} \sum_{i=1}^{n_j} x_{ij} = \frac{1}{n_j} \left(\sum_{i=1}^{n_j} (A + E_j + z_{ij}) \right) \tag{8.2}$$

$$\bar{x}._j = A + E_j + \bar{z}._j$$

and

$$\bar{x}.. = \frac{1}{N - \sum\limits_{j=1}^{k} n_j} \sum_{j=1}^{k} n_j \bar{x}._j = \frac{1}{N - \sum\limits_{j=1}^{k} n_j} \left(\sum_{j=1}^{k} n_j (A + E_j + \bar{z}._j) \right)$$

$$\bar{x}.. = A + \frac{\sum\limits_{j=1}^{k} n_j E_j}{N - \sum\limits_{j=1}^{k} n_j} + \bar{z}.. \tag{8.3}$$

where the terms $\bar{z}._j$ and $\bar{z}..$ are the appropriate means of the error terms z_{ij}. The term

$$\sum_{j=1}^{k} n_j E_j$$

is the total departure of all observations from the general level and it is clear that no restriction is placed on the model if the total departure of all observations from the general level A is taken to be zero. This gives the eminently reasonable result that the general level A is equal to $\bar{x}.. - \bar{z}...$ That is $x..$ differs from A by error terms only. Thus from expressions (8.1), (8.2) and (8.3) it is clear that

$$x_{ij} - \bar{x}._j = z_{ij} - \bar{z}._j \tag{8.4}$$

and

$$\bar{x}._j - \bar{x} = E_j + \bar{z}._j - \bar{z}.. \tag{8.5}$$

The difference between an observation and its sample mean involves only error terms whereas the difference between a sample mean and the grand mean involves the term E_j and error terms.

Consider the total sum of squares

$$S = \sum_{j=1}^{k} \sum_{i=1}^{n_j} (x_{ij} - \bar{x}..)^2.$$

If the E_j's are all zero the sum of squares S estimates $(N - 1)\sigma^2$, where N is the total number of observations. That is,

$$N = \sum_{j=1}^{k} n_j.$$

The analysis of variance technique consists of splitting the sum S into separate sums of squares with interesting properties. The way in which S is split up is indicated by expressions (8.4) and (8.5) and is as follows:

$$S = \sum_{j=1}^{k} \sum_{i=1}^{n_j} (x_{ij} - \bar{x}..)^2 = \sum_{j=1}^{k} \sum_{i=1}^{n_j} ((x_{ij} - \bar{x}._j) + (\bar{x}._j - \bar{x}..))^2$$

$$= \sum_{j=1}^{k} \sum_{i=1}^{n_j} (x_{ij} - \bar{x}._j)^2 + 2 \sum_{j=1}^{k} \sum_{i=1}^{n_j} (x_{ij} - \bar{x}._j)(\bar{x}._j - \bar{x}..) + \sum_{j=1}^{k} \sum_{i=1}^{n_j} (\bar{x}._j - \bar{x}..)^2$$

$$= \sum_{j=1}^{k} \sum_{i=1}^{n_j} (x_{ij} - \bar{x}._j)^2 + 2 \sum_{j=1}^{k} (\bar{x}._j - \bar{x}..) \sum_{i=1}^{n_j} (x_{ij} - \bar{x}._j) + \sum_{j=1}^{k} n_j(\bar{x}._j - \bar{x}..)^2$$

$$= \sum_{j=1}^{k} \sum_{i=1}^{n_j} (x_{ij} - \bar{x}._j)^2 + \sum_{j=1}^{k} n_j(\bar{x}._j - \bar{x}..)^2. \tag{8.6}$$

$\left(\text{Since } \sum_{i=1}^{n_j} (x_{ij} - \bar{x}._j) = 0 \text{ the cross-product term is zero.}\right)$ The first sum of squares in expression (8.6), namely:

$$\sum_{j=1}^{k} \sum_{i=1}^{n_j} (x_{ij} - \bar{x}._j)^2 = \sum_{j=1}^{k} \sum_{i=1}^{n_j} (z_{ij} - \bar{z}._j)^2$$

involves only error terms and may be shown to estimate

$$\sigma^2 \sum_{j=1}^{k} (n_j - 1)$$

whether the values of the E_j's are zero or not. The expression

$$\sum_{i=1}^{n_j} (z_{ij} - \bar{z}._j)^2$$

clearly estimates $\sigma^2(n_j - 1)$ so that

$$\sum_{j=1}^{k} \sum_{i=1}^{n_j} (z_{ij} - \bar{z}._j)^2$$

estimates

$$\sigma^2 \sum_{j=1}^{k} (n_j - 1)$$

and

$$\hat{\sigma}_2^2 = \frac{1}{\sum_{j=1}^{k} (n_j - 1)} \sum_{j=1}^{k} \sum_{i=1}^{n_j} (x_{ij} - \bar{x}._j)^2$$

estimates σ^2.

The second sum of squares in expression (8.6)

$$\sum_{j=1}^{k} n_j(\bar{x}._j - \bar{x}..)^2 = \sum_{j=1}^{k} (E_j + \bar{z}._j - \bar{z}..)^2$$

may be shown to estimate:

$$(k-1)\sigma^2 + \sum_{j=1}^{k} n_j E_j^2$$

so that the estimate $\hat{\sigma}_1^2$ of σ^2, given by

$$\hat{\sigma}_1^2 = \frac{1}{k-1} \sum_{j=1}^{k} n_j(\bar{x}._j - \bar{x}..)^2,$$

is increased by the presence of non-zero E_j's. The presence of non-zero E_j's may therefore be tested by the ratio of $\hat{\sigma}_1^2$ to $\hat{\sigma}_2^2$ using the F-test described in the previous chapter. It should be stressed that this comparison is possible because the two estimates $\hat{\sigma}_1^2$ and $\hat{\sigma}_2^2$ are independent estimates of σ^2.

The result of this comparison, and of others in the general application of analysis of variance, are normally presented in the form of an analysis of variance table which lists the sources of variation (only two in this example) that are being considered in the experiment. The analysis of variance table for this model is given in Table 8.1.

TABLE 8.1. ANALYSIS OF VARIANCE TABLE

Source of variation	Sum of squares	Degrees of freedom	Mean square	F-ratio
Between pop. means	$S_1 = \sum_{j=1}^{k} n_j(\bar{x}._j - \bar{x}..)^2$	$\nu_1 = k - 1$	$S_1/\nu_1 = M_1$	M_1/M_2
Residual variability	$S_2 = \sum_{j=1}^{k} \sum_{i=1}^{n_j} (x_{ij} - \bar{x}._j)^2$	$\nu_2 = \sum_{j=1}^{k} (n_j - 1)$	$S_2/\nu_2 = M_2$	
Total	$\sum_{j=1}^{k} \sum_{i=1}^{n_j} (x_{ij} - \bar{x}..)^2$	$\sum_{j=1}^{k} n_j - 1$		

8.1.2. Computation

The computation procedure using an electronic computer working to a fixed number of significant figures would be to compute the grand mean and the sample means and add up the sums of squares of the deviations as suggested by the sums S_1 and S_2 above. Reasons for the use of this procedure on an electronic computer were given in Section 4.5.2.

The computation procedure using a desk machine, however, may be shortened by expanding the above sums S_1 and S_2 as follows:

$$S_1 = \sum_{j=1}^{k} \frac{X_{.j}^2}{n_j} - \frac{X_{..}^2}{N}$$

$$S_2 = \sum_{j=1}^{k} \sum_{i=1}^{n_j} x_{ij}^2 - \sum_{j=1}^{k} \frac{X_{.j}^2}{n_j}$$

where $X_{.j}$ is the sum of the observations in the jth sample and $X_{..}$ is the sum of all observations. That is

$$X_{.j} = \sum_{i=1}^{n_j} x_{ij}$$

and

$$X_{..} = \sum_{j=1}^{k} \sum_{i=1}^{n_j} x_{ij}$$

and where

$$N = \sum_{j=1}^{k} n_j.$$

Two of the sums

$$\sum_{j=1}^{k} \sum_{i=1}^{n_j} x_{ij} \quad \text{and} \quad \sum_{j=1}^{k} \sum_{i=1}^{n_j} x_{ij}^2$$

needed in the calculation may be computed simultaneously on a desk machine so that the amount of computing is considerably reduced.

8.1.3. Example

We will use the following data as an example of the application of this analysis of variance model. Three experimentalists have each made a number of determinations of the yield of substance B from a certain reaction. The null hypothesis to be tested is that the three experimentalists do not differ in their determinations. The determinations are given in Table 8.2.

TABLE 8.2. DATA FOR EXAMPLE 8.1.3

	Determinations (in grams)	Totals
Experimentalist 1	304·2, 312·3, 309·8, 310·2, 306·5	1543·0
Experimentalist 2	284·1, 294·3, 288·3, 286·1	1152·8
Experimentalist 3	300·2, 301·3, 310·0, 304·1, 305·2, 308·6	1829·4

The study of experimental data normally begins, where possible, with a plot of the data such as that given in Fig. 8.1. In this plot the dotted lines represent the means for each experimentalist.

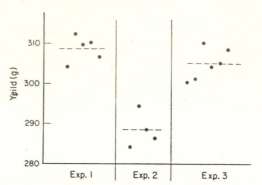

FIG. 8.1. Plot of data in Example 8.1.3.

The conclusions that most experimentalists would draw from this plot are that all three experimentalists are equally variable, that experimentalist 2 is consistently low and that experimentalists 1 and 3 differ only slightly if they differ at all. These conclusions are clear from the plot and the formal analysis is scarcely necessary. The analysis is performed, however, as an example of the computation. The analysis of variance method makes the assumption that the error terms z_{ij} have the same variance and the validity of this assumption should be checked before the analysis is performed. This assumption appears reasonable from the plot but the details of Bartlett's test and the maximum F ratio test are recorded below:

1. Experimentalist   Variance estimate   D. of F.   Log (var. est.)

j	$\hat{\sigma}_j^2$	ν_j	$\log_{10} \hat{\sigma}_j^2$
1	10·365	4	1·0156
2	19·480	3	1·2896
3	15·096	5	1·1789

2. *Combined estimate of variance $\hat{\sigma}^2$ and the first term*

$$\hat{\sigma}^2 = \frac{1}{12}(4 \times 10{\cdot}365 + 3 \times 19{\cdot}480 + 5 \times 15{\cdot}096)$$

$$= 14{\cdot}615$$

$$\log_{10} \hat{\sigma}^2 = 1{\cdot}1648$$

$$\sum_{j=1}^{k} \nu_j \log_{10} \hat{\sigma}^2 = 12 \times 1{\cdot}1648 = 13{\cdot}9776.$$

3. The second term

$$\sum_{j=1}^{k} v_i \log_{10} \hat{\sigma}_j^2 = 4 \times 1{\cdot}0156 + 3 \times 1{\cdot}2896 + 5 \times 1{\cdot}1789$$

$$= 13{\cdot}8257.$$

4. Bartlett's M

$$M = 2{\cdot}302585 \,(13{\cdot}9776 - 13{\cdot}8257)$$

$$= 0{\cdot}350.$$

5. First-order correction M'

$$\sum_{j=1}^{k} \frac{1}{v_j} = \frac{1}{4} + \frac{1}{3} + \frac{1}{5} = \frac{47}{60}$$

$$1 \Big/ \sum_{j=1}^{k} v_j = \frac{1}{12}$$

$$M' = M \Big/ \left\{ 1 + \left[\left(\frac{47}{60} - \frac{1}{12} \right) \Big/ (3 \times 2) \right] \right\} = \frac{60M}{67}$$

$$= 0{\cdot}273.$$

6. Pr $\{\chi^2 \geqq 0{\cdot}273 \mid v = 2\} = 0{\cdot}98.$

The agreement between the three estimates of variability is obviously good. The combined estimate of variance is 14·615 with 12 degrees of freedom.

7. The maximum F-ratio test

The ratio of the largest to the smallest variance estimate is $19{\cdot}480/10{\cdot}365 = 1{\cdot}88$. The tables for this test assume that each variance estimate has the same degrees of freedom but they may be used approximately for unequal degrees of freedom by entering the table with an average value. The tabulated upper 5 per cent point for the distribution of this ratio for $k = 3$ and $v = 4$ is 15·5 so that the observed value of 1·88 is well inside the upper 5 per cent point.

The conclusion from both these tests is that the experimenters are determining the yield equally precisely.

8. The analysis of variance computation

$$\sum_{j=1}^{k} \sum_{i=1}^{n_j} x_{ij}^2 = 304{\cdot}2^2 + 312{\cdot}3^2 + \cdots + 308{\cdot}6^2 = 1366366{\cdot}20$$

$$\sum_{j=1}^{k} \frac{X_{\cdot j}^2}{n_j} = \frac{1543{\cdot}0^2}{5} + \frac{1152{\cdot}8^2}{4} + \frac{1829{\cdot}4^2}{6} = 1366190{\cdot}82$$

$$\frac{X_{..}^2}{N} = \frac{4525 \cdot 2^2}{15} = 1365162 \cdot 34.$$

$$S_1 = 1366190 \cdot 82 - 1365162 \cdot 34 = 1028 \cdot 48,$$

$$S_2 = 1366366 \cdot 20 - 1366190 \cdot 82 = 175 \cdot 38.$$

The analysis of variance is given in Table 8.3.

TABLE 8.3. ANALYSIS OF VARIANCE OF EXAMPLE 8.1.3 DATA

Source of variation	S.S.	D.F.	M.S.	F.
Between experimentalists	1028·48	2	514·24	35·19
Residual	175·38	12	14·615	
Total	1203·86	14		

$$\text{Pr} \{F \geqq 35 \cdot 19 \mid \nu_1 = 2, \nu_2 = 12\} < 0 \cdot 001.$$

Having obtained a probability of less than 0·001 we feel confident in concluding that the three experimentalists differ in their determination of the yield.

The formal analysis of the data as outlined above has confirmed the main conclusions that were drawn from the plot in Fig. 8.1. It remains, however, to determine whether or not experimentalists 1 and 3 differ significantly in their determination of yield. The difference between the means for experimentalists 1 and 3 may be tested by using Student's t-test. The usual t-test described in Section 6.1.3 uses an estimate of variance obtained from the two samples involved. In the present situation, however, we have established that the *three* experimentalists are working with equal precision so that the variability estimate of 14·615 obtained in the analysis of variance is a better estimate of precision than would be obtained from the data for experimentalists 1 and 3 alone. The t-test computation is as follows:

$$t = \frac{308 \cdot 6 - 304 \cdot 9}{\sqrt{\left\{14 \cdot 615 \left(\frac{1}{5} + \frac{1}{6}\right)\right\}}} = \frac{3 \cdot 7}{2 \cdot 315} = 1 \cdot 60$$

$$\text{Pr} \{t \geqq 1 \cdot 60 \mid \nu = 12\} = 0 \cdot 136.$$

If we accept 0·05 as our criterion of significance we conclude that experimentalists 1 and 3 do not differ significantly although we might regard the probability of 0·136 as being close enough to the significance probability for the conclusion, one way or the other, to be doubtful.

8.1.4. Confidence Limits for Means

Having established that there are differences between the determinations made by the three experimentalists we may describe the determinations made by each experimentalist by calculating the mean determination and confidence limits for each experimentalist. The confidence limits may be calculated from the following expression obtained in Section 5.2.7:

$$\bar{x}_{.j} - t_{\nu,\alpha}\sqrt{(\hat{\sigma}^2/n_j)} \quad \text{to} \quad \bar{x}_{.j} + t_{\nu,\alpha}\sqrt{(\hat{\sigma}^2/n_j)}$$

where $\bar{x}_{.j}$ is the jth experimentalist mean,

$\quad n_j$ is the number of determinations made by the jth experimentalist,

$\quad \hat{\sigma}^2$ is the estimate of residual variability,

$\quad t_{\nu,\alpha}$ is the 100α significance point of the t-distribution with ν degrees of freedom,

$\quad \nu$ is the degrees of freedom associated with $\hat{\sigma}^2$.

Calculating the 95 per cent confidence limits for the experimentalist means we have

$$\hat{\sigma}^2 = 14.615; \quad \nu = 12$$

$$t_{\nu,\,0.025} = 2.179$$

and 95 per cent confidence limits are given by

$$\bar{x}_{.j} - 2.179\sqrt{(14.615/n_j)} \quad \text{to} \quad \bar{x}_{.j} + 2.179\sqrt{(14.615/n_j)}$$

substituting $\bar{x}_{.j} = 308.6,\ 288.2,\ 304.9$ and $n_j = 5,\ 4,\ 6$, respectively, for the three experimentalists we have the description given in Table 8.4.

TABLE 8.4. SUMMARY OF DATA OF EXAMPLE 8.1.3

Experi-mentalist	Sample size	Mean	95 per cent limits
1	5	308.6	304.9 to 312.3
2	4	288.2	284.0 to 292.4
3	6	304.9	301.5 to 308.3

8.1.5. The Analysis of Variance Approach (Random-effect Model)

The ith member of the jth sample may be represented by the following model:

$$x_{ij} = A + R_j + z_{ij}, \tag{8.7}$$

where A is a general level common to all observations

$\quad R_j$ is a random departure from A due to the jth population, is common to all members of the jth population, and is selected from a population with zero mean and variance σ_r^2.

z_{ij} is a random error having a normal distribution with zero mean and residual variance σ^2.

Using the fixed-effect model we attempt to establish differences between groups (for example experimenters) which are well defined and are such that a repeat experiment would represent the same groups. In the random-effect model, on the other hand, we seek to describe variation between groups which have occurred by chance so that a repeat experiment would not represent the same groups. An example of a random-effect model may be constructed as follows. A quantity of coal produced at one coal-mine is loaded into a number of railway trucks. If we now perform a number of determinations of the ash-content of the coal for each of the trucks the data obtained would be of the same form as that analysed above. We are interested in determining whether or not there is any difference between the truck contents. If such differences were to be found they would represent a lack of homogeneity in the coal produced by the coal-mine rather than any difference between railway trucks. If the trucks were to be loaded a second time and the experiment repeated any differences found again between trucks would almost certainly not be the same as those found in the first loading. That is, if the trucks were ordered separately in both loadings according to the ash-content of the coal they contain, the two orders so obtained would not be expected to be the same. If the three experimentalists, however, in the above fixed-effect example were to repeat their determinations we would not be surprised if the experimentalist order of 2, 3, 1 was repeated in the second set of determinations. That is we expect the experimentalists to be consistent in their determinations whether consistently high or low whereas we do not expect the same coal truck to always contain the coal with the highest ash-content.

The difference in terms of the models is that E_j is a fixed value for the jth population whereas R_j is a value selected from a population of values with expected value zero and variance σ_r^2. The hypothesis to be tested in the random model case is that σ_r^2 is zero. The derivation of the analysis of variance of this model is very similar to that of the fixed-effect model. In fact with this particular experimental design the formal analysis is the same. The between groups sum of squares

$$\sum_{j=1}^{k} n_j(\bar{x}_{.j} - \bar{x}_{..})^2$$

in the fixed-effect model estimates

$$(k-1)\sigma^2 + \sum_{i=1}^{k} n_j E_j^2$$

whereas in the random-effect model it estimates

$$(k - 1)\,\sigma^2 + \left(N - \frac{\sum\limits_{j=1}^{k} n_j^2}{N}\right)\sigma_r^2.$$

It is thus seen that the presence of a non-zero σ_r^2 may be tested for in the same way as that used in testing for non-zero E_j's. Although there is no difference between the two analyses for this experimental design, differences will be found in designs that are discussed below when a clear understanding of the different assumptions will become important.

8.1.6. The Three Types of Model

Models corresponding to experimental designs of a more complicated nature than that described above contain more than one term of the type E_j or R_j. A fixed-effect model is one in which all terms in the model are fixed type such as E_j. A random-effect model is one in which all terms in the model are of the random type. A mixed model is one in which some of the terms are of the fixed type and some of are of the random type.

8.2. Fixed-effect Model for a Hierarchical Design with Three Levels

Suppose that the three experimentalists discussed in the above example all work in the same laboratory and that several determinations of the yield of substance B have been made by each of two experimentalists working in a second laboratory. The data obtained could now be classified according to laboratories, experimentalists and determinations. This form of data is known as a hierarchical classification with three levels. The symbol x_{ijl} will be used to denote the ith determination made by the jth experimentalist in the lth laboratory. The fixed-effect analysis of variance model for this design is:

$$x_{ijl} = A + L_l + E_{jl} + z_{ijl}, \tag{8.8}$$

where A is a general level common to all observations,

L_l is a fixed-effect due to the lth laboratory,

E_{jl} is a fixed-effect due to the jth experimentalist in the lth laboratory,

z_{ijl} is a random error with a normal distribution with zero mean and variance σ^2.

The notation to be used in the following description will be:

m = the number of laboratories,

k_l = the number of experimentalists in the lth laboratory,

n_{jl} = the number of determinations made by the jth experimentalist in the lth laboratory,

$n_{.l}$ = the number of determinations made in the lth laboratory,

N = the total number of observations.

From the model (8.8) we may write down expressions for the experimentalists, laboratories, and grand means.

$$\bar{x}_{.jl} = A + L_l + E_{jl} + \bar{z}_{.jl}$$

$$\bar{x}_{..l} = A + L_l + \bar{z}_{..l} \qquad (8.9)$$

$$\bar{x}_{...} = A + \bar{z}_{...}$$

To obtain these expressions we have imposed the conditions

$$\sum_{j=1}^{k_l} n_{jl}E_{jl} = 0 \quad \text{and} \quad \sum_{l=1}^{m} n_{.l}L_l = 0.$$

These state that the total deviation of experimentalists from their laboratory mean is zero and that the total variation of laboratories from the grand mean is zero. It is necessary to make the assumption that these sums are constant. The assumption of a constant other than zero will simply alter the value of A correspondingly without affecting the sums of squares in the analysis so that the choice of the value zero, being the simplest value, therefore seems an obvious choice. These equations together with the original model enable us to determine how the total sum of squares is to be partitioned. We have:

$$x_{ijl} - \bar{x}_{.jl} = \bar{z}_{ijl} - \bar{z}_{.jl}$$

$$\bar{x}_{.jl} - \bar{x}_{..l} = \bar{z}_{.jl} - \bar{z}_{..l} + E_{jl} \qquad (8.10)$$

$$\bar{x}_{..l} - \bar{x}_{...} = \bar{z}_{..l} - \bar{z}_{...} + L_l .$$

So that the total sum of squares is split up in the following way:

$$S = \sum_{l=1}^{m}\sum_{j=1}^{k_l}\sum_{i=1}^{n_{jl}} (x_{ijl} - \bar{x}_{...})^2$$

$$= \sum_{l=1}^{m}\sum_{j=1}^{k_l}\sum_{i=1}^{n_{jl}} ((x_{ijl} - \bar{x}_{.jl}) + (\bar{x}_{.jl} - \bar{x}_{..l}) + (\bar{x}_{..l} - \bar{x}_{...}))^2$$

$$= \sum_{l=1}^{m}\sum_{j=1}^{k_l}\sum_{i=1}^{n_{jl}} (x_{ijl} - \bar{x}_{.jl})^2 + \sum_{l=1}^{m}\sum_{j=1}^{k_l}\sum_{i=1}^{n_{jl}} (\bar{x}_{.jl} - \bar{x}_{..l})^2 + \sum_{l=1}^{m}\sum_{j=1}^{k_l}\sum_{i=1}^{n_{jl}} (\bar{x}_{..l} - \bar{x}_{...})^2 .$$

(All cross-product terms being zero.)

$$= \sum_{l=1}^{m}\sum_{j=1}^{k_l}\sum_{i=1}^{n_{jl}} (x_{ijl} - \bar{x}_{.ll})^2 + \sum_{l=1}^{m}\sum_{j=1}^{k_l} n_{jl}(\bar{x}_{.jl} - \bar{x}_{..l})^2 + \sum_{l=1}^{m} n_{.l}(\bar{x}_{..l} - \bar{x}_{...})^2 .$$

$$(8.11)$$

The analysis of variance is given in Table 8.5.

TABLE 8.5. ANALYSIS OF VARIANCE OF A THREE-LEVEL HIERARCHICAL EXPERIMENT

Source	Sum of squares	D. of F.
Laboratories	$S_1 = \sum_{l=1}^{m} n_{.l}(\bar{x}_{..l} - \bar{x}_{...})^2$	$\nu_1 = m - 1$
Experimental-ists within labs.	$S_2 = \sum_{l=1}^{m} \sum_{j=1}^{k_l} n_{jl}(\bar{x}_{.jl} - \bar{x}_{..l})^2$	$\nu_2 = \sum_{l=1}^{m} (k_l - 1)$
Residual	$S_3 = \sum_{l=1}^{m} \sum_{j=1}^{k_l} \sum_{i=1}^{n_{jl}} (x_{ijl} - \bar{x}_{.jl})^2$	$\nu_3 = \sum_{l=1}^{m} \sum_{j=1}^{k_l} (n_{jl} - 1)$
Total	$\sum_{l=1}^{m} \sum_{j=1}^{k_l} \sum_{i=1}^{n_{jl}} (x_{ijl} - \bar{x}_{...})^2$	$\sum_{l=1}^{m} \sum_{j=1}^{k_l} (n_{jl}) - 1$

The computation will be illustrated by an example included in the next section.

The three mean squares corresponding to the three sums of squares listed in the above table may be shown in the case of the fixed-effect model to estimate respectively:

$$M_1 = S_1/\nu_1 \text{ estimates } \sigma^2 + \sum_{l=1}^{m} n_{.l}L_l^2/\nu_1$$

$$M_2 = S_2/\nu_2 \text{ estimates } \sigma^2 + \sum_{l=1}^{m} \sum_{j=1}^{k_l} n_{jl}E_{jl}^2/\nu_2$$

$$M_3 = S_3/\nu_3 \text{ estimates } \sigma^2.$$

The presence of non-zero values of L_l are detected by testing M_1 against M_3 and the presence of non-zero values of E_{jl} are detected by testing M_2 against M_3.

8.2.1. Random Model for a Hierarchical Design with Three Levels

The random model for a three-level hierarchical design is

$$x_{ijl} = A + R_l + S_{jl} + z_{ijl}, \tag{8.12}$$

where A is a general level,

R_l is a random effect due to the lth group and is drawn from a population with zero mean and variance σ_r^2,

S_{jl} is a random effect due to the jth subgroup in the lth group and is drawn from a population with zero mean and variance σ_s^2,

z_{ijl} is a random error drawn from a normal population with zero mean and variance σ^2.

The null hypotheses to be tested are that $\sigma_r^2 = 0$ and $\sigma_s^2 = 0$. That is, that there is no variation between the groups or between the subgroups within a group.

The analysis of variance for this model differs from that described for the fixed-effect model in Section 8.2 in the following important respect. The mean squares M_1 and M_2 obtained in the random model analysis are not independent unless $n_{jl} = n$ for all values of j and l. That is unless the same number of observations (determinations) are made in each subgroup (experimentalists). The mean squares M_1, M_2 and M_3 may be shown to estimate

$$M_1 = S_1/\nu_1 \text{ estimates } \sigma^2 + c_1\sigma_r^2 + c_2\sigma_s^2$$

$$M_2 = S_2/\nu_2 \text{ estimates } \sigma^2 + c_3\sigma_r^2$$

$$M_3 = S_3/\nu_3 \text{ estimates } \sigma^2.$$

If $n_{jl} \neq n$ for all values of j and l the constants c_1 and c_3 are not equal and the ratio M_1/M_2 does not have an F-distribution. If $n_{jl} = n$ for all values of j and l it may be shown that

$$c_1 = c_3 = n,$$

and

$$c_2 = \left(N - \frac{n^2 \sum_{l=1}^{m} k_l^2}{N}\right) \bigg/ (m - 1).$$

If, in addition $k_l = k$ for all values of l, that is if there are the same number of subgroups (experimentalists) in each group (laboratory), $c_2 = nk$ (= number of observations per group).

The presence of a non-zero σ_r^2 may be tested by comparing M_2 with M_3 as in the fixed-effect model. If, however, we test M_1 against M_3 as in the fixed-effect model and obtain a significant result it is not possible to attribute the significance to the presence of a non-zero value of σ_s^2 alone. The significance could be caused by non-zero values of σ_r^2 or σ_s^2 or by the presence of both. The testing procedure in the case of the random-effect model is therefore to test M_2 against M_3 to establish, or deny, the presence of a non-zero σ_r^2. If M_2 is not significantly different from M_3, M_1 may be tested against M_3 since the presence of σ_r^2 has been denied. If M_2 is significantly greater than M_3 and if $n_{jl} = n$, M_1 may be tested against M_2 since the ratio M_1/M_2 has unit expectation if σ_s^2 is zero.

It is usually the case that the degrees of freedom ν_3 are greater than ν_2 (it is just possible for ν_2 to be greater than ν_3 if most, but not all, values of n_{jl} are 1) so that the comparison of M_1 against M_3 is better from a power point of view than against M_2. Thus if in designing an experiment it is believed that σ_r^2 could exist the important factor in testing for a difference between groups is the number of subgroups rather than the number of observations made in each subgroup.

If the presence of σ_r^2 is established m_2 will be larger than M_3 so that the variability against which the presence of σ_s^2 must be tested is larger than if σ_r^2 is zero. This is another important factor to consider when determining the number of observations to be made in each subgroup and the number of subgroups.

8.2.2. A Mixed Model

The following mixedmodel is an interesting possibility.

$$x_{ijl} = A + L_l + S_{jl} + z_{ijl},$$

where L_l is a fixed laboratory effect and S_{jl} is a random effect due to an experimentalist chosen at random from a population of experimentalists.

The laboratory effect is regarded as a fixed effect which will be reproduced in repeated experiments but the variability about the fixed laboratory effect due to different experimentalists is random and the actual departures will not be reproduced in repeated experimentation. The expected values of the analysis of variance mean squares are:

$$M_1 \text{ estimates } \sigma^2 + c_1\sigma_r^2 + n \sum_{l=1}^{m} k_l L_l$$

$$M_2 \text{ estimates } \sigma^2 + c_3\sigma_r^2$$

$$M_3 \text{ estimates } \sigma^2,$$

where $c_1 = c_3 = n$ if $n_{jl} = n$.

The testing procedure for this model is then seen to be the same as for the random model described in Section 8.2.1.

8.2.3. Example

Three experimentalists in laboratory A and two experimentalists in laboratory B have each made four determinations of the yield of a substance B obtained in a chemical reaction. The yields obtained are recorded in Table 8.6 and plotted in Fig. 8.2.

It is reasonable to conclude from the plot of this data that all five experimentalists are working to the same precision but it is a little difficult to say whether or not there are differences between experimentalists within laboratory A. The experimentalists in laboratory B do appear to differ but the

TABLE 8.6. DATA FOR EXAMPLE 8.2.3

						Totals
Lab. A	Exp. 1	306·9	310·8	302·2	301·7	1221·6
	Exp. 2	297·8	300·1	299·8	306·3	1204·0
	Exp. 3	304·8	312·6	308·6	307·2	1233·2
				Laboratory A total		3658·8
Lab. B	Exp. 4	304·0	312·4	308·2	309·4	1234·0
	Exp. 5	301·1	304·4	302·0	298·9	1206·4
				Laboratory B total		2440·4

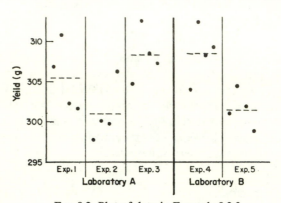

FIG. 8.2. Plot of data in Example 8.2.3.

two laboratories do not. The formal analysis would start by computing the mean and variance estimates for each experimentalist as recorded in Table 8.7.

TABLE 8.7. FIRST SUMMARY OF DATA OF EXAMPLE 8.2.3

	Mean	Var. est.	Log_{10} (var. est.)
Laboratory A			
Exp. 1	305·4	18·447	1·26593
Exp. 2	301·0	13·527	1·13120
Exp. 3	308·3	10·680	1·02857
Laboratory B			
Exp. 4	308·5	12·120	
Exp. 5	301·6	5·180	

The consistency of the variance estimates in laboratory A may be tested by computing Bartlett's statistic.

1. *Combined estimate of variance and the first term*

$$\hat{\sigma}^2 = \frac{1}{9}(3 \times 18\cdot447 + 3 \times 13\cdot527 + 3 \times 10\cdot680) = 14\cdot218$$

$$\sum_{j=1}^{k} v_j \log_{10}\hat{\sigma}^2 = 9 \times 1\cdot15284 = 10\cdot37556.$$

2. *The second term*

$$\sum_{j=1}^{k} v_j \log_{10}\hat{\sigma}_j^2 = 3 \times 1\cdot26593 + 3 \times 1\cdot13120 + 3 \times 1\cdot02857 = 10\cdot27710.$$

3. *Bartlett's statistic M*

$$M = (10\cdot37556 - 10\cdot27710) \times 2\cdot302585 = 0\cdot227.$$

4. *First order correction M'*

$$M' = 0\cdot227 \times 27/31 = 0\cdot198$$

$$\Pr\{\chi^2 \geqq 0\cdot198 \mid v = 2\} = 0\cdot9.$$

It seems reasonable, therefore, to believe that the experimentalists are working to the same precision.

The consistency of the variance estimates for laboratory B may be tested, since only two estimates are involved, by the F test.

$$F = 12\cdot120/5\cdot180 = 2\cdot34$$

$$\Pr\{F \geqq 2\cdot34 \mid v_1 = v_2 = 3\} \doteqdot 0\cdot25.$$

It is therefore reasonable to conclude that these two estimates are consistent. Having established that the experimentalists are equally precise within both laboratories we may now compare the mean determination by computing the two-level hierarchical analyses of variance for each laboratory.

Laboratory A

Between experimentalists S.S. $= \dfrac{1221\cdot6^2}{4} + \dfrac{1204\cdot0^2}{4} + \dfrac{1233\cdot2^2}{4} - \dfrac{3658\cdot8^2}{12}.$

$$= 108\cdot08$$

Total S.S. $= (306\cdot9^2 + 310\cdot8^2 + \cdots + 307\cdot2^2) - 3658\cdot8^2/12 = 236\cdot04.$

The analysis of variance for laboratory A is given in Table 8.8.

TABLE 8.8. ANALYSIS OF VARIANCE FOR LABORATORY A

Source	S.S.	D.F.	M.S.	F.
Experimentalists	108·08	2	54.04	3·80
Residual	127·96	9	14·218	
Total	236·04	11		

$$\Pr\{F \geqq 3\cdot80 \mid \nu_1 = 2,\ \nu_2 = 9\} > 0\cdot05$$
$$< 0\cdot10$$

Laboratory B

Between experimentalists S.S. $= \dfrac{1234\cdot0^2}{4} + \dfrac{1206\cdot4^2}{4} - \dfrac{2440\cdot4^2}{8} = 95\cdot22$.

Total S.S. $= (304\cdot1^2 + 312\cdot4^2 + \cdots + 298\cdot9^2) - \dfrac{2440\cdot4^2}{8} = 147\cdot12$.

The analysis of variance for laboratory B is given in Table 8.9.

TABLE 8.9. ANALYSIS OF VARIANCE FOR LABORATORY B

Source	S.S.	D.F.	M.S.	F.
Experimentalists	95·22	1	95·22	11·01
Residual	51·90	6	8·650	
Total	147·12	7		

$$\Pr\{F \geqq 11\cdot01 \mid \nu_1 = 1,\ \nu_2 = 6\} < 0\cdot025$$

The differences between the experimentalists in laboratory A produce an F-ratio that is close to the 5 per cent significance level. Although this evidence alone (assuming a 5 per cent level has been agreed on) is not sufficient for us to reject the null hypothesis of no difference it is sufficiently close to the chosen significance level for us to hesitate in accepting the null hypothesis. The decision in the case of laboratory B is clearer since the value of F is well outside the 95 per cent acceptance region and we feel confident in concluding that there are differences between the two experimentalists in laboratory B.

The residual mean squares for the two laboratories, 14·218 and 8·650, are not significantly different so that the combined analysis testing for laboratory differences may now be performed.

Between laboratories S.S. $= \dfrac{3658\cdot8^2}{12} + \dfrac{2440\cdot4^2}{8} - \dfrac{6099\cdot2^2}{20}$

$$= 1860012\cdot14 - 1860012\cdot03 = 0\cdot11.$$

Between experimentalists within laboratories

$$= \frac{1221 \cdot 6^2}{4} + \frac{1204 \cdot 0^2}{4} + \frac{1233 \cdot 2^2}{4} + \frac{1234 \cdot 0^2}{4} + \frac{1206 \cdot 4^2}{4} - \frac{3658 \cdot 8^2}{12}$$

$$- \frac{2440 \cdot 4^2}{8} = 1860215 \cdot 44 - 1860012 \cdot 14 = 203 \cdot 30.$$

Total S.S. $= 306 \cdot 9^2 + 310 \cdot 8^2 + \cdots + 298 \cdot 9^2 - 6099 \cdot 2^2/20$

$$= 1860395 \cdot 30 - 1860012 \cdot 03 = 383 \cdot 27.$$

The combined analysis of variance is given in Table 8.10.

TABLE 8.10. ANALYSIS OF VARIANCE FOR BOTH LABORATORIES

Source	S.S.	D.F.	M.S.	F.	Prob.
Laboratories	0·11	1	0·11	0·01	>0·5
Exp./laboratories	203·30	3	67·77	5·65	<0·001
Residual	179·86	15	11·99		
Total	383·27	19			

The F-ratios in the above analysis have been computed assuming a fixed-effect model and the conclusions are that there are significant differences between experimentalists within laboratories but no difference between laboratories.

It is possible to consider the experimentalist effect in this example to be a random effect. That is the experimentalists were chosen at random from a population of experimentalists. Repeated experimentation would be carried out with different experimentalists representing the two laboratories. The alternative assumption that the departure from the general level A due to each experimentalist is chosen at random from a population with zero mean and variance σ_r^2 (that is each experimentalist is behaving differently on different occasions) is a practical possibility. It is clear that if the experimentalist effect is assumed to be random the conclusions would be unchanged since the difference between the two laboratories is so small.

8.3. EXAMPLE OF A FOUR-LEVEL HIERARCHICAL DESIGN (MIXED MODEL)

The hierarchical design may be extended to contain any number of levels. The example experiment now described is an example of a hierarchy with four levels.

Four specimens of glass, two of each of two types of glass, were separately agitated for a period of a week in four containers each containing 10 cm^3 of distilled water. At the end of the week three determinations of the amount of glass dissolved in the water were made for each of the four 10 cm^3 "solutions". Each specimen was then further agitated in fresh water for a second week after which three determinations of the glass in solution were made for each of these solutions. The observation used in the analysis was the logarithm of the amount of a constituent of the glass in solution. The analysis of variance model for this experimental design is:

$$x_{ltij} = A + G_l + S_{lt} + W_{lti} + z_{ltij},$$

where A is a general level,

G_l is a fixedeffect due to the lth glass,

S_{lt} is a randomeffect due to the tth specimen for the lth glass,

W_{lti} is a randomeffect due to the ith week for specimen S_{lt},

z_{ltij} is a random error normally distributed with zero mean and variance σ^2.

The sums of squares are calculated in this analysis by a generalization of the above analyses. The analysis of variance table is given in Table 8.11.

TABLE 8.11. ANALYSIS OF VARIANCE OF A FOUR-LEVEL HIERARCHICAL EXPERIMENT

Source	S.S.	D.F.	M.S.	F.	Prob.
Glasses	11·8324	1	11·8324	6·2	>0·05
Specimens/glasses	3·7806	2	1·8903	10·28	0·05
Weeks/specimens/glasses	0·7357	4	0·1839	8·78	0·001
Determination/W/S/G	0·3349	16	0·02093		
Total	16·6836	23			

The conclusions are:

1. There is significantly more variability between weeks than between determinations. This may be due to either more variability between solutions than between determinations made on the same solution, or a genuine change in the leaching properties of the glass type from week to week. (Determinations do not contain solution differences and are therefore repeats with respect to weeks.)
2. There is significantly more variability between specimens than between weeks for the same specimen.
3. There is no significant difference between types of glasses as compared with different specimens of the same glass.

The F-ratio testing for glass differences has only 1 and 2 degrees of freedom in spite of the fact that twenty-four observations have been made. The glass comparison is made against a residual of $1\cdot8906$ whereas the weeks comparison is made against a residual of $0\cdot02093$. To improve the power of the glass comparison more specimens of each glass must be used. An increase in the number of determinations will have virtually no effect on the power of the glass comparison.

8.4. The General Method of Computation

The general method of computation of the analysis of variance of a hierarchical design may be described as follows. The computation involves entirely terms of the type

$$\sum_i \frac{T_i^2}{N_i},$$

where T_i is the total of N_i observations, and the summation is taken over all totals of the type T_i. For the lowest level (level k) the T_i's are the individual observations, the N_i are all unity and the summation is over all observations—that is the summation is k-dimensional. For level $k - 1$ the T_i are sums for all members of the second classification (for example, for all experimentalists) the N_i are the number of observations in each member of the second classification (the number of determinations made by each experimentalist) and the summation is over all members of the second classification (the number of experimentalists). If these sums are listed in order as in Table 8.12 the sums of squares are obtained as the differences between consecutive sums in this sequence.

TABLE 8.12. COMPUTATION OF A k-LEVEL HIERARCHICAL EXPERIMENT

Level	Sums	Differences	Sums of squares
k	$S_k = \sum_{i(k)} x_i^2$		
		$S_k - S_{k-1}$	Level k (Residual)
$k - 1$	$S_{k-1} = \sum_{i(k-1)} \frac{T_i^2(k-1)}{N_i(k-1)}$		
. . . .	. . . .	. . . .	. . . .
2	$S_2 = \sum_{i(2)} \frac{T_i^2(2)}{N_i(2)}$		
		$S_2 - S_1$	Level 2
1	$S_1 = \sum_{i(1)} \frac{T_i^2(1)}{N_i(1)}$		
		$S_1 - S_0$	Level 1
Total	$S_0 = \frac{T^2}{N}$		

Referring to Table 8.12 where the $T_i(j)$'s are the totals of $N_i(j)$ observations for all members of level j, where the summation is over all members of level j, and where T is the total of all N observations.

This table is illustrated by the computation for the three-level hierarchical experiment analysed in Section 8.2.3.

TABLE 8.13. EXAMPLE COMPUTATION BY THE GENERAL METHOD

Level		Differences	S.S.
1	$306 \cdot 9^2 + 310 \cdot 8^2 + \cdots + 298 \cdot 9^2 = 1860395 \cdot 30$		
2	$1221 \cdot 6^2/4 + \cdots + 1206 \cdot 4^2/4 = 1860215 \cdot 44$	179·86	Residual
		203·30	Experi-
			mentalists
3	$3658 \cdot 8^2/12 + 2440 \cdot 4^2/8 \quad = 1860012 \cdot 14$		Laboratories
Total	$6099 \cdot 2^2/20 \quad = 1860012.03$	0·11	

The comments on accuracy made in Section 4.5.2 apply to this analysis. The loss of accuracy due to cancellation is apparent in the subtractions in Table 8.13 where nine figures must be retained to obtain five figures in the results. Accuracy on a computer would be improved by subtraction of the grand mean from all observations prior to the analysis.

8.5. FURTHER READING ON ANALYSIS OF VARIANCE

Cochran and Cox (1957) contains a wealth of information on the practical aspects of experimentation both on the computation of analyses and on the choice of suitable experimental designs. Real practical situations are used as examples of the methods described. Kempthorne (1952) concentrates on the mathematical aspects of the subject but has much of practical interest to recommend it to the experimentalist. Cox (1958) is confined to the planning of experiments and it is to be strongly recommended as a book to be read from cover to cover. This book contains a wide range of carefully chosen examples of practical situations and discusses many interpretational problems that are avoided by other authors. Scheffe (1956) is a rather mathematical book but provided the experimentalist is prepared to ignore the mathematics much useful information can be extracted. This book is particularly useful in its descriptions and discussions of fixed-effect, random-effect and mixed models. Fisher (1935) and Quenouille (1953) are further books on the subject to which wide reference is made and which have a practical bias.

PROBLEMS

1. Three groups of ten animals were fed different diets and after a period on their particular diet all animals were weighed. The weights, in grams, are given below:

Diet 1: 78·4 81·5 79·7 74·0 83·1 76·9 78·8 80·6 77·4 81·0
Diet 2: 81·6 79·8 84·8 85·3 78·6 87·2 78·2 80·7 77·0 83·1
Diet 3: 87·5 81·6 79·9 85·3 86·2 78·9 86·4 88·0 82·7 85·1

(a) Place these weights in order within each diet and plot these on one sheet of probability paper with different symbols for each diet.

(b) On the basis of this plot would you conclude that the variability in weights is the same in all three diets?

(c) Would you conclude that the diets have affected the weights of the animals?

(d) Perform formal statistical tests to confirm, or deny, your conclusions in (b) and (c).

(e) What assumption have these tests made about the populations represented by these three samples? Is it reasonable to believe that these are true?

(f) Assuming that the conclusions have been that the diets have an effect on the average weight of animals but not on the variability, estimate the mean of the three diet populations and compute 95 per cent confidence limits for these estimates.

(g) What assumptions have been made in this question regarding the weights of the animals at the start of the experiment? How would the analysis have changed if these assumptions had proved unreasonable.

2. Repeat questions 1 (a) to 1 (f) to analyse the following data obtained from an experiment of the same form as that in question 1 but involving three diets numbered 4 to 6.

Diet 4: 76·9 78·2 77·0 81·9 80·5 78·8 73·6 80·2 82·3 76·6
Diet 5: 76·6 78·0 87·2 83·8 85·9 81·4 78·6 80·3 81·3 80·5
Diet 6: 88·8 86·5 80·9 78·4 81·9 83·7 84·2 81·6 85·2 83·8

Use the same scale for the probability plot as in question 1.

3. Place side by side the plots obtained in questions 1 and 2 and make comparisons between the two experiments.

(a) Is it reasonable to believe that the variation within diets is the same in both experiments?

(b) Do the two experiments differ in their conclusions bearing in mind that the three diets in each experiment are different?

(c) Combine the analyses obtained in question 1 and 2 and extend to test the difference between the two experiments.

(d) What assumptions regarding variability have been made by the extended analysis? How may these be tested formally?

(e) How would your testing procedure in the analysis of variance differ if you regarded the diets as a random selection from a distribution of diets?

CHAPTER 9

ANALYSIS OF VARIANCE—II.
FACTORIAL DESIGNS

THIS chapter continues the description of the technique of analysis of variance by considering the analysis of Factorial Experiments. The factorial experiment is sometimes referred to as a cross-classification experiment. The description begins by analysing a simple two-way factorial experiment, continues by considering more extensive experiments, and concludes with a discussion of an interesting experiment which introduces variants of the factorial experiment analysis.

The experiment described in Example 8.1.3 was performed to investigate differences between the yields obtained by three experimentalists. This experiment could be considered to be an example of a one-way factorial experiment with the "factor" experimentalists appearing at three "levels". If each experimentalist had performed a number of determinations for each of four solutions, the data obtained would be classifiable according to the two factors experimentalists and solutions appearing at three and four levels respectively. This experiment would be an example of a two-way factorial experiment. In general a factorial experiment is one in which observations are made for each combination of levels of the factors involved. The number of observations made for each combination may be equal or unequal and some combinations could even be omitted altogether. The analysis of variance of unequally replicated experiments is much more involved than that of the corresponding equally replicated experiment. The description given here will be confined to the equally replicated experiments. Analysis of unequally replicated experiments involving two or more factors may be performed by the general linear hypothesis method described in Kempthorne (1952). The computations involved are heavy and will persuade most experimentalists to replicate evenly where possible in their future experiments.

We will consider the analysis of a two-way experiment with one observation for each combination of the factor levels followed by the extension of the analysis to include experiments containing more than one observation for each combination.

9.1. Two-way Factorial Experiment without Replication (Fixedeffect Model)

The following two-factor experiment was carried out to investigate the effect of different heat treatments and of different doses of irradiation on the percentage inactivation of enzymes in milk. One observation was made of the percentage inactivation for each combination of four doses and six heat treatments. The data obtained is recorded in Table 9.1.

TABLE 9.1. EXAMPLE UNREPLICATED TWO-WAY DATA

Dose		A	B	C	D	Totals
Heat treatment	I	25·1	25·4	26·0	25·2	101·7
	II	24·4	24·9	25·2	27·1	101·6
	III	25·6	25·1	26·6	26·0	103·3
	IV	24·2	25·0	26·7	26·4	102·3
	V	26·1	26·0	25·9	26·9	104·9
	VI	23·9	24·5	25·6	26·4	100·4
Totals		149·3	150·9	156·0	158·0	614·2

The statistical model for this experiment represents the observation x_{ij} for the ith heat treatment and the jth dose by:

$$x_{ij} = A + H_i + D_j + I_{ij} + z_{ij}, \qquad (9.1)$$

where A is a general level,

H_i is a fixed effect due to the ith heat treatment,

D_j is a fixed effect due to the jth dose,

I_{ij} is a fixed change in the effect of the ith heat treatment due to the presence of the jth dose, or alternatively, the change in the effect of the jth dose due to the ith heat treatment. It is usually called the interaction between doses and heat treatments,

z_{ij} is a random variable distributed normally with zero mean and constant variance σ^2.

Effects involving one factor only such as H_i and D_j are referred to as main effects. Those involving more than one factor such as I_{ij} are referred to as interactions.

From this model we may obtain the following expressions for the heat treatment means, the dose means, and the grand mean:

$$\begin{aligned}
x_{ij} &= A + H_i + D_j + I_{ij} + z_{ij} \\
\bar{x}_{.i} &= A + H_i && + \bar{z}_{i.} \\
\bar{x}_{.j} &= A && + D_j && + \bar{z}_{.j} \\
\bar{x}_{..} &= A && && + \bar{z}_{..} \, .
\end{aligned} \qquad (9.2)$$

The restrictions placed on the parameters in this model are:

$$\sum_{i=1}^{h} H_i = 0, \quad \sum_{j=1}^{d} D_j = 0,$$

$$\sum_{i=1}^{h} I_{ij} = 0 \quad \text{for } j = 1, 2, \ldots, d, \tag{9.3}$$

$$\sum_{j=1}^{d} I_{ij} = 0 \quad \text{for } i = 1, 2, \ldots, h.$$

The differences

$$\bar{x}_{i.} - \bar{x}_{..} = H_i + \bar{z}_{i.} - \bar{z}_{..}$$

$$\bar{x}_{.j} - \bar{x}_{..} = D_j + \bar{z}_{.j} - \bar{z}_{..} \tag{9.4}$$

$$x_{ij} - \bar{x}_{i.} - \bar{x}_{.j} + \bar{x}_{..} = I_{ij} + z_{ij} - \bar{z}_{i.} - \bar{z}_{.j} + \bar{z}_{..}$$

separate the heating effect, the dose effect, and their interaction, but do not provide means of estimating the residual variability σ^2 independently of these effects. The three sources of variability in this experiment are heat treatments, doses and their interaction. To obtain an estimate of the residual variability we must make the further assumption that there is no interaction present so that the interaction terms $x_{ij} - \bar{x}_{i.} - \bar{x}_{.j} + \bar{x}_{..}$ may be used to provide an estimate of the residual variability.† The interaction term is often omitted altogether from the model (9.1) in descriptions of this experimental design. It has been included here, only to be ignored again, to make it clear that this term is assumed to be zero. It will be seen in a later section that the presence of an interaction term may be tested if more than one observation is made at each combination of the factor levels. It is therefore clear that if it is an important object of the experiment to test for the existence of interaction the experiment must be replicated. It is also clear that if it is at all possible that an interaction exists, even if the object of the experiment is to investigate the dose and heat treatment effects only, the experiment must again be replicated otherwise the estimate of residual variability will be inflated by the presence of interaction.

We see from expressions (9.4) that the total sum of squares is split up as follows:

$$\sum_{j=1}^{d} \sum_{i=1}^{h} (x_{ij} - \bar{x}_{..})^2$$

$$= \sum_{j=1}^{d} \sum_{i=1}^{h} [(x_{ij} - \bar{x}_{i.} - \bar{x}_{.j} + \bar{x}_{..}) + (\bar{x}_{i.} - \bar{x}_{..}) + (\bar{x}_{.j} - \bar{x}_{..})]^2$$

$$= \sum_{j=1}^{d} \sum_{i=1}^{h} (x_{ij} - \bar{x}_{i.} - \bar{x}_{.j} + \bar{x}_{..})^2 + d \sum_{i=1}^{h} (\bar{x}_{i.} - \bar{x}_{..})^2 + h \sum_{j=1}^{d} (\bar{x}_{.j} - \bar{x}_{..})^2$$

† It may be possible to define an interaction of a particular kind that is dependent on fewer parameters than the general form I_{ij} and to test for its existence, particularly if the factor levels take numeric values.

(all cross-product terms being zero). The analysis of variance table for this model is given in Table 9.2.

TABLE 9.2. ANALYSIS OF VARIANCE FOR AN UNREPLICATED TWO-WAY EXPERIMENT

Source of variation	S.S.	D. of F.
Doses	$h \sum\limits_{j=1}^{d} (\bar{x}_{.j} - \bar{x}..)^2$	$d - 1$
Heat treatments	$d \sum\limits_{i=1}^{h} (\bar{x}_{i.} - \bar{x}..)^2$	$h - 1$
Residual (Interaction)	$\sum\limits_{i=1}^{h} \sum\limits_{j=1}^{d} (x_{ij} - \bar{x}_{i.} - \bar{x}_{.j} + \bar{x}..)^2$	$(d - 1)(h - 1)$
Total	$\sum\limits_{i=1}^{h} \sum\limits_{j=1}^{d} (x_{ij} - \bar{x}..)^2$	$dh - 1$

The sums of squares may be computed on a desk machine by using the following more convenient expressions:

$$\text{Total sum of squares} = \sum_{i=1}^{h} \sum_{j=1}^{d} x_{ij}^2 - C.$$

$$\text{Dose sum of squares} = \frac{1}{h} \sum_{j=1}^{d} \left(\sum_{i=1}^{h} x_{ij} \right)^2 - C.$$

$$\text{Heat treatment sum of squares} = \frac{1}{d} \sum_{i=1}^{h} \left(\sum_{j=1}^{d} x_{ij} \right)^2 - C,$$

$$\text{where the correction factor } C = \frac{1}{dh} \left(\sum_{i=1}^{h} \sum_{j=1}^{d} x_{ij} \right)^2.$$

On a computer, however, the means $\bar{x}..$; $\bar{x}_{.j}$ for $j = 1, 2, ..., d$, and $\bar{x}_{i.}$ for $i = 1, 2, ..., h$ would be calculated and the expressions in Table 9.2 would be used to provide the sums of squares.

9.1.1. Example

Analysing the example data given in Table 9.1 we have:

The correction factor $C = 614 \cdot 2^2 / 24$	=	15718·4017
Total S.S. $= 25 \cdot 1^2 + 25 \cdot 4^2 + \cdots + 26 \cdot 4^2 - C$	=	17·4583
Doses S.S. $= \frac{1}{8} (149 \cdot 3^2 + 150 \cdot 9^2 + \cdots + 158 \cdot 0^2) - C$	=	8·4816
Heat treatment S.S.		
$= \frac{1}{4} (101 \cdot 7^2 + 101 \cdot 6^2 + \cdots + 100 \cdot 4^2) - C$	=	3·0483
Residual S.S. $= 17 \cdot 4583 - 8 \cdot 4816 - 3 \cdot 0483$	=	5·9284

The analysis of variance table is given in Table 9.3.

TABLE 9.3. ANALYSIS OF VARIANCE FOR DATA IN TABLE 9.1

Source	S.S.	D.F.	M.S.	F.	Prob.
Doses	8·4816	3	2·8272	7·15	0·005
Heat treatment	3·0483	5	0·6966	1·76	N.S.
Residual ($D \times M$)	5·9284	15	0·3952		
Total	17·4583	23			

We see from this analysis, testing doses and heat treatments mean squares against the residual mean squares, that the dose effect is significant at the 0·005 level and that there is no significant difference between heat treatments. The residual variance σ^2 is estimated by the residual mean square (assuming no interaction) as 0·3952 with 15 degrees of freedom.

9.1.2. Data Representation

The original model (9.1) represented x_{ij} by:

$$x_{ij} = A + H_i + D_j + I_{ij} + z_{ij} .$$

We have found it necessary to assume that the terms I_{ij} are zero and we have accepted the null hypothesis that the heat treatments have no effect, that is the terms H_i are zero. If we retain the dose effects D_j the individual observation x_{ij} is represented by

$$x_{ij} = A + D_j + z_{ij}. \tag{9.5}$$

It follows from equations (9.2) that the best estimates of A and the D_j are given by

$$\hat{A} = \bar{x}_{..}$$

$$\hat{D}_j = \bar{x}_{.j} - \bar{x}_{..}$$

and hence

$$\hat{A} + \hat{D}_j = \bar{x}_{.j} .$$

Thus the value predicted by the accepted model is the relevant dose mean, the actual heat treatment employed having no influence. The difference between the actual value and the predicted value is an estimate of the residual error term z_{ij}. That is

$$\text{residual } \hat{z}_{ij} = x_{ij} - \bar{x}_{.j} .$$

If the assumptions of normality made in the model are correct these residuals should show agreement with this assumption. The dose means are:

Dose Means	A	B	C	D
	24·8 (8)	25·1 (5)	26·0 (0)	26·3 (3)

The grand mean is 25·68 and the residuals are listed in Table 9.4.

TABLE 9.4. RESIDUALS AFTER ANALYSIS OF DATA IN TABLE 9.1

Dose		A	B	C	D
Heat treatment	I	0·2 (2)	0·2 (5)	0·0 (0)	−1·1 (3)
	II	−0·4 (8)	−0·2 (5)	−0·8 (0)	0·7 (7)
	III	0·7 (2)	0·0 (5)	0·6 (0)	−0·3 (3)
	IV	−0·6 (8)	−0·1 (5)	0·7 (0)	0·0 (7)
	V	1·2 (1)	0·8 (5)	−0·1 (0)	0·5 (7)
	VI	−0·9 (8)	−0·6 (5)	−0·4 (0)	0·0 (7)

The estimate of the variance of the distribution of the residuals is the residual mean square of 0·3952 with 15 degrees of freedom.

The normality of the residuals may be investigated by plotting on probability paper. Such a plot is made in Fig. 9.1 and it shows the normality assumption to be reasonable.

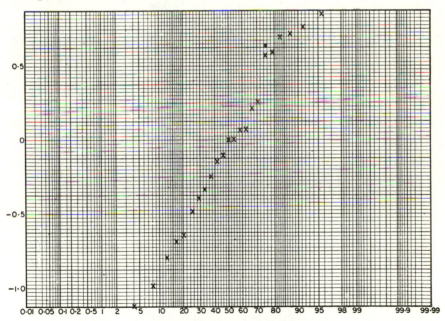

FIG. 9.1. Plot of residuals in Table 9.4.

It may be similarly shown that the best estimates of the heat treatment terms H_i, if these are not to be deleted from the model, are given by:

$$\hat{H}_i = \bar{x}_{i.} - \bar{x}_{..} \, .$$

If the heat treatment effect had been significant, and doses not significant, the data would have been represented by

$$\hat{A} + \hat{H}_i = \bar{x}_{i.} \, ,$$

that is, by the six heat treatment means.

If both factors had been significant the data would have been represented by

$$\hat{A} + \hat{H}_i + \hat{D}_j = \bar{x}_{i.} + \bar{x}_{.j} - \bar{x}_{..} \, .$$

If neither factor had been significant the model would include the general level A only and the data would be represented by $\bar{x}_{..}$.

9.1.3. Confidence Limits For Means

The means for the four dose levels are 24·88, 25·15, 26·00 and 26·33. These are estimates of the means of four populations corresponding to the four doses. We may compute confidence limits for these means by the methods described in Section 5.2.7. That is by solving the equations:

$$\frac{\bar{x}_{.j} - \mu_j}{\sqrt{(\hat{\sigma}_v^2/n)}} = \pm t_{\alpha, v} \, ,$$

where $\bar{x}_{.j}$ is the observed mean for the jth dose,

μ_j is the corresponding population mean,

$\hat{\sigma}_v^2$ is the estimate of the residual variability,

v is the degrees of freedom associated with $\hat{\sigma}_v^2$,

n is the number of observations of which $\bar{x}_{.j}$ is the mean,

$t_{\alpha, v}$ is the α significance level of the t-distribution with v degrees of freedom.

Limits for the dose population means are given by

$$\bar{x}_{.j} \pm t_{\alpha, v} \sqrt{(\hat{\sigma}_v^2/n)}$$

and for the first mean in particular the 95 per cent confidence limits are given by

$$24 \cdot 88 \pm 2 \cdot 131 \times \sqrt{(0 \cdot 3952/6)},$$

that is 24·33 to 25·43. The 95 per cent confidence limits for all four dose means are now given in Table 9.5.

TABLE 9.5. CONFIDENCE LIMITS (95 PER CENT) FOR DOSE MEANS FOR DATA IN TABLE 9.1

Dose	Lower limit	Mean	Upper limit
A	24·33	24·88	25·43
B	24·60	25·15	25·70
C	25·45	26·00	26·55
D	25·78	26·33	26·88

Similar confidence limits may be calculated for the heat treatment means using the almost identical expression

$$\bar{x}_{i\cdot} \pm t_{\alpha,\nu} \sqrt{(\hat{\sigma}_\nu^2/n)}$$

where $\bar{x}_{i\cdot}$ is the ith heat treatment mean and the other parameters are as defined above. Note that the value of n becomes four in the case of heat treatments.

9.1.4. The Investigation of Interaction

The interpretation of a two-factor interaction is given in Section 9.2.2 and the data plot suggested there may be used as a means of investigating the existence of an interaction term in the analysis of unreplicated data.

9.2. TWO-WAY EXPERIMENT WITH REPLICATION (FIXED-EFFECT MODEL)

The model for the replicated two-way experiment represents the tth observation for the jth dose and the ith heat treatment by:

$$x_{tij} = A + D_j + H_i + I_{ij} + z_{tij} , \tag{9.6}$$

where A, D_j, H_i and I_{ij} are as defined in Section 9.1 and z_{tij} is the usual error term. It is now possible to obtain from this model an estimate of the residual variance σ^2 which does not involve I_{ij} so that we may test for the existence of the interaction term I_{ij}.

The following functions of means isolate the parameters and the residuals in the model:

$$\bar{x}_{\cdot i\cdot} - \bar{x}_{\ldots} = H_i + \bar{z}_{\cdot i\cdot} - \bar{z}_{\ldots}$$

$$\bar{x}_{\cdot\cdot j} - \bar{x}_{\ldots} = D_j + \bar{z}_{\cdot\cdot j} - \bar{z}_{\ldots}$$

$$\bar{x}_{\cdot ij} - \bar{x}_{\cdot i\cdot} - \bar{x}_{\cdot\cdot j} + \bar{x}_{\ldots} = I_{ij} + \bar{z}_{\cdot ij} - \bar{z}_{\cdot i\cdot} - \bar{z}_{\cdot\cdot j} + \bar{z}_{\ldots}$$

$$x_{tij} - \bar{x}_{\cdot ij} = z_{tij} - \bar{z}_{\cdot ij}.$$

We see from these expressions that the total sums of squares is split up as follows:

$$\sum_{t=1}^{r} \sum_{i=1}^{d} \sum_{j=1}^{h} (x_{tij} - \bar{x}_{\ldots})^2 = \sum_{t=1}^{r} \sum_{i=1}^{d} \sum_{j=1}^{h} (x_{tij} - \bar{x}_{\cdot ij}) + rd \sum_{i=1}^{h} (\bar{x}_{\cdot i\cdot} - \bar{x}_{\ldots})^2$$

$$+ rh \sum_{j=1}^{d} (\bar{x}_{\cdot\cdot j} - \bar{x}_{\ldots})^2 + r \sum_{j=1}^{d} \sum_{i=1}^{h} (\bar{x}_{\cdot ij} - \bar{x}_{\cdot i\cdot} - \bar{x}_{\cdot\cdot j} + \bar{x}_{\ldots})^2,$$

where r is the number of observations for each dose–heat treatment combination. The analysis of variance table is given in Table 9.6.

TABLE 9.6. ANALYSIS OF VARIANCE FOR A REPLICATED TWO-WAY EXPERIMENT

Source	S.S.	D.F.
Doses	$rh \sum\limits_{j=1}^{d} (\bar{x}_{..j} - \bar{x}_{...})^2$	$d - 1$
Heat treatments	$rd \sum\limits_{i=1}^{h} (\bar{x}_{.i.} - \bar{x}_{...})^2$	$h - 1$
Interaction	$r \sum\limits_{i=1}^{d} \sum\limits_{j=1}^{h} (\bar{x}_{.ij} - \bar{x}_{.i.} - \bar{x}_{..j} + \bar{x}_{...})^2$	$(d-1)(h-1)$
Residual	$\sum\limits_{t=1}^{r} \sum\limits_{i=1}^{d} \sum\limits_{j=1}^{h} (x_{tij} - \bar{x}_{.ij})^2$	$dh(r-1)$
Total	$\sum\limits_{t=1}^{r} \sum\limits_{i=1}^{d} \sum\limits_{j=1}^{h} (x_{tij} - \bar{x}_{...})^2$	$rdh - 1$

9.2.1. Example

The example data given in Table 9.7 consists of two observations for each combination of four doses and six heat treatments.

TABLE 9.7. EXAMPLE REPLICATED TWO-WAY DATA

Dose	A		B		C		D		Totals
Heat treatment I	25·1	24·0	25·4	24·9	26·0	25·0	25·2	27·0	202·6
II	24·4	23·0	24·9	24·4	25·2	26·2	27·1	26·5	201·7
III	25·6	23·2	25·1	24·6	26·6	26·3	26·0	26·8	204·2
IV	24·2	23·1	25·0	25·5	26·7	25·8	26·4	27·8	204·5
V	26·1	24·2	26·0	24·7	25·9	26·3	26·9	27·7	207·8
VI	23·9	24·6	24·5	25·4	25·6	25·9	26·4	27·4	205·7
Totals	291·4		300·4		311·5		321·2		1224·5

The analysis of variance of this data is computed as follows:

Correction factor $C = 1224 \cdot 5^2 / 48$ $= 31237 \cdot 5052$

Dose S.S. $= \frac{1}{12} (291 \cdot 4^2 + 300 \cdot 4^2 + \cdots + 321 \cdot 2^2) - C$

                               $= \quad 42 \cdot 1456$

Heat treatment S.S. $= \frac{1}{8} (202 \cdot 6^2 + 201 \cdot 7^2 + \cdots + 205 \cdot 7^2) - C = \quad 2 \cdot 7535$

Interaction S.S. $= \frac{1}{2} (25{\cdot}1 + 24{\cdot}0)^2 + (25{\cdot}4 + 24{\cdot}9)^2 + \cdots + (26{\cdot}4 + 27{\cdot}4)^2)$

$$- C - \text{Dose S.S.} - \text{Heat treatment S.S.} \quad = 4{\cdot}4207$$

Total S.S. $= 25{\cdot}1^2 + 24{\cdot}0^2 + 25{\cdot}4^2 + \cdots + 27{\cdot}4^2 - C = 63{\cdot}6848$

The residual sum of squares is calculated by difference and all the sums of squares are collected together in the analysis of variance table given in Table 9.8.

TABLE 9.8. ANALYSIS OF VARIANCE FOR DATA IN TABLE 9.7

Source	S.S.	D.F.	M.S.	F.	Prob.
Doses	42·1456	3	14·0485	23·47	0·001
Heat treatment	2·7535	5	0·5507	0·92	N.S.
Interaction	4·4207	15	0·2947	0·49	N.S.
Residual	14·3650	24	0·5985		
Total	63·6848	47			

We see from Table 9.8 that the dose effect is highly significant and the heat-treatment effect and the interaction are not significant so that once again the data may be described by the model:

$$x_{tij} = A + D_j + z_{tij}$$

involving the general level A, the dose effect D_j and residual terms only.

9.2.2. The Interpretation of a Two-factor Interaction

The interpretation of a two-factor interaction may best be clarified by graphical illustration. Figure 9.2 shows plots of the percentage inactivation against doses for each heat treatment. The heavy line is the average dose effect measured by the dose sum of squares. The dose levels have been allocated to a scale at equal intervals. If the dose levels do have actual numerical values, these values would be used instead. We see from this plot that there is no difference between the heat treatments since the six dose response curves cross frequently and have the same general form. The overall relation between percentage inactivation and dose appears to be linear (assuming that the doses are allocated to the dose scale at equal intervals) with a positive (non-

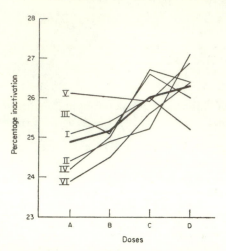

FIG. 9.2. Dose curves for each heat treatment for data in Table 9.7.

zero) slope. We may test the further hypothesis that the dose effect is linear by the methods described in Section 9.13. The conclusion is therefore that the dose effect is linear and the different heat treatments have no effect on the dose response. That is, there is no interaction between doses and heat treatments.

Let us now study the wide range of possible situations that can occur and the corresponding results of the analysis of variance. Figure 9.3 shows what the plot would look like ideally in different situations.

The eight plots included in Fig. 9.3, although not exhaustive, cover a wide range of situations. The first plot (a) shows that one straight line with zero slope will describe the data when no effects are significant. The next two plots (b) and (c) show that one function will describe the variation with dose so that points from all heat treatments lie on this line. Plots (d) and (e) show six separate parallel functions one for each heat treatment and the dose effect averaged over the six heat treatments is shown as a thick line. The significance of the dose effect means that the averaged dose effect is not a straight line with zero slope. It may be a straight line with a non-zero slope or a non-linear function. The significance of the heat treatment effect means that not all heat treatments have the same dose effect and the parallelism of these separate dose effects indicates a lack of interaction. Again the separate dose effects may be linear or non-linear. Two examples of the significance of all three effects are given in plots (f) and (g). The separate dose effects may each be linear or non-linear and some, but not all, may be coincident. The significance of the heat treatment effect and the interaction in plot (h) show that the average dose effect is horizontal but that separate dose effects still exist. These are shown to be linear in the plot but they may also be non-linear.

We have seen in Section 9.1 that it is not possible to test for an interaction term if the data is not replicated. However, in the unreplicated situation we may still plot the data in the manner described and investigate the possible existence of an interaction term. If the plot has the appearance of any one of the plots (a) to (e) we will feel confident that no interaction exists, whereas if the plot is similar to plots (f), (g) or (h) we will suspect that an interaction term does exist. While the plot is not a complete substitute for replication it is useful as additional evidence for, or against, an interaction term and in addition it will clarify the nature of the dose and heat-treatment effects. A

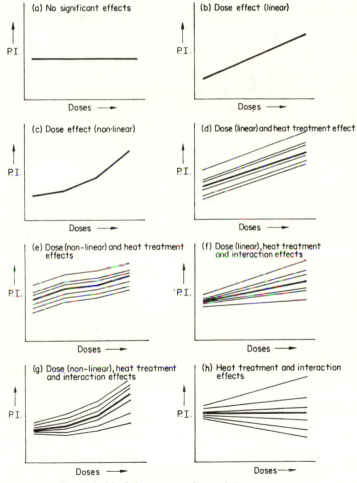

FIG. 9.3. Ideal dose curves for each heat treatment.

plot of the data in Table 9.1 is similar to, but more varied than, the plot in Fig. 9.2.

In these plots we have chosen to plot the percentage inactivation against dose for each heat treatment. In some experiments the choice of factor as abscissa is a natural one, especially if one factor has levels with numerical values. In other experiments where the choice is arbitrary it may be informative to plot the data twice choosing each factor in turn as abscissa.

9.3. Two-way Factorial Experiment (Random-effect Models)

Since there is no difference in the testing procedures between a fixed-effect model and a random-effect model in the unreplicated two-way case we pass on to consider the replicated case. The model is

$$x_{tij} = A + H_i + D_j + I_{ij} + z_{tij}, \tag{9.7}$$

where A is a general level, and where H_i, D_j, I_{ij} and z_{tij} are random variables independently drawn from normal populations with zero means and variances σ_h^2, σ_d^2, σ_I^2 and σ^2, respectively. The expected values of the mean squares in the analysis of variance are given in Table 9.9.

TABLE 9.9. EXPECTED VALUES OF MEAN SQUARES FOR A REPLICATED TWO-WAY EXPERIMENT (RANDOM MODEL)

Source	$\mathscr{E}$ (M.S.)
Doses	$\sigma^2 + r\sigma_I^2 + hr\sigma_d^2$
Heat treatments	$\sigma^2 + r\sigma_I^2 + dr\sigma_h^2$
Interaction	$\sigma^2 + r\sigma_I^2$
Residual	σ^2
Total	

It is clear from Table 9.9 that we must test first for the existence of the interaction variance σ_I^2 by comparing the interaction mean square with the residual mean square. If this is significant the dose and heat treatments mean squares must be compared with the interaction mean square since if we establish that the doses mean square, for example, is significantly greater than the residual mean square we would not be able to determine whether this was due to the interaction term, or to a genuine dose effect, or perhaps partly to both. If the significance of the interaction term is established the dose effect and the heat-treatment effect must now be compared with a variance estimate which is inflated by the interaction term and which also has fewer degrees of freedom than the residual estimate of variance. Thus we see that the power of the dose effect and the heat-treatment effect depends on the existence of an interaction term and, if it exists, on the size of the interaction term.

9.4. Two-way Factorial Experiment With Replication (Mixed Model)

Several mixed models are possible but the testing procedure is the same as the fixed-effect model analysis if the interaction term is a fixed effect, and the same as the random effect model analysis if the interaction term is a random effect.

9.5. Replicates or Repeats

The terms "Replicates" and "Repeats" have been introduced in Chapter 1 and briefly discussed there. It is now possible to consider their effect, on the analysis of the data given in Table 9.7. We consider the following two possible designs of the experiment.

(a) If the experimentalist had started with forty-eight separate containers filled with milk, irradiated them in random order, and then applied the relevant heat treatment to the container, again in random order, the duplicate pairs of observations would be replicates containing all sources of variation between pairs receiving the same dose–heat treatment combination as between those receiving different dose–heat treatment combinations.

(b) If, instead, the experimentalist had started with twenty-four separate containers, irradiated them in random order, applied the relevant heat treatment, then split each sample of milk into two subsamples and performed one determination on each subsample, the duplicate observations would be repeats since the duplicate observations would differ only because of determination errors and not because of irradiation or heating errors.

In this experimental procedure we are aware of two sources of error, firstly errors in the irradiation and heat-treatment processes, and secondly errors in the determination of the percentage inactivation. In the first design all forty-eight observations are independently subjected to both sources of variation, and these sources jointly make up the residual variability σ^2. In the second design the twenty-four pairs of observations are independently subjected to both sources of variation and pair members suffer the same process errors but different determination errors. The residual variability is therefore made up of determination errors only, whereas all other effects include process errors and determination errors. If we test for an interaction and obtain a non-significant result the difference between the two designs vanishes and we test the dose effect and heat-treatment effect against the residual mean square. If, however, the interaction proves significant we are faced with the problem of deciding whether there is a genuine interaction term, or whether the significant result is due to the variation introduced by the irradiation and heating processes. If we test one of the two main effects against the residual mean square and obtain a significant result we are again faced with the same difficult decision. The only chance of establishing the main effects is now to use the interaction mean square as the basis for comparison and so lose all the advantages of having more than one observation for each dose–heat treatment combination. The interaction term in the above analysis is not significant although it is close to the 10 per cent significance level. It we take, for the sake of example, the interaction as significant the F ratio for doses becomes 6·91 which, with 3 and 15 degrees of freedom, is significant at the 0·05 level. Thus we see that the dose effect is significant even if we assume the interaction term is significant.

It has already been pointed out that the comparison of the main effects with the interaction mean square is a less powerful comparison than with the residual mean square. The use of repeat observations instead of replicate observations thus leads to a less powerful comparison in addition to the other disadvantages already discussed. It is therefore an important consideration when designing an experiment to ensure, if at all possible, that duplicate observations are replicates rather than repeats.

9.6. Three-way Factorial Experiment (Fixed-effect Model)

If the first observation for each dose–heat treatment combination in Table 9.7 was obtained from a sample receiving the heat treatment before the irradiation, and the second observation was obtained from a sample receiving the heat treatment after the irradiation, the data could be analysed as a three-way factorial experiment without replication with factors doses, heat treatments, and a third factor which we will call experiments. This third factor has two levels, heating before irradiation, and heating after irradiation. Using the suffix t to denote the two levels of the third factor we have the following model:

$$x_{tij} = A + E_t + H_i + D_j + (EH)_{ti} + (ED)_{tj} + (HD)_{ij} + I_{tij} + z_{tij} , \quad (9.8)$$

where E_t, H_i and D_j are the three main effect terms for the tth experiment,
    the ith heat treatment, and the jth dose,
where $(EH)_{ti}$, $(ED)_{tj}$, and $(HD)_{ij}$ are interactions between pairs of factors,
where I_{tij} is the interaction between all three factors, and
where z_{tij} is the usual normally distributed error term.

It is necessary in this unreplicated three-way design, as in the unreplicated two-way design described in Section 9.1, to assume that the interaction between all factors is zero in order to obtain an estimate of the residual variance. The replicated form of this design provides an independent estimate of the residual variance allowing us to test for the existence of the final interaction term I_{tij}.

The total sum of squares

$$\sum_{t=1}^{e} \sum_{i=1}^{h} \sum_{j=1}^{d} (x_{tij} - \bar{x}...)^2$$

may be split up to give the analysis of variance shown in Table 9.10.

TABLE 9.10. ANALYSIS OF VARIANCE FOR AN UNREPLICATED THREE-WAY EXPERIMENT

Source of variation	S.S.	D.F.
Between doses	$\sum_{j=1}^{d} n_{..j}(\bar{x}_{..j} - \bar{x}...)^2$	$d-1$
Between heatings	$\sum_{i=1}^{h} n_{.i.}(\bar{x}_{.i.} - \bar{x}...)^2$	$h-1$
Between experiments	$\sum_{t=1}^{e} n_{t..}(\bar{x}_{t...} - \bar{x}..)^2$	$e-1$
Doses × heatings	$\sum_{i=1}^{h} \sum_{j=1}^{d} n_{.ij}(\bar{x}_{.ij} - \bar{x}_{.i.} - \bar{x}_{..j} + \bar{x}...)^2$	$(d-1)(h-1)$
Doses × experiments	$\sum_{t=1}^{e} \sum_{j=1}^{d} n_{t.j}(\bar{x}_{t.j} - \bar{x}_{t...} - \bar{x}_{..j} + \bar{x}...)^2$	$(d-1)(e-1)$
Heatings × experiments	$\sum_{t=1}^{e} \sum_{i=1}^{h} n_{ti.}(\bar{x}_{ti.} - \bar{x}_{t..} - \bar{x}_{.i.} + \bar{x}...)^2$	$(h-1)(e-1)$
Interaction and residual	$\sum_{t=1}^{e} \sum_{i=1}^{h} \sum_{j=1}^{d} (x_{tij} - \bar{x}_{ti.} - \bar{x}_{t.j} - \bar{x}_{.ij}$ $+ \bar{x}_{t..} + \bar{x}_{.i.} + \bar{x}_{..j} - \bar{x}...)^2$	$(h-1)(e-1)$ $(d-1)$
Totals	$\sum_{t=1}^{e} \sum_{i=1}^{h} \sum_{j=1}^{d} (x_{tij} + \bar{x}...)^2$	$ehd-1$

9.6.1. Example of an Unreplicated Three-way Experiment

We will analyse the data given above in Table 9.7 as a three-way experiment assuming that the first observation for each dose–heat treatment combination was heated before it was irradiated and that the second received the irradiation first. The rearranged data is given in Table 9.11.

TABLE 9.11. EXAMPLE UNREPLICATED THREE-WAY DATA

Heating Before Irradiation

Dose		A	B	C	D	Totals
Heat treatments	I	25·1	25·4	26·0	25·2	101·7
	II	24·4	24·9	25·2	27·1	101·6
	III	25·6	25·1	26·6	26·0	103·3
	IV	24·2	25·0	26·7	26·4	102·3
	V	26·1	26·0	25·9	26·9	104·9
	VI	23·9	24·5	25·6	26·4	100·4
Totals		149·3	150·9	156·0	158·0	614·2

Irradiation Before Heating

Dose		A	B	C	D	Totals
Heat treatments	I	24·0	24·9	25·0	27·0	100·9
	II	23·0	24·4	26·2	26·5	100·1
	III	23·2	24·6	26·3	26·8	100·9
	IV	23·1	25·5	25·8	27·8	102·2
	V	24·2	24·7	26·3	27·7	102·9
	VI	24·6	25·4	25·9	27·4	103·3
Totals		142·1	149·5	155·5	163·2	610·3

It is useful to compute the three two-way tables obtained by summing over each factor in turn, the tabular entry being the sum of all the observations for the particular combination of the levels of the two factors defining the table. These two-way tables are recorded in Table 9.12.

The sums of squares are now computed from these two-way tables and the raw data as follows:

Correction factor $C = 1224·5^2/48$ $= 31237·5052$

$$\text{Doses S.S.} = \frac{1}{12}(291·4^2 + 300·4^2 + \cdots + 321·2^2) - C = 42·1456$$

$$\text{Heat treatments S.S.} = \frac{1}{8}(202·6^2 + 201·7^2 + \cdots + 203·7^2)$$

$$- C = 2·7535$$

$$\text{Experiments S.S.} = \frac{1}{24}(614·2^2 + 610·3^2) - C = 0·3168$$

$$D \times H \text{ interaction} = \frac{1}{2}(49 \cdot 1^2 + 50 \cdot 3^2 + \cdots + 53 \cdot 8^2) - C$$

$$- \text{ Doses S.S.} - \text{ Heat treatments S.S.} \qquad = 4 \cdot 4207$$

$$D \times E \text{ interaction} = \frac{1}{6}(149 \cdot 3^2 + 150 \cdot 9^2 + \cdots + 163 \cdot 2^2) - C$$

$$- \text{ Doses S.S.} - \text{ Experiments S.S.} \qquad = 6 \cdot 4407$$

$$H \times E \text{ interaction} = \frac{1}{4}(101 \cdot 7^2 + 101 \cdot 6^2 + \cdots + 103 \cdot 3^2) - C$$

$$- \text{ Heat treatment S.S.} - \text{ Experiments S.S.} = 2 \cdot 3170$$

$$\text{Totals S.S.} = 25 \cdot 1^2 + 25 \cdot 4^2 + \cdots + 26 \cdot 4^2 + 24 \cdot 0^2 + \cdots$$

$$+ 27 \cdot 4^2 - C \qquad = 63 \cdot 6848$$

The complete analysis of variance table is given in Table 9.13.

TABLE 9.12. THREE TWO-WAY TABLES FOR DATA IN TABLE 9.11

Doses by Heat Treatments Two-way Table

Dose		A	B	C	D	Totals
Heat treatments	I	49·1	50·3	51·0	52·2	202·6
	II	47·4	49·3	51·4	53·6	201·7
	III	48·8	49·7	52·9	52·8	204·2
	IV	47·3	50·5	52·5	54·2	204·5
	V	50·3	50·7	52·2	54·6	207·8
	VI	48·5	49·9	51·5	53·8	203·7
Totals		291·4	300·4	311·5	321·2	1224·5

Doses by Experiments Two-way Table

Dose	A	B	C	D	Totals
Experiment 1	149·3	150·9	156·0	158·0	614·2
Experiment 2	142·1	149·5	155·5	163·2	610·3
Totals	291·4	300·4	311·5	321·2	1224·5

Heat Treatments by Experiments Two-way Table

Heat treatment	I	II	III	IV	V	VI	Totals
Experiment 1	101·7	101·6	103·3	102·3	104·9	100·4	614·2
Experiment 2	100·9	100·1	100·9	102·2	102·9	103·3	610·3
Totals	202·6	201·7	204·2	204·5	207·8	203·7	1224·5

TABLE 9.13. ANALYSIS OF VARIANCE TABLE FOR THREE-WAY DATA IN TABLE 9.11

Source	S.S.	D.F.	M.S.	F.	P.
Doses	42·1456	3	14·0485	39·83	0·001
Heat treatments	2·7535	5	0·5507	1·56	N.S.
Experiments	0·3168	1	0·3168	0·90	N.S.
D × H	4·4207	15	0·2947	0·84	N.S.
D × E	6·4407	3	2·1469	6·09	0·01
H × E	2·3170	5	0·4634	1·31	N.S.
Residual (D × H × E)	5·2905	15	0·3527		
Total	63·6848	47			

Using a 5 per cent significance level we establish from this analysis the dose effect and a dose × experiments interaction. The interaction term indicates that the dose effect differs in the two experiments. The nature of the interaction between doses and experiments may be studied by plotting the percentage inactivation against doses separately for each combination of experiments and heat treatments as in Fig. 9.4.

We see from Fig. 9.4 that the dose effect for experiment 1 is linear and that the dose effect for experiment 2 is also linear but different from that of experiment 1. This difference is measured by the doses × experiments interaction. There is also a slight suggestion that there is more variability present

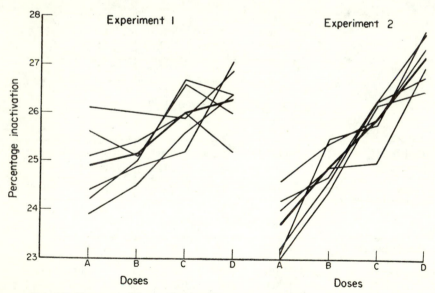

FIG. 9.4. Dose curves for each combination of heat treatments and experiments for data in Table 9.11.

in experiment 1 than in experiment 2. We may investigate this possibility, if we are prepared to assume no interaction between doses and heat treatments, by performing two-way analyses of variances for each experiment. These analyses are recorded in Table 9.14.

TABLE 9.14. TWO-WAY ANALYSES FOR EACH EXPERIMENT FOR DATA IN TABLE 9.11

Experiment 1 (heating first)

Source	S.S.	D.F.	M.S.	F.	Prob.
Doses	8·4816	3	2·8272	7·15	0·005
Heat treatments	3·0483	5	0·6966	1·76	N.S.
Residual (D × H)	5·9284	15	0·3952		
Totals	17·4583	23			

Experiment 2 (irradiation first)

Source	S.S.	D.F.	M.S.	F.	Prob.
Doses	40·1046	3	13·3682	53·01	0·001
Heat treatments	2·0221	5	0·4044	1·66	N.S.
Residual (D × H)	3·7829	15	0·2522		
Totals	45·9096	23			

We may test for equality of variance between the two experiments by the F-test as described in Section 6.2:

$$F = \frac{0 \cdot 3952}{0 \cdot 2522} = 1 \cdot 57, \qquad \nu_1 = \nu_2 = 15.$$

The value of F of $1 \cdot 57$ is not significant at the $0 \cdot 1$ level and we conclude that the variability in the two experiments is the same.

The comparison of these two analyses is also interesting. The dose effect in the second experiment is more marked than in the first. Although this is partly due to the decrease in residual mean square it does indicate an increase in the magnitude of the dose effect. This increase is clear in the plot in Fig. 9.4. The heat-treatment effect is very similar in the two separate analyses.

This example illustrates the usefulness of looking at the data in more than one way. It is frequently instructive to analyse the data separately for each level of one of the factors. If one of the factors is expected to be simply a means of replicating the experiment, such as two experiments performed at different times under hopefully identical conditions, or even under conditions known to be slightly different, it is a very useful consistency check on the experimental method to analyse separately. The extra information obtained by the separate analyses is a comparison of the variabilities in the two (or more in general) experiments as well as comparisons of the other factors. Factors such as experimentalists, or laboratories, should be analysed separately for each of their levels since it is by no means obvious that each experimentalist, or each laboratory, will work to the same precision. If different precisions are found then the combined analysis of variance with experimentalists, or laboratories, as a factor would not be valid because of the violation of the equal variance assumption made by the basic model (9.8).

The residual variance in this three-way analysis has been estimated as 0·3527 with 15 degrees of freedom and confidence limits for dose means (or interaction means also if required) may be calculated as in Section 9.1.3 using this estimate.

9.7. An Unsymmetrical Factor

Suppose in the above experiment that different heat treatments had been used in the two experiments, we would not have been able to calculate the sum of squares due to heating because of a lack of correspondence between the two experiments. Instead we must calculate the effect due to heat treatments, and the dose × heat treatments interaction, within experiments. That is, these effects are calculated separately for each experiment and then added together. The model becomes

$$x_{tij} = A + E_t + H_{ti} + D_j + (ED)_{tj} + (HD)_{tij} + z_{tij} \qquad (9.9)$$

and the analysis of variance table is given in Table 9.15.

TABLE 9.15. ANALYSIS OF VARIANCE WITH AN UNSYMMETRIC FACTOR

Source	S.S.	D.F.
Doses Experiments } D × E	as before	$\left\{ \begin{array}{c} d-1 \\ e-1 \\ (d-1)(e-1) \end{array} \right.$
Heat treatments within experiments	$d \sum\limits_{t=1}^{e} \sum\limits_{i=1}^{h} (\bar{x}_{ti.} - \bar{x}_{t..})^2$	$e(h-1)$
H × D within experiments	$\sum\limits_{t=1}^{e} \sum\limits_{i=1}^{h} \sum\limits_{j=1}^{d} (x_{tij} - \bar{x}_{t.j} - \bar{x}_{ti.} + \bar{x}_{t..})^2$	$e(h-1)(d-1)$
Total	$\sum\limits_{t=1}^{e} \sum\limits_{i=1}^{h} \sum\limits_{j=1}^{d} (x_{tij} - \bar{x}_{...})^2$	$ehd-1$

The analysis is best performed by computing separate two-way analyses of variance for each experiment. The heat treatments within experiments sum of squares may be calculated by summing the heat treatments sums of squares for each experiment. Heat treatments within experiments S.S. =

$$\tfrac{1}{6}(101 \cdot 7^2 + \cdots + 100 \cdot 4^2) - 614 \cdot 2^2/24 + \tfrac{1}{6}(100 \cdot 9^2 + \cdots + 103 \cdot 3^2)$$

$$- 610 \cdot 3^2/24 = 5 \cdot 0705.$$

The heat treatments × doses within experiments sum of squares may be calculated by difference and the new analysis of variance table is given in Table 9.16.

TABLE 9.16. ANALYSIS OF VARIANCE ILLUSTRATING AN UNSYMMETRIC FACTOR

Source	S.S.	D.F.	M.S.	F.	Prob.
Doses	42·1456	3	14·0485	43·30	0·001
Experiments	0·3168	1	0·3168	0·98	N.S.
D × E	6·4407	3	2·1469	6·63	0·005
Heat treatments/E	5·0705	10	0·5071	1·57	N.S.
H × D/E	9·7112	30	0·3237		
Total	63·6848	47			

The conclusions drawn from this analysis are consistent with those drawn from the usual analysis given in Table 9.13.

It should be noted that this analysis may be performed on data involving different number of levels for the unsymmetric factor (heat treatments).

9.8. REPLICATED THREE-WAY FACTORIAL EXPERIMENT (RANDOM-EFFECT MODEL)

The testing procedure for a random model may be understood by examination of the expected values of the mean squares for the three-way analysis of variance recorded in Table 9.17.

It will be seen from Table 9.17 that there is no difficulty in testing for the interaction terms but exact tests for the main effects may not exist. The main effects may only be tested exactly if certain of the interaction terms may be assumed to be zero. For example, if σ_{de}^2 may be assumed to be zero an exact test for the dose effect σ_d^2 may be made by comparing the doses mean square with the doses × heat treatments mean square.

The following approximate F-test is available for testing a main effect. We illustrate this by testing for the dose effect.

TABLE 9.17. EXPECTED VALUES OF MEAN SQUARES FOR RANDOM-EFFECT MODEL

Source	D.F.	$\mathscr{E}$ (M.S.)
Doses	$(d-1)$	$\sigma^2 + r\sigma_{dhe}^2 + er\sigma_{dh}^2 + hr\sigma_{de}^2 + her\sigma_d^2$
Heat treatments	$(h-1)$	$\sigma^2 + r\sigma_{dhe}^2 + er\sigma_{dh}^2 + dr\sigma_{he}^2 + der\sigma_h^2$
Experiments	$(e-1)$	$\sigma^2 + r\sigma_{dhe}^2 + dr\sigma_{he}^2 + hr\sigma_{de}^2 + dhr\sigma_e^2$
D × H	$(d-1)(h-1)$	$\sigma^2 + r\sigma_{dhe}^2 + er\sigma_{dh}^2$
D × E	$(d-1)(e-1)$	$\sigma^2 + r\sigma_{dhe}^2 + hr\sigma_{de}^2$
H × E	$(h-1)(e-1)$	$\sigma^2 + r\sigma_{dhe}^2 + dr\sigma_{he}^2$
D × H × E	$(d-1)(h-1)(e-1)$	$\sigma^2 + r\sigma_{dhe}^2$
Residual	$dhe\,(r-1)$	σ^2
Total	$dher - 1$	

Let

$$\tau = \sigma^2 + r\sigma_{dhe}^2 + er\sigma_{dh}^2 + hr\sigma_{de}^2$$

that is

$$\mathscr{E}(\text{dose M.S.}) = \tau + her\sigma_d^2$$

and

$$\tau = \mathscr{E}(\text{D} \times \text{H M.S.}) + \mathscr{E}(\text{D} \times \text{E M.S.}) - \mathscr{E}(\text{D} \times \text{H} \times \text{E M.S.}). \quad (9.10)$$

Thus τ may be estimated as a function of these three mean squares and the approximate F-test consists of testing the dose mean square against τ. The degrees of freedom for doses are $d - 1$, the degrees of freedom for τ, however, must be calculated. If we write the three mean squares on the right-hand side of expression (9.10) as $\hat{\tau}_1, \hat{\tau}_2, \hat{\tau}_3$, and their degrees of freedom as ν_1, ν_2, ν_3 so that

$$\hat{\tau} = \hat{\tau}_1 + \hat{\tau}_2 - \hat{\tau}_3,$$

then $\hat{\nu}$ is given by

$$\hat{\nu} = \hat{\tau}^2 \bigg/ \sum_{i=1}^{3} (\hat{\tau}_i^2/\nu_i).$$

Applying this test to the analysis above we have

$$\hat{\tau} = 0 \cdot 2947 + 2 \cdot 1469 - 0 \cdot 3527 = 2 \cdot 0889,$$

$$\hat{\nu} = 2 \cdot 0889^2/(0 \cdot 2947^2/15 + 2 \cdot 1469^2/3 + 0 \cdot 3527^2/15) = 3,$$

$$F = 14 \cdot 0485/2 \cdot 0889 = 6 \cdot 73,$$

$$\Pr\{F \geqq 6 \cdot 73 \mid \nu_1 = 3, \quad \nu_2 = 3\} < 0 \cdot 10.$$

We see that the conclusions drawn after the assumption of a random model can be very different from those drawn from a fixed-effect model. The failure to establish the significance of the dose effect is due to the presence of the doses × experiments interaction which has affected both the F-ratio and the degrees of freedom.

9.9. Main Effects and Interactions

Sums of squares for the main effects such as doses are computed by forming the totals of all observations for each level of the main effect, summing the squares of these totals, dividing by the number of observations in each total and subtracting the correction factor. The correction factor is the square of the grand total divided by the total number of observations.

The sums of squares for interactions are computed by forming the totals of all observations for each combination of the factor levels contained in the interaction, summing the squares of these totals, dividing by the number of observations in each total, subtracting the correction factor and subtracting the sums of squares for all main effects and interactions contained in this interaction.

These computational procedures apply to the analysis of all experiments containing an equal number of observations for each factor level combination.

9.10 UNEQUAL REPLICATION

The analyses described so far in this chapter assume that an equal number of observations were made for each combination of factor levels. The analysis in the case of unequal replication may still be performed although the analysis is involved and difficult. The analysis of a large experiment with unequal replication is a calculation best performed by an electronic computer.

9.11. VARIABILITY ESTIMATE FROM ANOTHER SOURCE

The analyses of experiments with no replication are forced to use the last interaction mean square as the estimate of residual variability. It is possible, however, that an estimate of variability is available from previous experiments, so that this estimate may be used to test for the presence of the last interaction term.

It is also possible that the residual variability is provided by a theoretical description of the experimental situation. If such a value is used as a basis for comparison with the experimental effects the resulting ratios multiplied by the degrees of freedom are distributed as χ^2 since the residual variability is known exactly.

9.12. VARIATIONS OF A FACTORIAL EXPERIMENT

An experiment which appears, at first sight, to be a four-way factorial experiment with replication was carried out as follows:

Four different batches of glass were prepared and six specimens of each batch of glass were separated. Two specimens of each type of glass received one of three doses of irradiation. After irradiation the twenty-four specimens of glass were placed in separate polythene pots together with 30 ml of distilled water. The pots were gently and continuously agitated for a period of 1 week after which the water was removed and dispatched for chemical analysis, the object of the analysis being to determine how much of a certain substance in the glass had been dissolved into the 30 ml of distilled water. As the solution was removed it was replaced with a further 30 ml of distilled water and agitated during the second week at the end of which the resulting solution was similarly analysed. This weekly process continued for a total of 10 weeks. The standardized amount of the particular glass constituent found in a solution was called the leach factor. The chemical analysis of each solution consisted of three determinations of the leach factor.

The facts which distinguish this experiment from a fixed-effect four-way factorial experiment with three replicates and involving the factors, glasses (four levels), specimens (two levels), doses (three levels) and weeks (ten levels) are:

1. Glass, dose and specimen comparisons involve variability between specimens and between solutions whereas week comparisons involve solution differences only.
2. The three determinations are repeats not involving solution differences.
3. The specimen factor is not a classification since there is no common treatment of one specimen of each glass. We cannot refer to specimen one and specimen two but only to two specimens.

The analysis of variance for this experiment is given in Table 9.18.

TABLE 9.18. EXAMPLE ANALYSIS FOR A VARIANT OF A FOUR-WAY EXPERIMENT

Source	S.S.	D.F.	M.S.	F_1	F_2	Prob.
Glasses	53·277	3	17·759		21·29	0·001
Doses	8·553	2	4·277		5·13	0·025
Glasses × doses	15·366	6	2·561		3·07	N.S.
III Specimens/glasses and doses	10·009	12	0·8341	5·41		0·001
Weeks	10·141	9	1·127	7·31		0·001
Weeks × glasses	9·973	27	0·369	2·39		0·005
Weeks × doses	19·262	18	1·070	6·94		0·001
Weeks × glasses × doses	13·453	54	0·249	1·62		0·025
II Weeks × specimens/ G and D	16·644	108	0·1541			
I Between determinations	10·555	480	0·02199			
Total	167·233	719				

The between determinations mean square provides an estimate of Variability between determinations for the same solution and does not contain Variability between determinations on different solutions. Since all other effects contain solution Variability this effect cannot be used as residual.

The "Week × specimens within glasses and doses" mean square is used as an estimate of Variability between solutions. Effects involving weeks contain solution Variability but not specimen Variability, and may be compared with this mean square.

Effects involving glasses and doses involve Variability between specimens and between solutions and must be compared with the Specimens within Glasses and Doses mean square since this is an estimate of specimen *and* solution variability.

The interesting points about this experiment are that there are three separate sources of residual in the experiment, and the testing procedure is modified to take this into account. The glass and dose comparisons are made with a residual with 12 degrees of freedom even though three determinations of each of 240 solutions were made. These degrees of freedom may only be increased by using more specimens of glass. The week effects are compared with a residual involving 108 degrees of freedom so that there was a good chance of establishing the week effects.

9.12.1. Confidence Limits

Confidence limits for glass and dose means may be calculated as described in Section 9.1.3, but must take the specimens within glasses and doses mean square as the estimate of variability.

Confidence limits for the week effects will use the weeks × specimens within glasses and doses mean square as the variability estimate if the limits are to describe week to week variability for a given specimen. If, however, the confidence limits are to include variability from week to week and between specimens then the specimens within glasses and doses mean square should be used.

9.13. POLYNOMIAL PARTITIONING

The factor levels in the analyses given so far in this chapter have been considered as defining groups into which observations are classed. The factor glasses in the experiment described in Section 9.12 is an example of this type of factor. Some factors such as doses, however, have levels which lie on a numerical scale so that the relationship between the observations and the factor levels may be considered. For example, with knowledge of the numerical values of the dose levels it is possible to split the dose sum of squares into two separate sums of squares. The first of these measures the linear response of the leach factor to dose and the second measures the quadratic response. Assuming, temporarily, that the factor levels are at equally spaced intervals the two "orthogonal" functions of the dose means

y_1, y_2, y_3 which measure the linear and quadratic parts of the dose effect are:

$$\text{Linear} = y_1 - y_3,$$

$$\text{Quadratic} = y_1 - 2y_2 + y_3. \tag{9.11}$$

The sums of squares which may be used to test the significance of the linear and quadratic parts are:

$$\text{Linear S.S.} = \frac{N}{2}(y_1 - y_3)^2,$$

$$\text{Quadratic S.S.} = \frac{N}{6}(y_1 - 2y_2 + y_3)^2, \tag{9.12}$$

where N is the number of observations contained in each of the means y_1, y_2 and y_3. The constant divisor of N is the sum of the squares of the coefficients of y_1, y_2 and y_3.

For factors at more than three levels it is possible to split the sum of squares into further partitions representing effects such as cubic, quartic or quintic. Coefficients of orthogonal polynomials assuming equally spaced factor levels are given in table 47 of Pearson and Hartley (1958). An explanation of the use of these tables is included in the introduction to these tables Sum of squares based on these tables for factors with four or five levels are given in Table 9.19.

TABLE 9.19. ORTHOGONAL PARTITIONS FOR FACTORS WITH FOUR OR FIVE EQUALLY SPACED LEVELS

	Four levels	Five levels
Linear	$\dfrac{N}{20}(3y_1 + y_2 - y_3 - 3y_4)^2$	$\dfrac{N}{10}(2y_1 + y_2 - y_4 - 2y_5)^2$
Quadratic	$\dfrac{N}{4}(y_1 - y_2 - y_3 + y_4)^2$	$\dfrac{N}{14}(2y_1 - y_2 - 2y_3 - y_4 + 2y_5)^2$
Cubic	$\dfrac{N}{20}(y_1 - 3y_2 + 3y_3 - y_4)^2$	$\dfrac{N}{10}(y_1 - 2y_2 + 2y_4 - y_5)^2$
Quartic		$\dfrac{N}{70}(y_1 - 4y_2 + 6y_3 - 4y_4 + y_5)^2$

9.13.1. Example

The three dose means for the above experiment were 5·6719, 5·4936 and 5·4107 and the doses were in equal logarithmic intervals. Each mean was based on 240 observations so that the linear and quadratic sums of squares are:

$$\text{Linear S.S.} = \frac{240}{2}(5{\cdot}6719 - 5{\cdot}4107)^2 = 8{\cdot}188,$$

$$\text{Quadratic S.S.} = \frac{240}{6}(5{\cdot}6719 - 2 \times 5{\cdot}4936 + 5{\cdot}4107)^2 = 0{\cdot}365.$$

The relevant part of the analysis of variance table is repeated in Table 9.20.

TABLE 9.20. EXAMPLE ANALYSIS ILLUSTRATING POLYNOMIAL PARTITIONING

Source		D.F.	M.S.	F.	Prob.
Glasses	53·277	3	17·759	21·29	0·001
Doses—linear	8·188	1	8·188	9·82	0·01
Doses—quadratic	0·365	1	0·365	0·44	N.S.
Glasses × doses	15·366	6	2·561	3·07	N.S.
Residual (specimens)	10·009	12	0·8341		

The conclusion from this analysis is that the dose effect is made up of a marked linear (logarithmic) response with no significant quadratic response.

9.13.2. Interactions between Partitions

It is also possible to compute the interactions between partitions and between other factors. The glasses × doses interaction in the above experiment may be split up into the interaction between glasses and the linear dose effect and the interaction between glasses and the quadratic dose effect. The significance of these sums of squares would establish that the linear dose effect or quadratic dose effect differed from glass to glass. The calculation of these sums of squares is now described.

The means for each glass–dose combination are given in Table 9.21

TABLE 9.21. GLASS × DOSE MEANS

Doses	1	2	3	Linear	Quadratic
Glasses A	5·3792	5·3320	5·4489	−0·0697	0·1641
B	5·1667	5·1106	5·2776	−0·1109	0·2231
C	5·9430	5·6042	5·2922	0·6508	0·0268
D	6·1988	5·9277	5·6239	0·5749	−0·0327

The computation begins by computing the linear and quadratic functions for each glass by using expressions (9.11). These are denoted by $L_1, Q_1, L_2, ..., Q_4$. For example:

$$L_1 = (5{\cdot}3792 - 5{\cdot}4489) = -0{\cdot}0697,$$
$$Q_2 = (5{\cdot}3792 - 2 \times 5{\cdot}3320 + 5{\cdot}4489) = 0{\cdot}1641.$$

The two interaction sums of squares are then calculated as

$$\text{Glasses} \times \text{doses (linear) S.S.} = \frac{N}{2} \sum_{i=1}^{4} (L_i - \bar{L})^2,$$

$$\text{Glasses} \times \text{doses (quadratic) S.S.} = \frac{N}{6} \sum_{i=1}^{4} (Q_i - \bar{Q})^2,$$

where $\bar{L}$ and $\bar{Q}$ are the means of the linear and quadratic functions. The constants outside the summations are defined as:

N is the number of observations contained in each mean,
the 2 or 6 is the same constant as in the linear and quadratic effects calculated above.

$$\text{Glasses} \times \text{doses (linear) S.S.} = \frac{60}{2} \times 0\cdot498149 = 14\cdot9445,$$

$$\text{Glasses} \times \text{doses (quadratic) S.S.} = \frac{60}{6} \times 0\cdot042143 = 0\cdot4214.$$

The analysis of variance table in its new form is given in Table 9.22.

TABLE 9.22. EXAMPLE ANALYSIS ILLUSTRATING INTERACTIONS BETWEEN PARTITIONS AND OTHER FACTORS

Source	S.S.	D.F.	M.S.	F.	Prob.
Glasses	53·277	3	17·759	21·29	0·001
Doses—linear	8·188	1	8·188	9·82	0·01
Doses—quadratic	0·365	1	0·365	0·44	N.S.
Glasses × doses—linear	14·945	3	4·982	5·97	0·01
Glasses × doses—quadratic	0·421	3	0·140	0·17	N.S.
Residual (specimens)	10·009	12	0·8341		

The conclusions from the analysis are that there is a significant glass effect, significant linear dose effect differing from glass to glass. Inspection of the glass × dose means reveals that for the first two glasses there is a slight increase in leach factor with dose but in the last two glasses there is a decrease in leach factor with increase dose.

Other interactions such as the interaction between the linear parts of two factors can be calculated by the same method as above. The interaction between the linear effects of doses and of glasses is calculated as an example. Firstly we calculate the linear effect of doses for each glass as described above. This gives the values

$$-0\cdot0697, \quad -0\cdot1109, \quad 0\cdot6508, \quad 0\cdot5749.$$

The further linear effect due to glasses is calculated on these values by using the linear function applicable to four levels. The required sum of squares is then given by

$$\text{Glasses (linear)} \times \text{doses (linear) S.S.} = \frac{N}{20} (3L_1 + L_2 - L_3 - 3L_4)^2,$$

$$= \frac{60}{20} (-3 \times 0\cdot0697 - 0\cdot1109 - 0\cdot6508 - 3 \times 0\cdot5749)^2,$$

$$= 21\cdot797.$$

9.13.3. Unequally Spaced Factor Levels and Other Functions

The coefficients given in the previous sections and in Pearson and Hartley apply to factors with equally spaced levels. It is possible to produce similar sets of coefficients for factors with unequally spaced levels by the methods described later in Section 12.10.3. Coefficients generated by these methods can be used to partition sums of squares into linear, quadratic and other parts.

It is also possible to develop sets of orthogonal coefficients that correspond to functions other than polynominals and to use these to partition sums of squares into parts with interesting properties. In fact, any set of coefficients can be used provided they satisfy the following conditions:

(a) The coefficients are mutually orthogonal (the scalar product of any two sets of coefficients is zero).

(b) The value obtained by applying one set of coefficients when squared is multiplied by the number of observations contained in the means and divided by the sum of squares of the coefficients (for example, expressions (9.12).

(c) Each partition measures something meaningful and its significance is interpretable.

Questions

1. Sixteen lengths of cotton have been chosen, two from each of eight rolls. One length from each roll was dyed with a newly developed red dye, and the other from each roll with a red dye that had been in use for some time. Each of the sixteen lengths of cotton was then washed and the amount of dye washed out was recorded. The results were:

Roll	1	2	3	4	5	6	7	8
New dye	12·5	14·3	16·8	14·9	17·4	11·4	15·6	15·2
Old dye	13·2	13·7	15·4	13·5	16·8	12·2	14·8	14·0

Assume a fixed effect model in the questions that follow. Plot the data where possible.

(a) Is the amount of dye washing out the same for both dyes?

(b) Is the amount of dye washing out the same for each roll of cotton?

(c) Compute the residuals from the analysis of variance model with both main effects included and plot these on probability paper to determine if the assumption of normality made for these residuals is acceptable.

(d) Compute the best estimate of the population means for each dye together with their 95 per cent confidence limits.

2. The experiment in question 1 was repeated with eight different rolls of cotton and the results were:

Roll	9	10	11	12	13	14	15	16
New dye	12·9	13·6	16·2	15·8	17·8	10·6	16·3	15·9
Old dye	13·6	13·9	15·1	12·3	17·8	11·5	14·2	14·7

(a) Is the amount of dye washing out the same for both dyes in this experiment?

(b) Is the amount of dye washing out the same for all rolls of cotton?

(c) Were both experiments equally precise?

(d) Combine these results with those in question 1 to obtain the following analysis of variance table:

Source	D.F.
Experiments	1
Dyes	1
Rolls × experiments	14
Dyes × experiments	1
Residual	14
Total	31

(e) Summarize the results of this analysis.

3. Imagine that the eight rolls of cotton used in question 2 were the same as those used in question 1. The results obtained were as given in questions 1 and 2.

(a) Perform the analysis appropriate to this situation and note the differences between it and that in question 2.

(b) Is there any evidence for believing that the difference between the two dyes differs in the two experiments?

(c) What means would you use to describe this data? Compute these means and their 95 per cent confidence limits.

(d) Compute the residuals and plot them on probability paper. Is it reasonable to believe that these are normally distributed?

4. Considering the same data used in previous questions assume that two lengths had been cut from each of the eight rolls but instead of conducting two separate experiments as in question 3 all thirty-two lengths had been processed in random order as one experiment. That is, the experimental design is a two-way factorial experiment with replication. The data rearranged in this way is given below:

New dye	12·5	14·3	16·8	14·9	17·4	11·4	15·6	15·2
	12·9	13·6	16·2	15·8	17·8	10·6	16·3	15·9
Old dye	13·2	13·7	15·4	13·5	16·8	12·2	14·8	14·0
	13·6	13·9	15·1	12·3	17·8	11·5	14·2	14·7

(a) Perform the formal analysis of variance of the data in this form.

(b) Compare the analysis with that in question 3. Could the analysis you have just performed been obtained from that in question 3?

5. (a) Do you feel that the assumption of a fixed-effect model made in the first four questions was justified?

(b) Under what conditions would you assume a random model?

6. (a) Repeat question 3 assuming a completely random model.

(b) How would you modify the testing procedure if only the factor "Rolls" were assumed random and the other factors fixed effects?

7. Assume now that the experiment in question 4 had been conducted by cutting two lengths from each of the eight rolls, dyeing one length from each roll with each dye, cutting each length in half, and determining the wash out. What effect would this have had on your analysis?

8. Consider the glass experiment discussed in the later part of this chapter. Treat the data given in Table 9.21 as though it contained the results of a simple two-way experiment without replication.

(a) Compute the two-way analysis of variance OMITTING glass D and separating the linear and quadratic glass effect, assuming the glasses to be equally spaced on a meaningful scale.

(b) Repeat the analysis, separating out the linear dose effect.

CHAPTER 10

EXPERIMENTAL DESIGN

THIS chapter will describe a wide range of experimental designs and their analyses of variance. The following experiment will be referred to in the description of many of the designs. It is the purpose of the experiment to compare the effect of four different doses of radiation on a material. It is decided that four observations for each dose will be made, making sixteen observations in all.

10.1. EXAMPLE EXPERIMENT

One possible design is to make all four observations for dose A first, followed by the four observations for dose B, and then similarly for doses C and D. It was pointed out in the first chapter that if an uncontrolled variable is varying with time this design could produce incorrect conclusions. The dose effect, if it exists, may be compensated for by the uncontrolled variable and its significance not established, or it would be possible for the uncontrolled variable to produce what looks like a dose effect when no dose effect really exists.

The main principle of experimental design is that the value of an uncontrolled variable is likely to change more between two points in time (or space) which are far apart than for two points which are close together. Applying this principle involves grouping the data so that comparisons between doses are made as close together as possible in time. Different types of grouping are discussed and the description starts by considering the example experiment with no grouping.

10.1.1. Example Data

The following data will be used in the description of these designs and each design will simply rearrange the data into the form required by the design. Each observation will remain a measurement for the same dose so that the dose effect in the analysis of variance will remain constant and the total sum of squares will remain constant also. The data is given in Table 10.1 and plotted in Fig. 10.1.

TABLE 10.1. EXAMPLE DATA

Dose					Totals
A	66·0	69·8	73·7	71·2	280·7
B	81·8	65·2	83·4	72·9	303·3
C	85·1	72·8	77·1	81·1	316·1
D	89·6	80·3	76·4	83·7	330·0
Total					1230·1

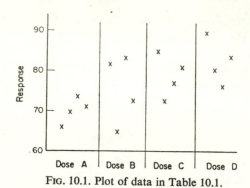

FIG. 10.1. Plot of data in Table 10.1.

The sum of squares due to doses is

$$\tfrac{1}{4}(280{\cdot}7^2 + 303{\cdot}3^2 + 316{\cdot}1^2 + 330{\cdot}0^2) - 1230{\cdot}1^2/16 = 329{\cdot}0219.$$

The total sum of squares is

$$66{\cdot}0^2 + 69{\cdot}8^2 + \cdots + 83{\cdot}7^2 - 1230{\cdot}1^2/16 = 752{\cdot}1644.$$

10.2. COMPLETE RANDOMIZATION

This is the simplest type of design and consists of making the sixteen observations in random order. The random order ensures that no treatment is favoured more than another in that each observation is equally likely to be made in each position in the order.

The model for this data represents the ith observation for the jth dose by

$$x_{ij} = A + D_j + z_{ij} \tag{10.1}$$

where A is a general level, D_j is an effect due the jth dose, z_{ij} is the usual normally distributed error term.

The analysis of variance of this type of design was discussed in Section 8.1 and the numerical results are given in Table 10.2.

TABLE 10.2. ANALYSIS OF VARIANCE FOR RANDOMIZED DESIGN

Source	S.S.	D.F.	M.S.	F.	Prob.
Doses	329·0219	3	109·6740	3·11	0·10
Residual	423·1425	12	35·2619		
Total	752·1644	15			

We conclude from this analysis that the variation observed in the dose means can be accounted for by the variability experienced between the experimental units, although the dose effect is significant at the 10 per cent level. The advantages of this design are:

1. Any number of treatments and replicates may be used. The number of replicates can be different for each dose if desired, although such variation should only be used when there is good reason.
2. The statistical analysis is simple even if the number of replicates differ from dose to dose.
3. The analysis is still simple if some observations are missing.

The main objection to the design is that the randomization is not restricted in any way to ensure that observations for different doses are made as close together as possible. The whole of the variation among the observations enters into the residual variability estimate. However, there is one compensation for the higher error in that this design has a higher number of degrees of freedom for the residual sum of squares than any other design. If the experimental conditions *are* well controlled this design is probably better than the others which follow.

10.3. RANDOMIZED BLOCKS

In this design the experimental observations are made in groups, or blocks, each consisting of an equal number of observations for each of the four doses. The order in which the observations are made within each block is chosen at random and the random order is chosen separately for each block.

The analysis of variance is different for the two following cases:

1. Each block contains more than one observation for each dose.
2. Each block contains one observation for each dose.

10.3.1. Each Dose Represented More than Once in Each Block

The above data will be rearranged so that two observations for each dose are included in each of two blocks. The order in which the eight observations for each block were made would normally be chosen at random. The model represents the ith observation (x_{tij}) for the jth dose in the tth block by

$$x_{tij} = A + D_j + B_t + (BD)_{tj} + z_{tij}, \qquad (10.2)$$

where A is a general level, D_j is an effect due to the jth dose, B_t is an effect due to the tth block, $(BD)_{tj}$ is the interaction between the jth dose and the tth block, and z_{tij} is the usual normally distributed error term.

The total sum of squares may be split up to give the analysis in Table 10.3.

TABLE 10.3. ANALYSIS OF VARIANCE OF A RANDOMIZED BLOCK DESIGN (CASE 1)

Source	D.F.
Doses	3
Blocks	1
Doses × blocks	3
Residual	8
Total	15

The rearranged data is given in Table 10.4.

TABLE 10.4. EXAMPLE DATA REARRANGED IN TWO RANDOMIZED BLOCKS

Block 1	66·0 (A)	89·6 (D)	85·1 (C)	81·8 (B)
	77·1 (C)	73·7 (A)	83·4 (B)	76·4 (D)
Block 2	83·7 (D)	69·8 (A)	71·2 (A)	65·2 (B)
	72·9 (B)	72·8 (C)	80·3 (D)	81·1 (C)

The block × dose totals are given in Table 10.5.

TABLE 10.5. BLOCK × DOSE TOTALS FOR REARRANGED DATA IN TABLE 10.4

Dose	A	B	C	D	Totals
Block 1	139·7	165·2	162·2	166·0	633·1
Block 2	141·0	138·1	153·9	164·0	597·0
Totals	280·7	303·3	316·1	330·0	1230·1

The blocks sum of squares and the doses × blocks interaction are:

$$\text{Blocks S.S.} = \tfrac{1}{8}(633 \cdot 1^2 + 597 \cdot 0^2) - 1230 \cdot 1^2/16 = 81 \cdot 4506$$

$$\text{Doses} \times \text{blocks S.S.} = \tfrac{1}{2}(139 \cdot 7^2 + 165 \cdot 2^2 + \cdots + 164 \cdot 0^2)$$

$$- 1230 \cdot 1^2/16 - \text{Dose S.S.} - \text{Block S.S.} = 120 \cdot 7969.$$

The complete analysis of variance is now given in Table 10.6.

TABLE 10.6. ANALYSIS OF EXAMPLE DATA REARRANGED AS TWO RANDOMIZED BLOCKS

Source	S.S.	D.F.	M.S.	F.	Prob.
Doses	329·0219	3	109·6740	3·97	0·10
Blocks	81·4506	1	81·4506	2·95	0·20
Doses × blocks	120·7969	3	40·2656	1·46	N.S.
Residual	220·8950	8	27·6119		
Total	752·1644	15			

We see from this analysis that the blocking has had the effect of reducing the residual mean square from 35·2619 to 27·6119 although the dose effect is still only significant at the 10 per cent level and not at the 5 per cent level.

The advantages of this design over the complete randomization design are:

1. By grouping observations more accurate results are obtained because observations making a complete replicate of the experiment are close together.
2. Any number of replicates and treatments may be used. If extra replication is desired for some treatments, each of these may be applied to two units in each block. This is useful for increasing the accuracy of the assessment of a standard or control treatment.
3. The statistical analysis is straightforward. It is still possible to analyse the data when some observations are missing, but if the gaps are numerous this design is less convenient than the completely randomized design.

The main disadvantage in this design is that the degrees of freedom of the residual term have been reduced from those for the residual term in the complete randomization design. If the blocking is effective, however, the residual mean square will have been reduced to more than compensate for the loss of degrees of freedom. The blocking has been effective in the example in that the residual mean square has been reduced but the reduction only marginally compensates for the loss of four degrees of freedom.

10.3.2. Each Dose Represented Once in Each Block

The above data will be rearranged so that one observation for each dose is included in each of four blocks. The order in which the four observations for each block were made would be chosen at random and the random orders for each block chosen separately.

The model for this design represents the jth dose in the tth block (x_{tj}) as:

$$x_{tj} = A + D_j + B_t + I_{tj} + Z_{tj}. \tag{10.3}$$

This design is a simple two-way factorial experiment with no replication and the analysis of variance is given in Table 10.7.

TABLE 10.7. ANALYSIS OF VARIANCE OF A RANDOMIZED BLOCK DESIGN (CASE 2)

Source	D.F.
Dose	3
Blocks	3
Interaction (residual)	9
Total	15

This design, as was discovered in Section 9.1, must make the assumption that there is no interaction between doses and blocks. The validity of this assumption depends on the nature of the experimental material, but since attempts are normally made to make the conditions under which the blocks are performed as identical as possible there is often some justification in making this assumption. The validity of the assumption must, however, be always considered when such an experiment is designed.

The rearranged data to be used as the example of this design is given in Table 10.8 with the doses in brackets.

TABLE 10.8. EXAMPLE DATA REARRANGED IN FOUR RANDOMIZED BLOCKS

Block					Totals
1	66·0 (A)	77·1 (C)	83·7 (D)	65·2 (B)	292·0
2	72·8 (C)	81·8 (B)	76·4 (D)	71·2 (A)	302·2
3	73·7 (A)	80·3 (D)	85·1 (C)	72·9 (B)	312·0
4	83·4 (B)	81·1 (C)	69·8 (A)	89·6 (D)	323·9

The block sum of squares is now

$$\text{Blocks S.S.} = \tfrac{1}{4}(292 \cdot 0^2 + 302 \cdot 0^2 + 312 \cdot 0^2 + 323 \cdot 9^2) - 1230 \cdot 1^2/16$$

$$= 139 \cdot 3869$$

and the complete analysis of variance is given in Table 10.9.

TABLE 10.9. ANALYSIS OF EXAMPLE DATA REARRANGED AS TWO RANDOMIZED BLOCKS

Source	S.S.	D.F.	M.S.	F.	Prob.
Doses	329·0219	3	109·6740	3·48	0·10
Blocks	139·3869	3	46·4623	1·47	N.S.
Interaction	283·7556	9	31·5284		
Total	752·1644	15			

The blocking has had the effect of reducing the residual mean square from 35·2619 to 31·5284 although in this example this reduction has only just compensated for the loss of residual degrees of freedom. The dose effect is not significant at the 5 per cent level but it is significant once again at the 10 per cent level.

The advantage of this design over the design in Section 10.3.1 is that the block size is the smallest possible (for the complete block designs) so that the doses are compared under as near identical conditions as possible. This design may be further improved by the extra balancing introduced in the designs described in the next sections.

10.4. THE LATIN SQUARE

The Latin square arrangement groups the treatments into blocks in two different ways simultaneously. In the example experiment the observations would be arranged so that each dose appears once in each block and once in each position within a block. Suppose that the example data had been obtained in the arrangement given in Table 10.10.

TABLE 10.10. EXAMPLE DATA REARRANGED AS A LATIN SQUARE

Block	Position				Totals
	1	2	3	4	
1	66·0 (A)	65·2 (B)	77·1 (C)	83·7 (D)	292·0
2	81·8 (B)	72·8 (C)	76·4 (D)	71·2 (A)	302·2
3	85·1 (C)	80·3 (D)	73·7 (A)	72·9 (B)	312·0
4	89·6 (D)	69·8 (A)	83·4 (B)	81·1 (C)	323·9
Total	322·5	288·1	310·6	308·9	1230·1

It will be seen from this arrangement that each dose appears in each row and in each column once. The rows have been labelled blocks and the columns positions within blocks. It is clear, however, that we could regard the columns as blocks and say that the treatments have been arranged in blocks in two ways.

The blocks in the example experiment may be performed on different days and the positions within blocks may correspond to different times of day. The latin square arrangement ensures that an experiment for each dose is performed once each day and once at each different time of day. The four dose means each include four observations one for each day (or block) and one for each time of day (or position) so that if one day produces high results or one time of day produces high results each dose mean will be equally affected.

The model for this design represents the jth dose in the tth block as:

$$x_{tij} = A + D_j + B_t + P_i + z_{tij}. \qquad (10.4)$$

It should be noted that although three subscripts are attached to an observation only two of these are necessary to define its position.

The Latin square arrangement is really an incomplete three-way factorial experiment in which the three factors all have the same number of levels, n, and in which only a particular choice of n^2 of the n^3 possible combinations are performed. A complete three-way experiment includes interaction terms as well as the three main effects. The Latin square sacrifices information about these interaction terms in order to reduce the number of observations from n^3 to n^2. Thus the analysis of variance table consists of the three main effects, days (or blocks), times of day (positions), and doses, and a residual term. The computations for the example data rearranged in Table 10.10 are given below:

Blocks S.S. $= \frac{1}{4}(292 \cdot 0^2 + 302 \cdot 2^2 + 312 \cdot 0^2 + 323 \cdot 9^2) - 1230 \cdot 1^2/16$

$= 139 \cdot 3869.$

Positions S.S. $= \frac{1}{4}(322 \cdot 5^2 + 288 \cdot 1^2 + 310 \cdot 6^2 + 308 \cdot 9^2) - 1230 \cdot 1^2/16$

$= 153 \cdot 2319.$

The analysis of variance table is given in Table 10.11.

TABLE 10.11. ANALYSIS OF EXAMPLE DATA REARRANGED AS A LATIN SQUARE

Source	S.S.	D.F.	M.S.	F.	Prob.
Doses	329·0219	3	109·6740	5·04	0·05
Blocks (days)	139·3869	3	46·4623	2·14	N.S.
Positions (times)	153·2319	3	51·0773	2·35	N.S.
Residual	130·5237	6	21·7539		
Total	752·1644	15			

The extra blocking has reduced the residual mean square in this example to 21·7539 and this reduction has more than compensated for the loss of residual degrees of freedom. The dose effect is now significant at the 5 per cent level.

The advantages of this design are:

1. Two possible sources of variation, blocks and positions, are eliminated from the experimental error. The double grouping is likely to increase the precision of the experiment.
2. The statistical analysis is straightforward and is still possible if some observations are missing.

The disadvantages of this design are:

1. The number of replicates must equal the number of treatments and for large numbers of treatments the experiment size becomes very large.
2. The number of degrees of freedom for the residual mean square is reduced from that of the comparable randomized block design, although the extra grouping often more than compensates for this loss.
3. The necessary assumption that there are no interaction terms present.

10.4.1. Randomization of Treatments in a Latin Square

The random allocation of treatments to blocks is made more difficult in the case of a Latin square because of the additional requirement that each treatment must appear once in each position within a block. The possible method of repeating the randomized block procedure of random allocation within a block, until a Latin square arrangement is the result, is likely to be a lengthy process. Tables of certain Latin squares from which it is possible to derive a random Latin square are given in Fisher and Yates (1958) and description of the use of these tables is given in the introduction to these tables. The method is briefly described here. There are 12 3 × 3 squares, 576 4 × 4 squares, 161,280 5 × 5 squares and 812,851,200 6 × 6 squares. While it is possible to tabulate all the 3 × 3 squares and possibly all the 4 × 4 squares it soon becomes apparent that tabulation of all possible squares for sizes larger than 4 × 4 is impracticable. However, it is clear that the rows and columns of all squares may be rearranged to form a standard square containing the first row and the first column in a standard order. It is also clear that there are far fewer possible standard squares so that tabulation of standard squares is more practicable than tabulation of all squares. If we choose a standard square at random we may randomly permute the rows, columns and letters to obtain a random Latin square. It has been found that it is possible to divide all possible squares of a given size into transformation sets each consisting of squares that are obtainable from each other by permutation of rows, columns and letters. Each transformation set is independent of any other set in that the squares in one set are all different from squares in another set. Thus if one standard square is available from each transformation set we may choose at random a transformation set and then randomly permute rows, columns and letters. However, the $(6!)^3$ squares generated from one square are not all different and the number of different squares obtained varies from set to set. Fortunately within one set each different square appears equally often so that we may choose the transformation set with probability proportional to the number of different squares in the set and then randomly permute rows, columns and letters. This is the procedure adopted in Fisher and Yates (1953) where their Table 15 contains one standard square from each transformation set and the number of different squares in each set is recorded beneath each square. Space is saved in these tables by observing that the squares in one set may be obtained from the squares in another set by interchanging rows for columns and where this is possible only one square is recorded. Two squares are called conjugate if the rows of one are the columns of the other, and a square is self-conjugate if the process of replacing rows by columns produces the same square. Table 15 in Fisher and Yates (1953) records three standard 4×4 squares in the first transformation set and one in the second. There are three times as many different squares in the first set as in the second so that we may select one square from the four recorded with equal probability. For 5×5 squares there are two transformation sets and the number of different squares in the two sets are in the ratio 50 to 6. The table records twenty-five conjugate pairs belonging to the first set and six self-conjugate squares belonging to the second set. The selection of a square thus consists of selecting a random number in the range 1 to 56 and selecting the square corresponding to the chosen number. The twenty-five conjugate pairs in the first set are numbered with two numbers per square from 1 to 50 and the six squares from the second set are numbered from 51 to 56. If the random number falls between 1 and 50 and if it corresponds to the first number recorded below a square then that square is chosen as it stands; if it corresponds to the second number recorded then the conjugate of that square is chosen. If the random number falls between 51 and 56 the relevant square from the second set is chosen.

For 6×6 squares only one square from each set is recorded and five of the squares given represent two transformation sets one set being obtained from the square as it is given and the other being obtained from the conjugate square. The number of squares in each set is recorded below each square in the form of a range such as 9241–9280; that is forty squares. The five squares representing two sets have two ranges recorded, the first corresponding to the square as given and the second corresponding to the conjugate. The selection process for a particular transformation set consists of selecting a random number in the range 1 to 9408 and selecting the set in whose range the random number falls. For example, if the random number is 6512 square VI is chosen and if the random number is 6001 the conjugate of square V is chosen.

Having chosen a square from the tables the rows, columns and letters are randomly permuted to obtain the randomly chosen Latin square. Random numbers are available in Fisher and Yates (1953) for this purpose.

For squares larger than 6 × 6 the tables do not list squares from all transformation sets but only four 7 × 7 squares and one each of 8 × 8 to 12 × 12 squares. The suggested approximate process for randomly selecting a Latin square of size 7 × 7 is to choose, with equal probability, one of the squares given and to randomly permute rows, columns and letters. The approximate process for 8 × 8 to 12 × 12 squares consists simply of randomly permuting rows, columns and letters of the square tabulated.

10.4.2. Uses of Latin Squares

The variables used above to define the double grouping were both time variables. The Latin square has many uses in which the groupings are determined by variables such as experiments, machines, rows and columns of a field, or a layout of items on a tray. The above example experiment could have been arranged as follows. Four different experimenters were to make an observation of the effect of each of four doses on a material and the four observations for each experimenter were to be performed one on each of four days. The columns in the above Latin square design would represent experimenters and the rows would represent days. Thus one observation for each dose would be made by each experimenter and on each day. If one day's results or one experimenter's results were different from the others the dose means would be equally effected and the comparison between the means would still be valid.

10.4.3. An Interesting Use of a Latin Square

The following experiment is an interesting example of the use of a Latin square and shows that the treatments themselves could form a further classification. The purpose of the experiment was to test the preference to rats of two different foods and each rat was to be given a free choice of foods. It had been observed previously that each rat had a favourite corner for eating and would drag a movable container to this corner. It was necessary to balance the positions to be occupied by the two containers and the containers would be fixed in these positions. There are six ways of choosing two corners from four and since each container may contain either food there are twelve combinations. We will denote the foods by the numbers 1 and 2 and the four positions by LB, LF, RB, RF. The twelve combinations form the treatments A, B, ..., L to be used as the treatments in a 12 × 12 Latin square. The rows of the Latin square were twelve different rats and the columns were twelve different feeding period of 2 days each. The rats were presented with equal amounts of the two foods and the observations collected were the intakes of food 1 and food 2 for each rat. The twelve combinations are given in Table 10.12.

TABLE 10.12. TWELVE POSITION–FOOD COMBINATIONS

Treatment		A	B	C	D	E	F	G	H	I	J	K	L
Corners	LB	1	1	1				2	2	2			
	LF	2			1	1		1			2	2	
	RB		2		2		1		1		1		2
	RF			2		2	2			1		1	1

The analysis of variance for this experiment is different from that of a normal Latin square in that two measurements are made for each combination and interest is concentrated on the difference between pairs of observations. The analysis of variance is given in Table 10.13.

TABLE 10.13. ANALYSIS OF VARIANCE USING BOTH FOOD INTAKES

Source	D.F.
Foods	1
Positions	11
Positions × foods	11
Periods	11
Periods × foods	11
Rats	11
Rats × foods	11
Residual	220
Total	287

An interesting alternative method of analysis is worth study. If the observation used for the analysis of variance is the difference between the two food intakes we would be interested in testing the null hypothesis that the difference between the food intakes is zero. The analysis of variance is that of a 12 × 12 Latin square with the extra sum of squares $144\bar{x}^2...$ (where $\bar{x}...$ is the grand mean of the differences). This sum of squares enables us to test the null hypothesis of no difference between the food intakes. The analysis is given in Table 10.14.

TABLE 10.14. ANALYSIS OF VARIANCE USING THE FOOD INTAKES DIFFERENCES

Source	D.F.
Foods difference	1
Positions	11
Periods	11
Rats	11
Residual	110
Total	144

The total sum of squares in this analysis is computed about the hypothetical population mean of zero. That is the total sum of squares is

$$\sum_{t=1}^{12} \sum_{j=1}^{12} x_{tij}^2$$

and has 144 degrees of freedom. The extra degree of freedom is taken up in the sum of squares $144\bar{x}_{..}^2$ measuring the difference between the observed mean and the population mean of zero. In both these analyses the position effect can be split up into various sums of squares measuring particular aspects of the position effect.

10.5. THE GRAECO-LATIN SQUARE

In this design the treatments are grouped into replicates in three different ways with the consequence that the effects of three different sources of variation are removed from the residual. We will denote the extra classification by Greek letters when the arrangement for a 4 × 4 square might be as given in Table 10.15.

TABLE 10.15. A GRAECO-LATIN SQUARE ARRANGEMENT

Aα	Cβ	Dγ	Bδ
Bβ	Dα	Cδ	Aγ
Cγ	Aδ	Bα	Dβ
Dδ	Bγ	Aβ	Cα

Both the Greek letters and the Latin letters form latin squares on their own and they jointly show the further property that each Greek letter appears with each Latin letter once and once only.

The model for this design represents the observation in the ith row and tth column as

$$x_{tijk} = A + D_j + R_i + C_t + V_k + z_{tijk}, \qquad (10.5)$$

where A is a general level, D_j is an effect due to the jth dose, R_i is an effect due to the ith row, C_t is an effect due to the tth column, V_k is an effect due to the kth value of the extra classification, z_{tijk} is the usual normally distributed error term.

The analysis of variance is given in Table 10.16.

TABLE 10.16. ANALYSIS OF VARIANCE OF A GRAECO-LATIN SQUARE

Source	D.F.
Doses	3
Rows	3
Columns	3
Variable	3
Residual	3
Total	15

The advantages of this design are:

1. Three possible sources of variation are eliminated from the residual.
2. The statistical analysis is straightforward and missing data can be dealt with provided only a small amount is missing.

The disadvantages of this design are:

1. The number of replicates is fixed.
2. The triple balancing is sometimes difficult to achieve in practice.
3. The number of degrees of freedom for the residual are further reduced and are small for the smaller squares.

We now apply this analysis to the example data as rearranged in Table 10.17.

TABLE 10.17. EXAMPLE DATA REARRANGED AS A GRAECO-LATIN SQUARE

66·0 (Aα)	77·1 (Cβ)	83·7 (Dγ)	65·2 (Bδ)	292·0
81·8 (Bβ)	76·4 (Dα)	72·8 (Cδ)	71·2 (Aγ)	302·2
85·1 (Cγ)	73·7 (Aδ)	72·9 (Bα)	80·3 (Dβ)	312·0
89·6 (Dδ)	83·4 (Bγ)	69·8 (Aβ)	81·1 (Cα)	323·9
322·5	310·6	299·2	297·8	1230·1

The totals for the Greek letters are:

α	β	γ	δ	Total
296·4	309·0	323·4	301·3	1230·1

and the main effects are:

Blocks S.S. $= \frac{1}{4}(292 \cdot 0^2 + \cdots + 323 \cdot 9^2) - 1230 \cdot 1^2/16 = 139 \cdot 3869,$

Positions S.S. $= \frac{1}{4}(322 \cdot 5^2 + \cdots + 297 \cdot 8^2) - 1230 \cdot 1^2/16 = 99 \cdot 3969$,

Greek letters S.S. $= \frac{1}{4}(296 \cdot 4^2 + \cdots + 301 \cdot 3^2) - 1230 \cdot 1^2/16 = 104 \cdot 1769$.

These sums of squares are collected together in Table 10.18.

TABLE 10.18. ANALYSIS OF EXAMPLE DATA REARRANGED AS A GRAECO-LATIN SQUARE

Source	S.S.	D.F.	M.S.	F.	Prob.
Doses	329·0219	3	109·6740	4·10	0·20
Blocks (rows)	139·3869	3	46·4623	1·74	N.S.
Positions (columns)	99·3969	3	33·1323	1·24	N.S.
Greek letters	104·1769	3	34·7256	1·30	N.S.
Residual	80·1818	3	26·7272		
Total	752·1644	15			

The Latin square formed by the Latin letters in this Graeco-Latin square is not the same as the Latin square analysed in Section 10.4 above. (It is not in fact possible to form a Latin square by superimposing a second Latin square on to the square analysed in Section 10.4.) So that this above analysis differs from that in Section 10.4 by more than the extra variable. The positions within blocks totals are different in the two analyses. Comparison with the randomized blocks analysis in Section 10.3.2 shows that the two extra classifications have reduced the residual mean square from 31·5284 to 26·7272 but this reduction has not compensated for the loss of residual degrees of freedom. The three degrees of freedom remaining in the Graeco-Latin square analysis is very low and the dose effect is now only significant at the 20 per cent level.

10.5.1. Random Selection of a Graeco-Latin Square

Table 16 in Fisher and Yates contains complete sets of orthogonal Latin squares for square sizes 3, 4, 5, 7, 8 and 9. Any two of these squares, of a given size, will jointly form a Graeco-Latin square if the treatment numbers in one are replaced by Latin letters and by Greek letters in the other. The random selection of a Graeco-Latin square consists of selecting two Latin squares at random, combining to form a Graeco-Latin square, and randomly permuting rows, columns, Latin letters and Greek letters.

10.6. FURTHER ALPHABETS

It is possible to extend this type of design by adding further alphabets in the same fashion. The maximum numer of possible alphabets is equal to one less than the size of the square. Although this is possible and the analysis is straightforward these more complicated designs are not frequently useful.

1. Five treatments are compared and four observations for each treatment made. The experiment is arranged in a randomized block design with four blocks. The data obtained and the design are:

```
Block 1   54·6 (A)   63·3 (D)   64·8 (E)   60·6 (C)   56·0 (B)
Block 2   65·5 (E)   61·9 (D)   57·1 (B)   53·6 (A)   59·7 (C)
Block 3   57·0 (C)   54·0 (A)   59·9 (D)   63·4 (E)   55·5 (B)
Block 4   59·3 (D)   62·8 (E)   57·2 (C)   55·8 (B)   54·0 (A)
```

(a) Plot this data assuming that the treatments are equally spaced on some scale. Join points for the same block so that separate treatment responses can be seen for each block. What conclusions would you make on the basis of this plot and which points would you seek to confirm by the analysis of variance?

(b) Perform the formal analysis of variance for this design and make your conclusions.

(c) Repeat the analysis separating out the linear treatment effect and its interaction with blocks. What are your conclusions now?

2. Sixteen animals are to be fed on four different foods. They are placed in cages one animal per cage and the cages arranged in a four by four square. The foods are allocated to the animals in the form of the following Latin square:

```
A   B   C   D       23·4   26·5   25·9   27·8
C   D   A   B       26·8   28·3   23·0   24·7
B   C   D   A       25·3   27·9   28·9   22·9
D   A   B   C       29·4   23·0   25·3   26·4
```

The data given above is the gain in weight for the first week for each animal.

(a) Perform the analysis of variance appropriate to this design and make your conclusions.

(b) For the second week of this experiment the cages were rearranged to form a second Latin square as given below. The weight gains are also given.

```
C   B   A   D       26·8   25·8   22·5   28·7
A   D   C   B       23·4   28·1   27·6   25·0
B   A   D   C       24·2   22·5   29·4   27·1
D   C   B   A       28·3   26·0   26·2   23·8
```

Perform the analysis of variance for this square and make your conclusions.

(c) Suppose now that the animals in the first week had been allocated to the 16 positions in the order 1 to 16 and in the second week in the order 16 to 1. That is animal 1 occupied row 1 column 1 in week 1, but row 4 column 4 in week 2. Ignoring row, column, and treatment effects for the moment the experiment can be regarded as a simple two-way design without replication with factors weeks and treatments. Analyse the data in this way.

(d) To include treatments (foods) in the analysis performed in the part (c) we would split the animals sum of squares into "Treatments" and "Remainder"—representing variability between animals which is not attributable to treatments. The "Weeks × Animals" sum of squares would be similarly split into "Weeks × Treatments" and "Weeks × Remaining animal variability". Thus the analysis would consist of the following sums of squares:

Source	D.F.	D.F.	Source
Treatments	3 ⎫	15	Animals
Remainder (animals)	12 ⎭		
Weeks	1	1	Weeks
Weeks × Treatments	3 ⎫	15	Weeks × Animals
Weeks × Remainder (animals)	12 ⎭		
Total	31	31	

Perform the analysis in this form. Why is the testing procedure in this analysis a little different from normal? Explain the reasons for choosing your testing procedure.

(e) Which sum or sums of squares in the analysis in part (d) contain the row and column effects and their interactions with weeks?

CHAPTER 11

BALANCED INCOMPLETE RANDOMIZED BLOCK DESIGNS AND THE ANALYSIS OF COVARIANCE

THIS chapter is devoted to two separate subjects. The first part describes the use of balanced incomplete randomized block designs in experimental situations where the number of treatments that can be placed in one block is not equal to the number of treatments to be tested. The second part describes the use of the technique of analysis of covariance as a means of adjusting the usual analysis of variance for the measured values of a concomitant variable.

11.1. BALANCED INCOMPLETE RANDOMIZED BLOCK DESIGNS

The blocked experimental designs considered so far have assumed that it is possible to include all treatments an equal number of times in each block. It is possible, however, to analyse experiments in which the blocks contain incomplete replication, but unless special attention is given to the allocation of the treatments to the blocks the analysis is involved and laborious. The allocation of treatments to blocks in incomplete block designs is made to achieve a certain amount of balance between the treatments, and a design may be "balanced" or "partially balanced". The number of observations made in each block is usually less than the number of treatments but it is possible to achieve balance with blocks containing more observations than the number of treatments. Each block then contains one complete replicate and an incomplete replicate.

Suppose that in the example experiment described in the previous chapter it had been possible to perform only three experiments during one day. The randomized blocks and Latin square designs would require four experiments to be performed during one day so that these designs could not be used. A convenient design for this situation would be the balanced incomplete block design, consisting of four blocks (days) each containing three treatments, given in Table 11.1. For this design the order of performing the three experiments on a given day would be chosen at random.

194

TABLE 11.1. EXAMPLE BALANCED INCOMPLETE RANDOMIZED BLOCK DESIGN

Day	1	2	3	4
Doses	A	C	A	B
	D	B	C	C
	B	A	D	D

The three properties shown by this design which are necessary for a balanced design are:

1. Each dose appears equally often.
2. Each block contains the same number of doses.
3. Each dose appears with each other dose equally often in the same block.

The following notation will be used for five important parameters:

t = the number of treatments (doses),
r = the number of times each treatment appears,
b = the number of blocks,
k = the number of observations per block,
λ = the number of times pairs of treatments appear together in the same block.

It is clear that the following relations exist between these parameters:

$tr = bk$ = the total number of observations,

$r(k - 1) = \lambda(t - 1)$ = the number of experimental units occurring together with any one treatment.

In the above design $t = 4$, $r = 3$, $b = 4$, $k = 3$, and $\lambda = 2$ so that

$$tr = bk = 12$$

and

$$r(k - 1) = \lambda(t - 1) = 6. \tag{11.1}$$

The existence of these relations between the parameters means that we are not free to choose all parameters independently and some compromise may be necessary. The number of treatments and the block size are normally decided by the experimental situation. The values of the other parameters are then chosen to satisfy the relations (11.1) above. It may be found, however, that a balanced design of a convenient size does not exist. Cox (1958) includes a list of the values of t, r, b and k for which balanced designs exist.

The analyses of these types of experiments are more involved than for complete experiments. If the results for one block (or day) are high, only three of the four doses are affected so that the observed dose means are not equally affected by block differences. The observed dose means must therefore be adjusted for block differences. The third necessary condition given above that each treatment (or dose) appears with each of the other treatments in the same block the same number of times enables this to be done.

The jth adjusted treatment mean is given by

$$\bar{x}.. + A_j = \bar{x}.. + \frac{kT_j - \Sigma B_i}{\lambda t}, \qquad (11.2)$$

where T_j is the jth treatment total, B_i is the ith block total, and the sum ΣB_i is taken over those blocks containing the jth treatment.

The ith adjusted block mean is given by

$$\bar{x}.. + B_i = \bar{x}.. + \frac{rB_i - \Sigma T_j}{\lambda t}, \qquad (11.3)$$

where the sum ΣT_j is taken over those treatments contained in the ith block.

The analysis of variance of these experiments is arranged differently from that of the complete randomized block experiment. This is because the sums of squares due to treatments, due to blocks, and due to residual variability do not add up to the total sum of squares. The terms adjusted and unadjusted used in Table 11.2 refer to sums of squares calculated using the adjusted and unadjusted means respectively. The analysis of variance is given in Table 11.2 and the letter G denotes the grand total of all observations.

TABLE 11.2. ANALYSIS OF VARIANCE OF A BALANCED INCOMPLETE RANDOMIZED BLOCK DESIGN

Sources	S.S.	D.F.
Treatments (adjusted)	$S_1 = \dfrac{t\lambda}{k} \sum\limits_{j=1}^{t} A_j^2$	$t - 1$
Blocks (unadjusted)	$S_2 = \dfrac{1}{k} \sum\limits_{i=1}^{b} B_i^2 - \dfrac{G^2}{kr}$	
Blocks (adjusted)	$S_3 = S_1 + S_2 - S_4$	$b - 1$
Treatments (unadjusted)	$S_4 = \dfrac{1}{r} \sum\limits_{j=1}^{t} T_j^2 - \dfrac{G^2}{kr}$	
Residual	$S_5 = S_6 - (S_1 + S_2)$	$tr - t - b + 1$
Total	$S_6 = \sum\limits_{i=1}^{t} \sum\limits_{j=1}^{b} (\bar{x}_{ij} - \bar{x}..)^2$	$tr - 1$

The unadjusted sums of squares are calculated in the normal way as main effects on the unadjusted treatment and block totals.

11.1.1. Example

We will apply this analysis to the data given in Table 11.3.

TABLE 11.3. EXAMPLE BALANCED INCOMPLETE RANDOMIZED BLOCK DATA

	Day			
	1	2	3	4
	84·8 (A)	78·2 (C)	82·1 (A)	78·2 (B)
	85·7 (B)	79·4 (B)	85·0 (C)	70·2 (C)
	86·6 (D)	78·7 (A)	81·1 (D)	82·8 (D)
Totals	257·1	236·3	248·2	231·2

The treatment totals are

$$
\begin{array}{cccc}
\text{A} & \text{B} & \text{C} & \text{D} \\
245\cdot6 & 243\cdot3 & 233\cdot4 & 250\cdot5
\end{array}
$$

and the grand total G is 972·8.

The corrections for treatment means (A_j in expression (11.2)) are

$$\tfrac{1}{8}(3 \times 245\cdot6 - 257\cdot1 - 236\cdot3 - 248\cdot2) = -0\cdot6000$$

$$\tfrac{1}{8}(3 \times 243\cdot3 - 257\cdot1 - 236\cdot3 - 231\cdot2) = 0\cdot6625$$

$$\tfrac{1}{8}(3 \times 233\cdot4 - 236\cdot3 - 248\cdot2 - 231\cdot2) = -1\cdot9375$$

$$\tfrac{1}{8}(3 \times 250\cdot5 - 257\cdot1 - 248\cdot2 - 231\cdot2) = 1\cdot8750$$

Adjusted treatment S.S. $= \dfrac{8}{3}[(-0\cdot6000)^2 + \cdots + (1\cdot8750)^2] = 21\cdot516.$

Unadjusted treatment S.S. $= \dfrac{1}{3}(245\cdot6^2 + \cdots + 250\cdot5^2) - \dfrac{972\cdot8^2}{12}$

$$= 51\cdot700.$$

Unadjusted block S.S. $= \dfrac{1}{3}(257\cdot1^2 + \cdots + 231\cdot2^2) - \dfrac{972\cdot8^2}{12}$

$$= 136\cdot607.$$

Adjusted block S.S. $= 21 \cdot 516 + 136 \cdot 607 - 51 \cdot 700 = 106 \cdot 423.$

Total S.S. $\qquad = 84 \cdot 8^2 + \cdots + 82 \cdot 8^2 - \dfrac{972 \cdot 8^2}{12} = 228 \cdot 467.$

Residual S.S. $\qquad = 228 \cdot 467 - 21 \cdot 516 - 136 \cdot 607 = 70 \cdot 344.$

TABLE 11.4. ANALYSIS OF VARIANCE OF EXAMPLE DATA

Source	S.S.	D.F.	M.S.	F.	Prob.
Treatments (adjusted)	21·516	3	7·172	0·51	N.S.
Blocks (unadjusted)	136·607				
Blocks (adjusted)	106·423	3	35·474	2·43	N.S.
Treatments (unadjusted)	51·700				
Residual	70·344	5	14·067		
Total	228·467	11			

The complete analysis of variance given in Table 11.4 shows that the treatments (doses) effect may be explained entirely by the variability observed between experimental units.

11.2. A YOUDEN SQUARE

The Youden square (Youden, 1937) is a balanced incomplete block design which balances, in addition, the order of units within a block. It is thus a design of the Latin square type. The Youden square may be regarded as an incomplete Latin square or as an incomplete square with one, or more, complete Latin squares added. We may rearrange the above data to form a Youden square as shown in Table 11.5.

TABLE 11.5. EXAMPLE DATA REARRANGED AS A YOUDEN SQUARE

Day	1	2	3	4	Totals
Period 1	84·8 (A)	79·4 (B)	85·0 (C)	82·8 (D)	332·0
2	86·6 (D)	78·2 (C)	82·1 (A)	78·2 (B)	325·1
3	85·7 (B)	78·7 (A)	81·1 (D)	70·2 (C)	315·7
Totals	257·1	236·3	248·2	231·2	972·8

This arrangement is still a balanced incomplete block design but in addition each dose appears once in each position in the block (day).

The analysis of this data differs from that of the balanced incomplete block experiments given above in that the residual sum of squares in the above analysis is divided into two sums of squares. One of these represents the position effect calculated as an ordinary main effect and the other is the new residual.

The position sum of squares is

$$\frac{1}{4}(332 \cdot 0^2 + 325 \cdot 1^2 + 315 \cdot 7^2) - \frac{972 \cdot 8^2}{12} = 33 \cdot 472$$

and the complete analysis of variance is shown in Table 11.6.

TABLE 11.6. ANALYSIS OF VARIANCE OF THE EXAMPLE YOUDEN SQUARE

Source	S.S.	D.F.	M.S.	F.	Prob.
Treatments (adjusted)	21·516	3	7·172	0·58	N.S.
Blocks (unadjusted)	136·607				
Blocks (adjusted)	106·423	3	35·474	2·89	N.S.
Treatments (unadjusted)	51·700				
Positions	33·472	2	16·736	1·36	N.S.
Residual	36·872	3	12·291		
Total	228·467	11			

The results of this analysis are very similar to those of Section 11.1.1 although the residual mean square has been reduced slightly from 14·067 to 12·291. This reduction has barely compensated for the loss of two degrees of freedom.

11.3. OTHER INCOMPLETE BLOCK DESIGNS

Other incomplete blocks designs exist and these are described in the various textbooks on experimental design such as Cox (1958), Cochran and Cox (1957) and Kempthorne (1952). The number of replicates necessary to balance an incomplete block design is sometimes too large for the experiment to be practically possible and the partially balanced incomplete block designs have been introduced to enable a smaller but still useful experiment to be carried out. The lattice designs are useful designs for experiments designed to compare n^2 treatments, that is four, nine, and sixteen treatments.

11.4. ANALYSIS OF COVARIANCE (CONCOMITANT OBSERVATIONS)

The use of concomitant observations as a method of reducing the variability in an experiment has been briefly discussed in Chapter 1. The method consists of observing, in addition to the main variable, the value of one or more uncontrolled variables and using this extra information in the analysis. The method may be applied to any design and more than one such variable may be measured and used in the analysis. We will consider the use of one concomitant variable in a simple randomized block design.

11.4.1. *Analysis of a Two-way Factorial with One Concomitant Variable*

The purpose of the example experiment was to investigate the effect of the addition of three different quantities of substance A on the yield of substance B from a chemical reaction. It was believed that the temperature of the reaction was an important variable and this could not be controlled. The temperature could, however, be measured and its effect allowed for in the analysis. Four randomized blocks make up the design and the data is recorded in Table 11.7.

TABLE 11.7. EXAMPLE DATA FOR A TWO-WAY FACTORIAL WITH ONE CONCOMITANT OBSERVATION

Block		1	2	3	4	Totals
A_1	Yield	248·6	263·2	277·1	256·3	1045·2
	Temp.	341·5	351·3	371·4	354·3	1418·5
A_2	Yield	267·6	278·6	275·6	284·8	1106·6
	Temp.	349·4	360·2	347·2	364·9	1421·7
A_3	Yield	284·4	289·6	272·4	284·9	1131·3
	Temp.	355·0	340·2	345·0	356·9	1397·1
Total yield		800·6	831·4	825·1	826·0	3283·1
Total temp.		1045·9	1051·7	1063·6	1076·1	4237·3

The analysis of variance model represents the yield y_{jt} in the tth block for the jth treatment (quantity of A) as

$$y_{jt} = A + T_j + B_t + C(x_{jt} - \bar{x}..) + Z_{jt}, \qquad (11.4)$$

where A is a general level, T_j is an effect due to the jth treatment, B_t is an effect due to the tth block, x_{jt} is the temperature of the reaction for yield y_{jt}, C is a coefficient to be estimated, Z_{jt} is the usual normally distributed error term.

This model contains the treatment and block effects of a normal randomized block experiment and the total sum of squares is split up into these two effects and a residual. Each of these sums of squares needs adjustment for the different temperature values observed in each block and treatment.

The treatment means similarly require adjustment. The analysis is carried out in the following stages.

Firstly analyses of variance are performed on each variable separately. For the main variable (yield) we have:

Total S.S. $\quad = 248 \cdot 6^2 + 263 \cdot 2^2 + \cdots + 284 \cdot 9^2 - \dfrac{3283 \cdot 1^2}{12} = 1736 \cdot 91.$

Treatment S.S. $= \dfrac{1}{4}(1045 \cdot 2^2 + 1106 \cdot 6^2 + 1131 \cdot 3^2) - \dfrac{3283 \cdot 1^2}{12} = 982 \cdot 77.$

Block S.S. $\quad = \dfrac{1}{3}(800 \cdot 6^2 + \cdots + 826 \cdot 0^2) - \dfrac{3283 \cdot 1^2}{12} = 188 \cdot 64.$

Residual S.S. $\quad = 1736 \cdot 91 - 982 \cdot 77 - 188 \cdot 64 = 565 \cdot 50.$

For the concomitant variable we have:

Total S.S. $\quad = 341 \cdot 5^2 + 351 \cdot 3^2 + \cdots + 356 \cdot 9^2 - \dfrac{4237 \cdot 3^2}{12} = 962 \cdot 35.$

Treatment S.S. $= \dfrac{1}{4}(1418 \cdot 5^2 + 1421 \cdot 7^2 + 1397 \cdot 1^2) - \dfrac{4237 \cdot 3^2}{12} = 89 \cdot 45.$

Block S.S. $\quad = \dfrac{1}{3}(1045 \cdot 9^2 + \cdots + 1076 \cdot 1^2) - \dfrac{4237 \cdot 3^2}{12} = 179 \cdot 35.$

Residual S.S. $\quad = 962 \cdot 35 - 89 \cdot 45 - 179 \cdot 35 = 693 \cdot 55.$

The next stage of the computation is to calculate an analysis of cross-products of the two variables in the same way as the ordinary analysis of variance. That is the above computation is performed again using the cross-products between the two variables instead of the squares of one of them. It should be noted that a sum of cross-products may be negative. The analysis of cross-products is:

Total S.C.P. $\quad = 248 \cdot 6 \times 341 \cdot 5 + \cdots + 284 \cdot 9 \times 356 \cdot 9$

$$- \frac{3283 \cdot 1 \times 4237 \cdot 3}{12} = 396 \cdot 85.$$

Treatments S.C.P. $= \dfrac{1}{4}(1045 \cdot 2 \times 1418 \cdot 5 + \cdots + 1131 \cdot 3 \times 1397 \cdot 1)$

$$- \frac{3283 \cdot 1 \times 4237 \cdot 3}{12} = -187 \cdot 81.$$

Blocks S.C.P. $\quad = \dfrac{1}{3}(800\cdot6 \times 1045\cdot9 + \cdots + 826\cdot0 \times 1076\cdot1)$

$$-\frac{3283\cdot1 \times 4237\cdot3}{12} = 96\cdot66.$$

Residual S.C.P. $= 396\cdot85 + 187\cdot81 - 98\cdot66 = 686\cdot00.$

The sums of squares and cross-products are summarized in Table 11.8.

TABLE 11.8. SUMS OF SQUARES AND CROSS-PRODUCTS IN THE EXAMPLE ANALYSIS OF COVARIANCE

Source	S.S. (yield)	S.C.P.	S.S. (temperature)
Treatments	982·77	−187·81	89·45
Blocks	188·64	98·66	179·35
Residual	565·50	486·00	693·55
Total	1736·91	396·85	962·35

We will denote the sums of squares and cross-products for the residual by R_{xx}, R_{yy} and R_{xy}. The coefficient C is estimated by:

$$\hat{C} = \frac{R_{xy}}{R_{xx}} = \frac{486\cdot00}{693\cdot55} = 0\cdot7007.$$

The corrected residual sum of squares is given by:

$$R_{yy} - \frac{R_{xy}^2}{R_{xx}} = 565\cdot50 - \frac{486\cdot00^2}{693\cdot55} = 224\cdot94.$$

The other sums of squares in the corrected analysis of variance are obtained in the following way. We form a new table, Table 11.9, by omitting the block effect from Table 11.8.

TABLE 11.9. SUMS OF SQUARES AND CROSS-PRODUCTS FOR TREATMENTS AND RESIDUAL ONLY

	S.S. (yield)	S.C.P.	S.S. (temperature)
Treatments	982·77	−187·81	89·45
Residual	565·50	486·00	693·55
Total	1548·27	298·19	783·00

We now correct the total sum of squares (excluding the block effect) by subtracting the contribution due to the concomitant variable, that is:

$$1548 \cdot 27 - \frac{298 \cdot 19^2}{783 \cdot 00} = 1434 \cdot 71.$$

The corrected treatment sum of squares is now calculated by subtracting from this total the corrected residual sum of squares calculated earlier. This gives the corrected analysis shown in Table 11.10.

TABLE 11.10. CORRECTED ANALYSIS OF COVARIANCE

Source	S.S.	D.F.	M.S.	F.	Prob.
Treatments	1209·77	2	604·885	13·45	0·01
Residual	224·94	5	44·988		
Total	1434·71	7			

It should be noticed that the degrees of freedom of the residual term is reduced by one from the usual analysis of variance because of the estimation of the constant C. The usual analysis of variance is given in Table 11.11.

TABLE 11.11. ANALYSIS OF VARIANCE OF THE MAIN VARIABLE

Source	S.S.	D.F.	M.S.	F.	Prob.
Treatments	982·77	2	491·385	5·21	0·05
Blocks	188·64	3	62·880		
Residual	565·50	6	94·250		
Total	1736·91	11			

The usual analysis of variance establishes the significance of the treatment effect but the analysis of covariance establishes the significance of the treatment effect more strongly. The residual mean square for the covariance analysis is smaller than for the usual analysis so that the observation of the concomitant variable has improved the precision of the experiment.

It is possible to obtain a significant treatment effect in the analysis of variance and not in the analysis of covariance. The interpretation of this situation would be that the treatment differences were due to the differing values of the concomitant variable rather than due to the treatments themselves.

QUESTIONS

1. Twenty automobile tyres were used in an experiment to compare five tyre treatments believed to have an effect on the wear of the tyre. The same automobile was used for a total time of 15 months. It is clear that only four treatments could be included on the same car during one period. The treatments were allocated to wheel positions so that each treatment appears in each of the four positions once and only once. The results and treatment allocation are given below:

	Wheel position			
	1	2	3	4
Period 1	0·429 (A)	0·517 (B)	0·471 (C)	0·444 (D)
Period 2	0·502 (B)	0·458 (C)	0·432 (D)	0·530 (E)
Period 3	0·463 (C)	0·440 (D)	0·524 (E)	0·425 (A)
Period 4	0·429 (D)	0·518 (E)	0·429 (A)	0·530 (B)
Period 5	0·506 (E)	0·405 (A)	0·527 (B)	0·461 (C)

(a) Analyse this data and assess the evidence for the belief that the treatments have different effects on the wear of the tyres.

(b) Do the four wheel positions have an equal effect on the wear of the tyres?

(c) What effect on the results would you have expected if instead of using one car for five different periods we had used five different cars?

(d) Compute the adjusted treatment means.

2. The experiment in question 1 was repeated with a second automobile and the results obtained are as follows:

	Wheel position			
	1	2	3	4
Period 1	0·415 (A)	0·528 (B)	0·486 (C)	0·460 (D)
Period 2	0·498 (B)	0·465 (C)	0·448 (D)	0·548 (E)
Period 3	0·455 (C)	0·456 (D)	0·529 (E)	0·430 (A)
Period 4	0·410 (D)	0·534 (E)	0·451 (A)	0·539 (B)
Period 5	0·486 (E)	0·421 (A)	0·536 (B)	0·477 (C)

(a) Analyse this data and test treatment differences.

(b) Do the four wheel positions have an equal effect on the wear of the tyres?

(c) Compute the adjusted treatment means.

(d) Compare this data with that in question 1. Do you feel that the two automobiles have produced the same comparisons between the treatments?

(e) Compare the two automobiles.

(f) If the fourth wheel position in the above design had not been used would this new design with three blocks (wheel positions) and five treatments have been a Youden square? If not, which condition for a Youden square is not satisfied?

3. The data given in question 1 after Chapter 10 is given again below together with the values of a concomitant variable that was measured at the time of each observation. The data is rearranged according to treatments and blocks and may appear at first sight to be different from that in question 1 after Chapter 10.

Treatments	A	B	C	D	E
Block 1	54·6, 186	56·0, 168	60·6, 162	63·3, 150	64·8, 153
Block 2	53·6, 184	57·1, 172	59·7, 168	61·9, 158	65·5, 142
Block 3	54·0, 184	55·5, 164	57·0, 165	59·9, 161	63·4, 148
Block 4	54·0, 179	55·8, 160	57·2, 157	59·3, 159	62·8, 144

(a) Perform an analysis of variance of the main (first) variable if this was not done after Chapter 10.

(b) Perform an analysis of covariance, taking into account the values of the concomitant variable, and compare your conclusions with those obtained from the analysis of variance.

(c) Has the measurement of the concomitant variable had an effect on the precision of the experiment?

(d) Are the main variable and the concomitant variable related?

CHAPTER 12

CORRELATION AND FUNCTION FITTING

THE estimation procedures and tests of significance described in previous chapters, with the exception of the analysis of covariance, have been confined to the consideration of one variable. This chapter introduces, firstly, the use of the correlation coefficient as a measure of association between two variables, and secondly the description of one variable as a function of another variable or more than one other variable. The aim in correlation problems is usually to determine whether or not two variables are associated and if so, to measure the degree of association. To describe one variable as a function of one or more variables we choose the values of unknown parameters of the function (for example, the slope of a straight line) to provide the "best" description. The criterion by which the best description may be chosen will be that of least squares as briefly introduced in Section 4.11.

12.1. THE CORRELATION COEFFICIENT

We have n pairs of observations (x_i, y_i) of two variables x and y and we seek to determine if knowledge of the value of one variable gives any information about the value of the other. If we denote by $\bar{x}$ and $\bar{y}$ the means of the observations of x and y respectively, the correlation coefficient, which measures the association between the two variables, is defined as:

$$r = \frac{\sum_{i=1}^{n} (x_i - \bar{x})(y_i - \bar{y})}{\left\{ \sum_{i=1}^{n} (x_i - \bar{x})^2 \sum_{i=1}^{n} (y_i - \bar{y})^2 \right\}^{\frac{1}{2}}} . \tag{12.1}$$

It may be shown that the value of the correlation coefficient r cannot be less than -1 or greater than $+1$. To appreciate what is measured by the correlation coefficient we study a plot of one variable against the other.

Figure 12.1 shows a plot of y against x with axes drawn through the point with coordinates $(\bar{x}, \bar{y})$. The four quadrants so formed are numbered 1 to 4 and a typical point (x_i, y_i) is shown in the first quadrant. Let us consider the

206

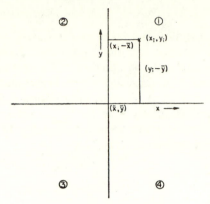

Fɪɢ. 12.1. Plot of y against x.

contribution made by typical points in each quadrant to the cross-product term:

$$C = \sum_{i=1}^{n} (x_i - \bar{x}) (y_i - \bar{y}) \qquad (12.2)$$

in the expression (12.1) for the correlation coefficient. If an observation falls in the first quadrant $(x_i - \bar{x})$ and $(y_i - \bar{y})$ are both positive so that the point makes a positive contribution to C. If an observation falls in the third quadrant $(x_i - \bar{x})$ and $(y_i - \bar{y})$ are both negative so that the point again makes a positive contribution to C. If an observation falls in the second or fourth quadrants one of the expressions $(x_i - \bar{x})$ and $(y_i - \bar{y})$ is positive and one is negative so that the point makes a negative contribution to C. The sum C is the main part of the correlation coefficient (the denominator being always positive and simply a scale factor) so that if r is positive we may conclude that there is a tendency for observations to lie in the first and third quadrants, and if r is negative we may conclude that there is a tendency for observations to lie in the second and fourth quadrants. If r is near zero we may conclude that there is an approximately equal distribution of points in the even quadrants as in the odd quadrants. Values of r near $+1$ indicate positively correlated variables, values of r near -1 indicate negatively correlated variables and values of r near zero indicate uncorrelated variables when the value of one variable gives little information about the value of the other.† The correlation coefficient r is a sample statistic and may be shown to estimate the population correlation coefficient which is usually denoted by ϱ. Tests of significance testing hypotheses about ϱ are described in Section 12.1.3 after discussion of the Normal Bivariate Distribution.

† This statement is qualified in Section 12.1.2.

12.1.1. The Normal Bivariate Distribution

Tests of significance and estimation procedures associated with the correlation coefficient are usually, but not necessarily, based on the normal bivariate distribution. It should be noted that it is theoretically possible to derive procedures similar to those described below starting from a non-normal bivariate distribution. A bivariate distribution provides the probability that the two variables jointly take particular values (or more strictly values within infinitesimal intervals such as dx and dy). The normal bivariate distribution has the mathematical form:

$p(x, y)\, dxdy$

$$= K\exp\left\{\frac{-1}{2(1 - \varrho^2)}\left[\frac{(x - \mu_x)^2}{\sigma_x^2} - \frac{2\varrho(x - \mu_x)\,(y - \mu_y)}{\sigma_x\sigma_y} + \frac{(y - \mu_y)^2}{\sigma_y^2}\right]\right\} dxdy$$

$$(12.3)$$

where

$$K = (2\pi\sigma_x\sigma_y\sqrt{(1 - \varrho^2)})^{-1}$$

μ_x and μ_y are the population x and y means,
σ_x^2 and σ_y^2 are the population x and y variances, and
ϱ is the population correlation coefficient as defined in Section 4.3.1.

This distribution may be visualized by imagining the x-scale and the y-scale at right angles in a horizontal plane and the probability of observing a particular combination of x and y values represented by a vertical line the length of which is proportional to the probability. The tops of these lines will form the bivariate normal surface. The general shape of the bivariate normal surface is shown in Fig. 12.2.

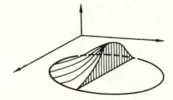

FIG. 12.2. The normal bivariate distribution.

The exact shape of the bivariate normal distribution depends on the values of three parameters σ_x^2, σ_y^2 and ϱ. If $\sigma_x^2 > \sigma_y^2$ the distribution is more spread out in the x-direction than the y-direction. The best way of appreciating the effect of the value of ϱ is to consider, for a standardized distribution with $\sigma_x^2 = \sigma_y^2 = 1$ and $\mu_x = \mu_y = 0$, the contours of the surface. That is the contours formed by joining points with equal probability. The shape of these contours is, in general, elliptical. If ϱ is large and positive an ellipse has a long major axis on the line $y = x$ and a short minor axis on the line $y = -x$. As ϱ is reduced the major axis shortens and the minor axis lengthens until when $\varrho = 0$ both axes are equal in length and the contours are circular. As ϱ is further reduced to approach -1 the major axis, now lying on the line $y = -x$, increases in length and the minor axis, now lying on $y = x$, decreases.

The following properties are shown by the normal bivariate surface:

1. The distribution of y for a given value of x is normal.
2. The distribution of x for a given value of y is normal.
3. The distribution of x for all values of y (the x-marginal distribution) is normal.
4. The distribution of y for all values of x (the y-marginal distribution) is normal.
5. The x-means for given values of y lie on a straight line and the y-means for given values of x also lie on a straight line. These two lines are not the same unless ϱ is $+1$ or -1.
6. If the variance of the x-marginal distribution is σ_x^2 the variance of the distribution of x for a given value of y is $\sigma_x^2 \sqrt{(1 - \varrho^2)}$. A similar relation holds for the y-distributions.

12.1.2. Correlation and Independence

If two variables x and y are found to be uncorrelated—that is, if

$$\mathscr{E}(x - \bar{x})(y - \bar{y}) = 0,$$

it does not necessarily follow that x and y are independent. It is possible for two variables to be directly related and for the population coefficient to be zero. If, for example, the relationship between x and y was any one of the following:

$$y = ax^2 + b$$

$$y = ax^4 + bx^2 + c$$

$$y^2 = x^2 + c^2 \text{ (a circle)}$$

the population correlation coefficient would be zero.

12.1.3. A Test for Correlation

It may be shown that the sample correlation coefficient r, defined above, is an unbiased estimator of the population correlation coefficient ϱ. That is:

$$\mathscr{E}(r) = \varrho.$$

To answer questions about the population correlation coefficient ϱ on the basis of the sample coefficient r we require the distribution of r for given values of ϱ. The most common null hypothesis to be tested is that $\varrho = 0$. If we assume that the population distribution is bivariate normal with $\varrho = 0$ the distribution of r for a given sample size n is:

$$p(r \mid \varrho = 0, n) \, dr = \frac{\Gamma\{(n-1)/2\}}{\sqrt{\pi}\,\Gamma\{(n-2)/2\}} \, (1 - r^2)^{\frac{1}{2}(n-4)}. \qquad (12.4)$$

This distribution is symmetric about zero, its mean is zero, its variance is $1/(n-1)$ and it tends to normality as n increases. Six significance levels of this distribution are given in Table 13 of Pearson and Hartley (1958) and Table 7 in the Tables Section for $\nu\,(=n-2)$ less than 100. For values of ν greater than 100 the statistic

$$r(n-1)^{\frac{1}{2}}$$

may be assumed to be normally distributed with zero mean and unit variance. The 5 per cent and 1 per cent points for some sample sizes are given in Table 12.1 to show that the correlation coefficient estimated from small samples must be high to be significant.

TABLE 12.1. SELECTED 5 PER CENT AND 1 PER CENT SIGNIFICANCE POINTS OF r, GIVEN $\varrho = 0$

Sample size n	Degrees of freedom $\nu = n - 2$	5 per cent level	1 per cent level
3	1	0·9877	0·999507
5	3	0·805	0·9343
12	10	0·497	0·658
22	20	0·360	0·492
42	40	0·257	0·358

12.1.4. Example

The data given in Table 12.2 consists of measurements of the heights and waists of twenty men and we wish to test the hypothesis that heights and waists are uncorrelated. This data is plotted in Fig. 12.3.

TABLE 12.2. EXAMPLE CORRELATION DATA

Heights	(x)	64·8	72·9	60·8	68·9	67·4	63·8	73·0	61·2	69·2	66·2
Waists	(y)	32·3	35·1	35·3	36·0	33·4	29·7	37·9	32·1	31·7	30·7
Heights	(x)	63·1	68·2	72·0	67·2	61·2	67·9	65·1	66·6	70·9	62·3
Waists	(y)	33·9	34·3	36·2	30·8	28·3	38·3	36·2	35·4	31·9	30·6

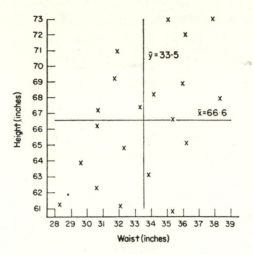

Fig. 12.3. Plot of bivariate data in Table 12.2.

From this plot we see that there is a suggestion of an increase in waist measurement with increase in height but it is not obvious that this association is significant. To test for significant correlation between height and waist we compute:

$$n = 20 \qquad \sum_{i=1}^{n} x_i = 1332 \cdot 7 \qquad \sum_{i=1}^{n} y_i = 669 \cdot 9$$

$$\sum_{i=1}^{n} x_i^2 = 89086 \cdot 23 \qquad \sum_{i=1}^{n} x_i y_i = 44738 \cdot 62 \qquad \sum_{i=1}^{n} y_i^2 = 22587 \cdot 17.$$

From these sums we compute

$$\bar{x} = 66 \cdot 635 \qquad \bar{y} = 33 \cdot 495$$

$$\sum_{i=1}^{n} (x_i - \bar{x})^2 = \sum_{i=1}^{n} x_i^2 - \frac{1}{n} \left(\sum_{i=1}^{n} x_i \right)^2 = 89086 \cdot 23 - 88804 \cdot 46 = 281 \cdot 77$$

$$\sum_{i=1}^{n} (y_i - \bar{y})^2 = \sum_{i=1}^{n} y_i^2 - \frac{1}{n} \left(\sum_{i=1}^{n} y_i \right)^2 = 22587 \cdot 17 - 22438 \cdot 30 = 148 \cdot 87$$

$$\sum_{i=1}^{n} (x_i - \bar{x})(y_1 - \bar{y}) = \sum_{i=1}^{n} x_i y_i - \frac{1}{n} \sum_{i=1}^{n} x_i \sum_{i=1}^{n} y_i$$

$$= 44738 \cdot 62 - 44638 \cdot 79 = 99 \cdot 83.$$

From these we obtain the sample correlation coefficient.

$$r = \frac{99 \cdot 83}{(281 \cdot 77 \times 148 \cdot 87)^{\frac{1}{2}}} = \frac{99 \cdot 83}{204 \cdot 8099} = 0 \cdot 487$$

$$\Pr\{r \geq 0 \cdot 487 \mid \nu = 18, \quad \varrho = 0\} \leq 0 \cdot 025$$

or

$$\Pr\{|r| \geq 0 \cdot 487 \mid \nu = 18, \quad \varrho = 0\} \leq 0 \cdot 05.$$

Accepting a 5 per cent level as our significance criterion we conclude that heights and weights are correlated although the relation is not well marked.

12.1.5. *An Approximation to the Distribution of r*

The following approximation to the distribution of r for a given sample size n assuming the population distribution to be bivariate normal with correlation coefficient ϱ (not necessarily zero) was given by R. A. Fisher.

Let

$$z = \frac{1}{2} \log_e \left(\frac{1 + r}{1 - r} \right) = \tanh^{-1} r$$

and

$$Z = \frac{1}{2} \log_e \left(\frac{1 + \varrho}{1 - \varrho} \right) = \tanh^{-1} \varrho,$$

then

$$z \approx N(Z, 1/(n - 3)),$$

that is

$$(z - Z)(n - 3)^{\frac{1}{2}} \approx N(0, 1).$$

This approximation may be used to test the hypothesis that the population correlation coefficient is a particular value. It may also be used to obtain confidence limits for the population correlation coefficient by finding the range of values of ϱ with which r is consistent.

12.2. FUNCTION FITTING

The need to describe one variable as a function of others arises in a wide variety of situations. In the first type of situation we require to know the value of a variable such as the life of a piece of electrical apparatus, or the breaking strength of a crash helmet or of a mechanical part. Measurement of such a variable leaves the article useless, and we seek means of predicting the value of the variable without destroying the article. One means of doing this is to express breaking strength as a function of variables which may be measured without destruction. Another reason for a prediction of this type might be that although the variable can be measured without destruction it is very expensive to do so. Expense may be saved therefore if we can predict, to acceptable accuracy, the value of the expensive variable by measuring several other variables cheaply. Thus we see that the decisions in function fitting problems may not always be statistical. This situation will require means of assessing the precision with which the variable is predicted.

A second type of situation requires the fitting of a theoretical function to data. In this situation we not only wish to estimate the parameters of the function but we must determine whether or not the theoretical function provides a good description of the experimental data. We may even have more than one possible theoretical function to describe the practical situation and we may require to choose that function which provides the best description of the experimental data. This situation will require goodness-of-fit tests.

A third type of situation requires the description of a variable as a function of another simply as a means of reproducing the values in subsequent computation. For example, the function relating absorption cross-sections to energy for a given nucleus is often complex and nuclear calculations require cross-sections for a wide range of energies. Calculation of a cross-section for a particular energy is lengthy and the large number required may make the overall calculation too long on a computer for it to be practical. If we can approximate the cross-section function with a simpler function which provides the cross-section to acceptable accuracy the overall calculation may be reduced in length sufficiently to make it possible. A variant of this situation, again arising particularly with the use of computers, is the approximation to an analytic function, such as an integral, which is difficult to compute. We may calculate the function at a number of points and use the method of least squares to fit a simpler function such as a polynomial. In this situation the ultimate goodness-of-fit criterion is whether, or not, the fitted function produces values close enough to those of the original function. The statistical tests of goodness-of-fit are of little interest in this situation.

In all three situations described above it is possible that we may wish to restrict the solution in some way. We may, for example, require the fitted function to pass through a given point such as the origin, or perhaps through more than one point. An example of this type of restriction is considered in Section 12.3.4. In this section a straight line is fitted through a fixed point (x_0, y_0) and this restriction is used to reduce the problem to estimating the slope only. Restrictions of the comparative type such as the function must not take negative values or some of the parameters must take values within certain ranges are more difficult to deal with. If the function is fitted ignoring the restrictions and the fit obtained does satisfy the imposed restrictions then this fit is the best solution and the problem is solved. If the restrictions are not satisfied, then other methods of obtaining the best fit must be employed. Restricted function fitting problems of this type are now being formulated as Linear Programming or Quadratic Programming problems and are not considered here.

12.2.1. The Method of Least Squares

The method of least squares, briefly introduced in Section 4.11, will be described and illustrated by the fitting of the straight line

$$y = a + bx.$$

The extension of the method to higher degree polynomials and to functions of more than one variable will be given. It is the main concern of the next sections to illustrate five different practical situations which may arise in function fitting problems and to consider the difficulties that arise in determining goodness-of-fit.

12.3. FUNCTION FITTING (SITUATION 1)

The data available in this first situation consists of one observation of the dependent variable y for each value of the independent variable x. The x-values are assumed to be known exactly and the variance of the y-values is unknown but is assumed to be constant for all values of x. Thus the data consists of samples of n pairs of values:

$$\{(x_1, y_1), \quad (x_2, y_2), \ldots, \quad (x_i, y_i), \ldots, \quad (x_n, y_n)\}.$$

The statistical model to be fitted, in general terms, is

$$y_i = f(x_i) + z_i,$$

where $f(x_i)$ is the function of x to be fitted and z_i is a random error with zero mean and variance σ^2. It is usually assumed further that the z_i's have a normal distribution although this assumption is only used when goodness-of-fit tests are performed or when confidence limits are calculated.

We will illustrate the process of fitting by the method of least squares by fitting the straight line $y = a + bx$.

The statistical model is

$$y_i = a + bx_i + z_i \tag{12.5}$$

and we obtain estimates of the parameters a and b by minimizing the sum of squares of the residual terms:

$$S = \sum_{i=1}^{n} z_i^2 = \sum_{i=1}^{n} (y_i - a - bx_i)^2 \tag{12.6}$$

with respect to a and b.

This gives the equations:

$$\frac{\partial S}{\partial a} = 0 = -2 \sum_{i=1}^{n} (y_i - \hat{a} - \hat{b}x_i)$$

$$\frac{\partial S}{\partial b} = 0 = -2 \sum_{i=1}^{n} x_i(y_i - \hat{a} - \hat{b}x_i)$$

and hence

$$\sum_{i=1}^{n} y_i = \hat{a}n + \hat{b} \sum_{i=1}^{n} x_i$$

$$\sum_{i=1}^{n} x_iy_i = \hat{a} \sum_{i=1}^{n} x_i + \hat{b} \sum_{i=1}^{n} x_i^2.$$

(12.7)

Solving these equations for $\hat{a}$ and $\hat{b}$ we have

$$\hat{b} = \frac{n \sum_{i=1}^{n} x_iy_i - \sum_{i=1}^{n} x_i \sum_{i=1}^{n} y_i}{n \sum_{i=1}^{n} x_i^2 - \left(\sum_{i=1}^{n} x_i\right)^2} = \frac{\sum_{i=1}^{n} (x_i - \bar{x})(y_i - \bar{y})}{\sum_{i=1}^{n} (x_i - \bar{x})^2}$$

(12.8)

and

$$\hat{a} = \bar{y} - \hat{b}\bar{x}$$

where

$$\bar{x} = \frac{1}{n} \sum_{i=1}^{n} x_i \quad \text{and} \quad \bar{y} = \frac{1}{n} \sum_{i=1}^{n} y_i.$$

12.3.1. The Analysis of Variance for Situation 1

The total sum of squares

$$\sum_{i=1}^{n} (y_i - \bar{y})^2$$

represents the variability observed among the y-values when the x-values are not considered. This may be split, in this situation, into two sums of squares to give the analysis of variance shown in Table 12.3.

The "about line" sum of squares is the minimized sum of squares of the errors z_i in expression 12.6. The "due to slope" sum of squares may be written in many forms:

$$\hat{b}^2 \sum_{i=1}^{n} (x_i - \bar{x})^2 = \hat{b} \sum_{i=1}^{n} (x_i - \bar{x})(y_i - \bar{y})$$

$$= \left[\sum_{i=1}^{n} (x_i - \bar{x})(y_i - \bar{y})\right]^2 \bigg/ \sum_{i=1}^{n} (x_i - \bar{x})^2.$$

TABLE 12.3. ANALYSIS OF VARIANCE FOR SITUATION 1

Source	S.S.	D.F.
Due to slope	$\hat{b}^2 \sum\limits_{i=1}^{n} (x_i - \bar{x})^2$	1
About line	$\sum\limits_{i=1}^{n} (y_i - \hat{a} - \hat{b}x_i)^2$	$n - 2$
Total	$\sum\limits_{i=1}^{n} (y_i - \bar{y})^2$	$n - 1$

The "due to slope" sum of sqares measures the amount of variation accounted for by the fitted function. The remaining variability about the function is contained in the "about line" sum of squares and the corresponding mean square provides an estimate $\hat{\sigma}_2^2$ of the variance σ^2 of the error terms z_i. If the first mean square is significantly greater than the second we conclude that a significant amount of variability is accounted for by the fitted function, or, in other words, that the slope of the straight line is significantly greater than zero. The application of the F-test here brings in the normality assumption discussed in Section 12.3.

It may be shown that the correlation significance test described in Section 12.1.3 is exactly equivalent to the F-test testing for a significant slope in this section.

12.3.2. Confidence Limits

Having established that there is a significant linear relation between the two variables we frequently wish to describe the variability of the data about the fitted line in terms of confidence limits placed about the line. Two types of confidence limits may be computed. The first describing the variability of a point on the fitted line, and the second describing the variability of an individual observation y_i for a given value of x_i. The general expression for confidence limits is

$$Y_i \pm t_{v,\alpha/2}[\text{var}]^{\frac{1}{2}} \qquad (12.9)$$

where Y_i denotes the point on the fitted line corresponding to x_i, that is, $Y_i = \hat{a} + \hat{b}x_i$, where $t_{v,\alpha/2}$ is the value of Student's t with $v(=n-2)$ degrees of freedom corresponding to the double-tailed probability level α, and where "var" is the appropriate variance. To compute confidence limits describing the variability of a point on the fitted line the variance is given by

$$\text{var} = \text{var}(Y_i) = \frac{\sigma^2}{n} + (x_i - \bar{x})^2 \, \text{var}(\hat{b}),$$

where the variance of the estimated slope is given by

$$\text{var}(\hat{b}) = \frac{\sigma^2}{\sum\limits_{i=1}^{n} (x_i - \bar{x})^2}.$$

That is, the confidence limits applying to points on the fitted line are given by

$$Y_i \pm t_{v,\alpha/2} \left[\frac{\sigma^2}{n} + (x_i - \bar{x})^2 \, \text{var}(\hat{b}) \right]^{\frac{1}{2}}. \tag{12.10}$$

To compute confidence limits describing the variability of an individual observation y_i the variance is given by:

$$\text{var} = \text{var}(y_i) = \sigma^2 + \frac{\sigma^2}{n} + (x_i - \bar{x})^2 \, \text{var}(\hat{b}).$$

That is, the confidence limits describing the variability of an individual point y_i are given by

$$Y_i \pm t_{v,\alpha/2} \left[\frac{n+1}{n} \sigma^2 + (x_i - \bar{x})^2 \, \text{var}(\hat{b}) \right]^{\frac{1}{2}}. \tag{12.11}$$

12.3.3. Example

Six samples of material were irradiated one at each of six doses and a response y was measured. The data shown in Table 12.4 was observed.

TABLE 12.4. EXAMPLE DATA FOR SITUATION 1

Variable x (= $\log_{10}$ (dose))	4·0	4·2	4·4	4·6	4·8	5·0
Variable y (= response)	5·10	9·74	20·80	27·68	41·96	49·48

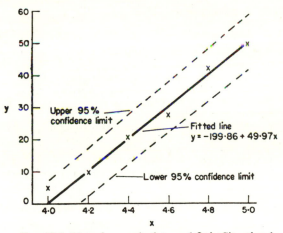

FIG. 12.4. Plot of example data and fit in Situation 1.

The plot of this data in Fig. 12.4 shows a marked linear response. To fit the least squares straight line we compute:

$$n = 6 \quad \sum_{i=1}^{n} x_i = 27 \cdot 0 \quad \sum_{i=1}^{n} y_i = 154 \cdot 76$$

$$\sum_{i=1}^{n} x_i^2 = 122 \cdot 20 \quad \sum_{i=1}^{n} x_i y_i = 728 \cdot 9640 \quad \sum_{i=1}^{n} y_i^2 = 5528 \cdot 6120.$$

From these totals we now compute:

$$\sum_{i=1}^{n} (x_i - \bar{x})^2 = 0 \cdot 70 \qquad \bar{x} = 4 \cdot 5$$

$$\sum_{i=1}^{n} (y_i - \bar{y})^2 = 1536 \cdot 8358 \qquad \bar{y} = 25 \cdot 7933$$

$$\sum_{i=1}^{n} (x_i - \bar{x})(y_i - \bar{y}) = 32 \cdot 5440,$$

and hence obtain:

$$\hat{b} = \frac{32 \cdot 5440}{0 \cdot 70} = 46 \cdot 4914,$$

$$\hat{a} = 25 \cdot 7933 - 46 \cdot 4914 \times 4 \cdot 5 = -183 \cdot 4180.$$

The fitted straight line is therefore:

$$Y_i = -183 \cdot 4180 + 46 \cdot 4914 x_i.$$

The analysis of variance sums of squares are:

Due to slope S.S. $= 46 \cdot 4914^2 \times 0 \cdot 70 = 1513 \cdot 0171$

Total S.S. $= 1536 \cdot 8358.$

TABLE 12.5. ANALYSIS OF VARIANCE OF EXAMPLE DATA FOR SITUATION 1

Source	S.S.	D.F.	M.S.	F.	Prob.
Due to slope	1513·0171	1	1513·0171	254·09	<0·001
Residual	23·8187	4	5·9547		
Total	1536·8358	5			

The analysis of variance shown in Table 12.5 (unnecessary as a testing procedure in this clear-cut example) shows that the slope is significant and provides the estimate 5·9547 of σ^2 with four degrees of freedom.

The variance of the estimated slope is estimated as

$$\text{vâr}(\hat{b}) = 5\cdot9547/0\cdot70 = 8\cdot5067$$

and

$$\text{vâr}(y_i) = \frac{7}{6} \times 5\cdot9547 + (x_i - \bar{x})^2 \times 8\cdot5067.$$

This variance for the six values of x are recorded in Table 12.6 with the 95 per cent confidence limits describing the variability of individual observations ($t_{4,\,0.025} = 2\cdot776$).

TABLE 12.6. GRADUATION WITH CONFIDENCE LIMITS OF THE FITTED LINE IN SITUATION 1

x	Vâr(y_i)	$[\text{Vâr}(y_i)]^{1/2}$	$Y_i = \hat{a} + \hat{b}x_i$ (fitted)	y_i (observed)	95 per cent confidence limits
4·0	9·0738	3·012	2·548	5·10	−5·81 to 10·91
4·2	7·7126	2·777	11·859	9·74	4·15 to 19·57
4·4	7·0322	2·652	21·144	20·80	13·78 to 28·51
4·6	7·0322	2·652	30·442	27·68	23·08 to 37·80
4·8	7·7126	2·777	39·741	41·96	32·03 to 47·65
5·0	9·0738	3·012	49·039	49·48	40·68 to 57·40

The data, fitted straight line, and 95 per cent confidence limits for individual observations are plotted in Fig. 12.4.

12.3.4. Fitting a Straight Line Through the Point (x_0, y_0)

It is not uncommon that we require the fitted function to pass through a fixed point (x_0, y_0). The fixed point is frequently the origin.

Since the point (x_0, y_0) lies on the straight line we have:

$$y_0 = a + bx_0$$

so that we may write the statistical model as:

$$y_i = y_0 + b(x_i - x_0) + z_i \qquad (12.12)$$

and obtain a model involving only one parameter b. We obtain an estimate of b by minimizing the sum of squares:

$$S = \sum_{i=1}^{n} z_i^2 = \sum_{i=1}^{n} (y_i - y_0 - b(x_i - x_0))^2$$

with respect to b. Thus

$$\frac{\partial S}{\partial b} = 0 = -2 \sum_{i=1}^{n} (x_i - x_0)(y_i - y_0 - \hat{b}(x_i - x_0))$$

and hence

$$b = \frac{\sum\limits_{i=1}^{n} (x_i - x_0)(y_i - y_0)}{\sum\limits_{i=1}^{n} (x_i - \bar{x}_0)^2}.$$

(12.13)

The analysis of variance table is given in Table 12.7.

TABLE 12.7. ANALYSIS OF VARIANCE FOR A RESTRICTED FIT FOR SITUATION 1

Source	S.S.	D.F.
Due to slope	$\hat{b}^2 \sum\limits_{i=1}^{n} (x_i - x_0)^2$	1
About line	$\sum\limits_{i=1}^{n} (y_i - y_0 - b(x_i - x_0))^2$	$n - 1$
Total	$\sum\limits_{i=1}^{n} (y_i - y_0)^2$	n

It should be noted that the degrees of freedom for the "about line" mean square are $n - 1$. The variances required in confidence limits calculations are as follows:

$$\text{var}(\hat{b}) = \frac{\hat{\sigma}^2}{\sum\limits_{i=1}^{n} (x_i - \bar{x}_0)^2}$$

$$\text{var}(Y_i) = \frac{\hat{\sigma}^2}{n} + (x_i - x_0)^2 \, \text{var}(\hat{b})$$

(12.14)

$$\text{var}(y_i) = \frac{n + 1}{n} \hat{\sigma}^2 + (x_i - x_0)^2 \, \text{var}(\hat{b}).$$

12.3.5. Example

We will illustrate this computation by fitting a straight line passing through (4,0) to the data given in Table 12.4. The sums of squares about x_0 and y_0 do not expand to give forms which are particularly convenient computationally and they are best computed as follows:

$$\sum\limits_{i=1}^{n} (x_i - x_0)^2 = 0^2 + 0 \cdot 2^2 + 0 \cdot 4^2 + \cdots + 1 \cdot 0^2 = \quad 2 \cdot 2000$$

$$\sum\limits_{i=1}^{n} (y_i - y_0)^2 = 5 \cdot 10^2 + 9 \cdot 74^2 + \cdots + 49 \cdot 48^2 = 5528 \cdot 6120$$

$$\sum\limits_{i=1}^{n} (x_i - x_0)(y_i - y_0) = 0 \times 5 \cdot 10 + \cdots + 1 \cdot 0 \times 49 \cdot 48 = \quad 109 \cdot 9240.$$

Thus

$$b = \frac{109 \cdot 9240}{2 \cdot 2000} = 49 \cdot 9655$$

and the fitted line is:

$$Y_i = -199 \cdot 8618 + 49 \cdot 9655 x_i.$$

The slope sum of squares is

$$49 \cdot 9655^2 \times 2 \cdot 2000 = 5492 \cdot 4026$$

and the analysis of variance table is given in Table 12.8.

TABLE 12.8. ANALYSIS OF VARIANCE OF RESTRICTED FIT TO EXAMPLE DATA FOR SITUATION 1

Source	S.S.	D.F.	M.S.	F.	Prob.
Due to slope	5492·4026	1	5492·4026	758·42	≪0·001
About line	36·2094	5	7·2419		
Total	5528·6120	6			

The slope of the fitted line is clearly seen to be significant without performing the significance test. However, the estimate of the variance of the error terms z_i is obtained from this analysis as $\hat{\sigma}^2 = 7 \cdot 2419$ with 5 degrees of freedom. The variances necessary for the calculation of confidence limits describing individual points are

$$\text{var}(\hat{b}) = \frac{7 \cdot 2419}{2 \cdot 2000} = 3 \cdot 2918,$$

$$\text{var}(y_i) = \left[\frac{7}{6} \times 7 \cdot 2419 + (x - 4 \cdot 0)^2 \, \text{var}(\hat{b}) \right]$$

$$(t_{5,0.025} = 2 \cdot 571).$$

The data summary and plot are given in Table 12.9 and Fig. 12.5.

12.3.6. Does a Particular Straight Line Fit the Data?

The question "Does a particular straight line describe the data?" may be conveniently investigated by an analysis of variance splitting the total sum of squares into the three sources of variation as described below. The particular straight line is rearranged into the following form:

$$y = \alpha + \beta(x - \bar{x}) \tag{12.15}$$

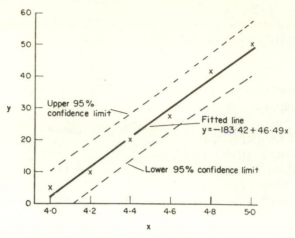

FIG. 12.5. Plot of example data and restricted fit in Situation 1.

TABLE 12.9. GRADUATION WITH CONFIDENCE LIMITS OF RESTRICTED FIT IN SITUATION 1

x	$\text{Var}(y_i)$	$[\text{Var}(y_i)]^{1/2}$	Y_i (fitted)	y_i (observed)	95 per cent confidence limits
4·0	8·4489	2·907	0·000	5·10	7·47 to 7·47
4·2	8·5806	2·929	9·993	9·74	12·46 to 17·52
4·4	8·9755	2·996	19·986	20·80	22·28 to 27·69
4·6	9·6339	3·104	29·979	27·68	2·00 to 37·96
4·8	10·5557	3·249	39·972	41·96	31·62 to 48·33
5·0	11·7407	3·426	49·966	49·48	41·16 to 58·77

TABLE 12.10. ANALYSIS OF VARIANCE TESTING THE FIT OF A GIVEN STRAIGHT LINE FOR SITUATION 1

Source	S.S.	D.F.
Slope	$(\hat{b} - \beta)^2 \sum_{i=1}^{n} (x_i - \bar{x})^2$	1
Constant	$n(\bar{y} - \alpha)^2$	1
About line	$\sum_{i=1}^{n} (y_i - \bar{y} - \hat{b}(x_i - \bar{x}))^2$	$n-2$
Total	$\sum_{i=1}^{n} (y_i - \alpha - \beta(x_i - \bar{x}))^2$	n

and compared with the best (in the least squares sense) straight line fitted to the data similarly rearranged as follows:

$$y = \bar{y} + \hat{b}(x - \bar{x}). \tag{12.16}$$

The analysis of variance given in Table 12.10 has three sources of variation.

The significance of the "slope" mean square when compared with the "about line" mean square establishes that it is not reasonable to believe that the slope of the population straight line is β. Similarly significance of the "constant" mean square when compared with the "about line" mean square establishes that it is not reasonable to believe that the constant term of the population straight line is α.

12.3.7. Example

We will test the hypothesis that the straight line

$$y = 200 + 50x$$

describes the data given in Table 12.4. The mean of the x-values is 4·5 so that this line may be rewritten

$$y = 25 + 50(x - 4·5).$$

That is:

$$\beta = 50 \quad \text{and} \quad \alpha = 25.$$

The computations are as follows:

Slope S.S. $= (46·4914 - 50)^2 \times 0·70 = 8·6172,$

Constant S.S. $= b \times (25·7933 - 25)^2 = 3·7763,$

and the complete analysis of variance table is given in Table 12.11.

TABLE 12.11. EXAMPLE ANALYSIS OF VARIANCE TESTING THE FIT OF A GIVEN STRAIGHT LINE FOR SITUATION 1

Source	S.S.	D.F.	M.S.	F.	Prob.
Slope	8·6172	1	8·6172	1·45	>0·25
Constant	3·7763	1	3·7763	0·63	>0·50
About line	23·8187	4	5·9547		
Total	36·2122	6			

The fitted straight line does not differ significantly from the test line.

12.4. FUNCTION FITTING (SITUATION 2)

In the second function fitting situation to be considered the variance of the error terms z_i are assumed to be equal to $c_i\sigma^2$, where the c_i's are known constants. That is we know the relative variability for each x_i but not the absolute variability. A not infrequent practical situation is one in which the variability in y increases as x increases. If we restrict the fitted function to pass through the origin we will require the variability to be zero at the origin and to gradually increase as x is increased. To achieve this in the statistical model we chose steadily increasing values for the c_i's.

The least squares procedure in this situation minimizes the weighted sum of squares:

$$S = \sum_{i=1}^{n} w_i z_i^2 = \sum_{i=1}^{n} w_i(y_i - a - bx_i)^2,$$

where

$$w_i = 1/c_i.$$

The individual terms contained in the sum S, namely $\sqrt{(w_i z_i^2)}$, have constant variance σ^2. The mean square $S/(n-2)$ provides an estimate of σ^2 as in the first situation. Each of the cases considered above may be analysed in situation 2 by including the weight w_i in each sum or sum of squares and by replacing n where this occurs by

$$\sum_{i=1}^{n} w_i.$$

Modified formulae fitting a straight line without restriction are:

$$b = \frac{\sum_{i=1}^{n} w_i(x_i - \bar{x})(y_i - \bar{y})}{\sum_{i=1}^{n} w_i(x_i - \bar{x})^2} \qquad \hat{a} = \bar{y} - \hat{b}\bar{x}$$

$$\bar{y} = \frac{\sum_{i=1}^{n} w_i y_i}{\sum_{i=1}^{n} w_i} \qquad\qquad \bar{x} = \frac{\sum_{i=1}^{n} w_i x_i}{\sum_{i=1}^{n} w_i} \qquad (12.17)$$

and the analysis of variance table is given in Table 12.12.

The estimate of σ^2 is given by:

$$\hat{\sigma}^2 = \frac{\sum_{i=1}^{n} w_i(y_i - \hat{a} - \hat{b}x_i)^2}{(n-2)} \qquad (12.18)$$

TABLE 12.12. ANALYSIS OF VARIANCE FOR SITUATION 2

Source	S.S.	D.F.
Due to slope	$\hat{b}^2 \sum_{i=1}^{n} w_i(x_i - \bar{x})^2$	1
About line	$\sum_{i=1}^{n} w_i(y_i - \hat{a} - \hat{b}x_i)^2$	$n - 2$
Total	$\sum_{i=1}^{n} w_i(y_i - \bar{y})^2$	$n - 1$

and the variance of the slope by:

$$\text{var}(\hat{b}) = \frac{\hat{\sigma}^2}{\sum_{i=1}^{n} w_i(x_i - \bar{x})^2}. \tag{12.19}$$

Confidence limits applying to individual observations in this situation are given by:

$$y \pm t_{\nu,\alpha/2}[\text{var}]^{\frac{1}{2}},$$

where $t_{\nu,\alpha/2}$ is as defined in Section 12.3.2, and where

$$\text{var} = \text{var}(y_i) = \frac{(n+1)\,c_i\hat{\sigma}^2}{n} + (x_i - \bar{x})^2\,\text{var}(\hat{b}). \tag{12.20}$$

12.5. FUNCTION FITTING (SITUATION 3)

The third situation to be considered differs from the first situation in that the variances of the z_i's are different but known. That is:

$$\text{var}(z_i) = \sigma_i^2.$$

The least squares procedure minimizes the sum of squares

$$S = \sum_{i=1}^{n} w_i z_i^2 = \sum_{i=1}^{n} w_i(y_i - a - bx_i)^2,$$

where $w_i = 1/\sigma_i^2$.

The individual terms in the sum S, namely $\sqrt{(w_i z_i)}$, have constant variance equal to unity. The residual mean square $S/(n-2)$ provides an estimate of the variances of the terms $\sqrt{(w_i z_i)}$ and should be an estimate of unity. The value of S obtained in this situation has a χ^2-distribution (if the errors z_i are normally distributed) with $n - 2$ degrees of freedom if the function describes the data. We have, therefore, a test for the goodness-of-fit of the fitted function.

The formulae (12.17) and the analysis of variance table (Table 12.12) given above apply to this situation but confidence limits applying to individual observations are given by:

$$y \pm X_{\alpha/2}[\text{var}]^{\frac{1}{2}},$$

where

$$\text{var} = \text{var}(y_i) = \frac{(n+1)\,\sigma_i^2}{n} + (x_i - \bar{x})^2\,\text{var}(\hat{b}) \qquad (12.21)$$

and where $X_{\alpha/2}$ is the upper $\alpha/2$ significance level of a standardized normal distribution.

12.5.1. Example

The data recorded in Table 12.13 will be used as an example of the fitting procedure for situation 3.

TABLE 12.13. EXAMPLE DATA FOR SITUATION 3

y_i	11·38	18·12	28·11	41·48	39·60	72·90
x_i	1	2	3	4	5	6
c_i	1	4	9	16	25	36
w_i	1	0·2500	0·1111	0·0625	0·0400	0·0278

The computation is as follows:

$$\sum_{i=1}^{n} w_i = 1\cdot4914 \qquad \sum_{i=1}^{n} w_i x_i = 2\cdot4500 \qquad \sum_{i=1}^{n} w_i y_i = 25\cdot2348$$

$$\bar{x} = 1\cdot6428 \qquad \bar{y} = 16\cdot9202$$

$$\sum_{i=1}^{n} w_i(x_i - \bar{x})^2 = \sum_{i=1}^{n} w_i x_i^2 - \left[\sum_{i=1}^{n} w_i x_i\right]^2 \Big/ \sum_{i=1}^{n} w_i$$

$$= 6\cdot0000 - 2\cdot45^2/1\cdot4914$$

$$= 1\cdot9753$$

$$\sum_{i=1}^{n} w_i(y_i - \bar{y})^2 = \sum_{i=1}^{n} w_i y_i^2 - \left[\sum_{i=1}^{n} w_i y_i\right]^2 \Big/ \sum_{i=1}^{n} w_i$$

$$= 617\cdot2718 - 25\cdot2348^2/1\cdot4914$$

$$= 190\cdot2937$$

$$\sum_{i=1}^{n} w_i(x_i - \bar{x})(y_i - \bar{y}) = \sum_{i=1}^{n} w_i x_i y_i - \sum_{i=1}^{n} w_i x_i \sum_{i=1}^{n} w_i y_i \bigg/ \sum_{i=1}^{n} w_i$$

$$= 60 \cdot 2500 - 2 \cdot 45 \times 25 \cdot 2348/1 \cdot 4914$$

$$= 18 \cdot 7956.$$

The slope and constant are therefore estimated to be:

$$\hat{b} = \frac{18 \cdot 7956}{1 \cdot 9753} = 9 \cdot 5153,$$

$$\hat{a} = 16 \cdot 9202 - 9 \cdot 5153 \times 1 \cdot 6428 = 1 \cdot 2885.$$

The fitted straight line is $y = 1 \cdot 2885 + 9 \cdot 5153x$, the sum of squares due to slope is

$$9 \cdot 5153^2 \times 1 \cdot 9753 = 178 \cdot 8460$$

and the analysis of variance table is given in Table 12.14.

TABLE 12.14. EXAMPLE ANALYSIS OF VARIANCE FOR SITUATION 3

Source	S.S.	D.F.	M.S.	F.	Prob.
Due to slope	178·8460	1	178·8460	62·49	<0·005
About line	11·4477	4	2·8619		
Total	190·2937	5			

The weighted sum of squares S of the errors z_i is 11·4477 with 4 degrees of freedom, and if the function fits, S has a χ^2-distribution. Thus, since

$$\Pr \{\chi^2 \geq 11 \cdot 4477 \mid \nu = 4\} = 0 \cdot 022,$$

the variability as estimated by the "about line" mean square is significantly greater than unity so that there is significantly more variation of the points about the line than would be expected by chance. We therefore conclude that although the straight line fitted to the data is the best straight line that can be fitted it does not describe the data adequately.

12.6. FUNCTION FITTING (SITUATION 4)

In the fourth situation to be considered the variances of the errors z_i are known as a function of the fitted y-values. That is, the variances depend on the function to be fitted. The formulae given for situation 2 above apply to this situation but since the weights to be used in fitting the function depend on the function itself the solution must be obtained by iteration. In the case of a straight line this normally consists of the following steps:

1. Fit a straight line either without weighting or by using approximate weights which are sometimes available.
2. Obtain from this line the fitted y-values and hence another set of weights.
3. Fit another straight line using the weights obtained.
4. Repeat steps 2 and 3 until the final line is obtained, that is when tolerably little change in the line is observed from one iteration to the next.

One type of data to which this situation would apply is one in which the observation y_i is the proportion of success; for example, the number of plants showing a certain characteristic. Probit analysis described in Finney (1952a, 1952b) is an example of this fitting situation in which the observation y_i is usually the proportion of times a certain character is observed in a group of animals at each of a number of doses x_i (of a chemical or of radiation) and the function to be fitted is the cumulative normal distribution. In these examples the number of successes has a binomial distribution with the probability of a success at a single trial given by the point P_i on the fitted function. If r_i successes are observed in n_i trials at dose x_i the observed proportion of successes is $p_i = r_i/n_i$, and the variance of p_i is given by

$$\text{var}(p_i) = \frac{P_i(1 - P_i)}{n_i},$$

where P_i lies on the fitted function.

The fitting process in this case would start by using the *observed* proportions p_i to obtain approximate weights w_{i1} and hence the first estimate of the fitted function. That is, the first weights would be

$$w_{i1} = \frac{n_i}{p_i(1 - p_i)}.$$

Having obtained the first fitted function using these weights we may obtain the first set of fitted values of the proportions of successes P_{i1} by substituting the values x_i of x in the first fitted function. From these we obtain the second set of weights

$$w_{i2} = \frac{n_i}{P_{i1}(1 - P_{i1})}$$

and hence obtain a second fitted function using these weights. If the second fitted function is found to be sufficiently close to the first fitted function we would accept the later fit as final, otherwise we would continue the iteration process until an acceptable agreement has been obtained. There is no automatic guarantee that the process described will converge to a final fit. Experience has shown, however, that if the first fitted function is obtained by using weights calculated from the observed proportions the process usually converges. If the first fitted function is obtained without weights the process fails to converge more often.

The formulae given in Section 12.4 apply to the finally fitted line and the goodness-of-fit test based on the "about line" sum of squares described in Section 12.5 also applies to this situation. An example of this situation will be given in Section 13.4.

12.7. FUNCTION FITTING (SITUATION 5)

The data for this fifth situation consists of more than one observation of variable y for each (or most) of the values of x and the variances of the errors z_i are constant but unknown. The data thus consists of r_i observations y_{ij} for each of n values x_i of variable x. In this situation it is possible to obtain a goodness-of-fit test if the several observations for each observation are replicate measurements rather than repeat measurements (see Sections 1.2.2 and 9.5). The statistical model represents the jth observation for the ith value of y as:

$$y_{ij} = a + bx_i + d_i + z_{ij}, \qquad (12.22)$$

where a and b are parameters to be estimated, d_i is a departure from linearity common to all observations made for $x = x_i$ and z_{ij} is a normally distributed error term with zero mean and variance σ^2.

The formulae for this situation may be obtained from those given in Section 12.4 by taking weights $w_i = r_i$ and replacing y_i by $\bar{y}_i$.—the mean of the r_i observations $y_{i1}, \ldots, y_{ir_i}$ for $x = x_i$. That is we have:

$$\hat{b} = \frac{\sum_{i=1}^{n} r_i (x_i - \bar{x})(\bar{y}_i. - \bar{y}..)}{\sum_{i=1}^{n} r_i (x_i - \bar{x})^2} = \frac{\sum_{i=1}^{n} \sum_{j=1}^{r_i} x_i y_{ij} - \sum_{i=1}^{n} r_i \bar{x}\bar{y}..}{\sum_{i=1}^{n} r_i (x_i - \bar{x})^2}$$

$$\hat{a} = \bar{y}.. - \hat{b}\bar{x},$$

where

$$\bar{y}_i. = \frac{1}{r_i} \sum_{j=1}^{r_i} y_{ij} \quad \text{(the mean } y\text{-value for the } i\text{th value of } x\text{)}$$

$$\bar{y}.. = \frac{\sum_{i=1}^{n} \sum_{j=1}^{r_i} y_{ij}}{\sum_{i=1}^{n} r_i} = \frac{\sum_{i=1}^{n} r_i \bar{y}_i.}{\sum_{i=1}^{n} r_i} \quad \text{(the grand } y\text{-mean)}$$

and

$$\bar{x} = \frac{\sum\limits_{i=1}^{n} r_i x_i}{\sum\limits_{i=1}^{n} r_i}.$$

The total sum of squares

$$\sum_{i=1}^{n} \sum_{j=1}^{r_i} (y_{ij} - \bar{y}..)^2$$

may be split into three sums of squares to give the analysis of variance table given in Table 12.15.

TABLE 12.15. ANALYSIS OF VARIANCE FOR SITUATION 5

Source	S.S.	D.F.
Slope	$b^2 \sum\limits_{i=1}^{n} r_i(x_i - \bar{x})^2$	1
About line	$\sum\limits_{}^{n} r_i(\bar{y}_i. - \hat{a} - bx_i)^2$	$n - 2$
Residual	$\sum\limits_{i=1}^{n} \sum\limits_{j=1}^{r_i} (y_{ij} - \bar{y}_i.)^2$	$\sum\limits_{i=1}^{n} r_i - n$
Total	$\sum\limits_{i=1}^{n} \sum\limits_{j=1}^{r_i} (y_{ij} - \bar{y}..)^2$	$\sum\limits_{i=1}^{n} r_i - 1$

The residual mean square provides an estimate of the residual variability σ^2 whether the function fits or not. The "about line" mean square provides an estimate of σ^2 if the function fits so that a comparison between these two estimates provides a means of testing the adequacy of the fit provided that the several y-measurements $y_{i1}, \ldots, y_{ir_i}$ for the value x_i of x are replicate rather than repeat measurements.

12.7.1. Example

The data given in Table 12.16 consists of the logarithms of the surviving fractions of organisms for each of six doses of radiation (six has been added to each logarithm to avoid negative values).

The assumption of equal variance made in the model (12.22) could be tested by performing Bartlett's test on this data. This gives a non-significant result in fact.

TABLE 12.16. EXAMPLE DATA FOR SITUATION 5

Dose (M-rads)	Log (surviving fraction) + 6								Number of replicates r_i
6	4·96	4·79	4·29	3·96	4·08				5
9	2·87	3·37	3·15	3·17					4
12	3·00	2·88	2·16	2·85	2·10	2·36			6
15	2·15	1·08	1·46	1·49	1·52	1·48	0·79	0·48	11
	1·79	1·45	1·48						
18	0·48								1

The calculation commences by computing the totals given in Table 12.17.

TABLE 12.17. BASIC y-SUMS FOR SITUATION 5—EXAMPLE

Dose	$\sum_{j=1}^{r_i} y_{ij}^2$	$\sum_{j=1}^{r_i} y_{ij}$	$\sum_{j=1}^{r_i} (y_{ij} - \bar{y}_{i\cdot})^2$ $\left(=\sum_{j=1}^{r_i} y_{ij}^2 - (\sum_{j=1}^{r_i} y_{ij})^2/r_i\right)$	
6	98·2778	22·08	0·7725	$(= 98·2778 - 1/5 \ (22·08)^2)$
9	39·5652	12·56	0·1268	$(= 39·5652 - 1/4 \ (12·56)^2)$
12	40·0621	15·35	0·7917	$(= 40·0621 - 1/6 \ (15·35)^2)$
15	22·9929	15·17	2·0720	$(= 22·9929 - 1/11 \ (15·17)^2)$
18	0·2304	0·48	0·0000	
Totals	201·1284	65·64	3·7631	

From Table 12.17 we have:

Residual S.S. = 3·7631.

$$\text{Total S.S.} = 201·1284 - \frac{(65·64)^2}{5 + 4 + 6 + 11 + 1}$$

$$= 41·5503.$$

$$\sum_{i=1}^{n} r_i x_i^2 = 5 \times 6^2 + 4 \times 9^2 + \cdots + 1 \times 18^2 = 4167.$$

$$\sum_{i=1}^{n} r_i x_i = 5 \times 6 + 4 \times 9 + \cdots + 1 \times 18 \quad = 321.$$

$$\sum_{i=1}^{n} r_i (x_i - \bar{x})^2 = 4167 - \frac{1}{27} \times 321^2 = 350·6667.$$

$$\sum_{i=1}^{n} \sum_{j=1}^{r_i} x_i y_{ij} = 6 \times (4·96 + 4·79 + \cdots + 4·08) + \cdots$$

$$+ 15 \times (2·15 + 1·08 + \cdots + 1·48) + 18 \times 0·48.$$

$$= 665·91.$$

$$\sum_{i=1}^{n} r_i(x_i - \bar{x})(\bar{y}_i. - \bar{y}..) = 665 \cdot 91 - 321 \times 65 \cdot 64/27$$

$$= -114 \cdot 4767.$$

Hence the slope b is estimated by

$$\hat{b} = \frac{-114 \cdot 4767}{350 \cdot 6667} = -0 \cdot 3265.$$

The slope sum of squares is $(-0 \cdot 3265)^2 \times 350 \cdot 6667 = 37 \cdot 366$ and the analysis of variance table is given in Table 12.18.

TABLE 12.18. EXAMPLE ANALYSIS OF VARIANCE FOR SITUATION 5

Source	S.S.	D.F.	M.S.	F.	P.
Slope	37·3766	1	37·3766	219·88	≪0·001
Departures from linearity	0·4106	3	0·1369	0·82	>0·50
Residual	3·7631	22	0·1711		
Total	41·5503	26			

We conclude from this analysis that the slope is significantly less than zero and that the straight line is an acceptable description of the data.

12.8. FITTING OTHER FUNCTIONS

Each of the above situations can occur when functions other than a straight line are to be fitted to data. Similar analyses of variance may be computed by replacing $a - bx_i$ in the "about line" sums of squares by $f(x_i)$. That is the "about line" sum of squares becomes:

$$\sum_{i=1}^{n} (y_i - f(x_i))^2 \quad \text{in situation 1}$$

or

$$\sum_{i=1}^{n} w_i(y_i - f(x_i))^2 \quad \text{in situations 2, 3, 4} \tag{12.23}$$

or

$$\sum_{i=1}^{n} r_i(\bar{y}_i. - f(x_i))^2 \quad \text{in situation 5}.$$

The "about line" sum of squares has $n - p$ degrees of freedom, where p is the number of parameters estimated in fitting the function $f(x)$.

12.9. MULTIPLE REGRESSION

Multiple regression is the description of a variable y by a function of more than one variable. For example, we may seek to describe a man's weight as a function of his height, waist measurement, and back length. That is, we seek to fit the model:

$$(WT)_i = A + H_i + (WM)_i + B_i + z_i. \qquad (12.24)$$

The general form of the multiple regression function is:

$$y_i = a_0 + a_1 x_{1i} + a_2 x_{2i} + \cdots + a_k x_{ki} + z_i, \qquad (12.25)$$

where y is the variable to be described (the dependent variable) as a function of the variables $x_1, x_2, ..., x_k$ (the independent variables). It may be shown that the point with coordinates $(\bar{x}_1, \bar{x}_2, ..., \bar{x}_k, \bar{y})$ lies on the fitted function so that the model (12.25) may be rewritten with one less parameter as:

$$y_i - \bar{y} = a_1(x_{1i} - \bar{x}_1) + a_2(x_{2i} - \bar{x}_2) + \cdots + a_k(x_{ki} - \bar{x}_k) + z_i. \qquad (12.26)$$

The coefficients are estimated by minimizing

$$S = \sum_{i=1}^{n} z_i^2 = \sum_{i=1}^{n} (y_i - \bar{y} - a_1(x_{1i} - \bar{x}_1) - a_2(x_{2i} - \bar{x}_2) - \cdots - a_k(x_{ki} - \bar{x}_k))^2$$

with respect to $a_1, a_2, ..., a_k$ to obtain equations:

$$\Sigma(y_i - \bar{y})(x_{1i} - \bar{x}_1) = \hat{a}_1 \Sigma(x_{1i} - \bar{x}_1)^2$$
$$+ \hat{a}_2 \Sigma(x_{1i} - \bar{x}_1)(x_{2i} - \bar{x}_2) + \cdots + \hat{a}_k \Sigma(x_{1i} - \bar{x}_1)(x_{ki} - \bar{x}_k)$$

$$\Sigma(y_i - \bar{y})(x_{2i} - \bar{x}_2) = \hat{a}_1 \Sigma(x_{1i} - \bar{x}_1)(x_{2i} - \bar{x}_2)$$
$$+ \hat{a}_2 \Sigma(x_{2i} - \bar{x}_2)^2 + \cdots + \hat{a}_k \Sigma(x_{2i} - \bar{x}_2)(x_{ki} - \bar{x}_k)$$

. . . .

$$\Sigma(y_i - \bar{y})(x_{ki} - \bar{x}_k) = \hat{a}_1 \Sigma(x_{1i} - \bar{x}_1)(x_{ki} - \bar{x}_k)$$
$$+ \hat{a}_2 \Sigma(x_{2i} - \bar{x}_2)(x_{ki} - \bar{x}_k) + \cdots + \hat{a}_k \Sigma(x_{ki} - \bar{x}_k)^2, \qquad (12.27)$$

where all summations are for $i = 1, 2, ..., n$.

The solution of these simultaneous equations provides the estimates of the coefficients $a_1, a_2, ..., a_k$ which minimize S. There are many descriptions of methods of solving simultaneous equations to be found in the mathematical, statistical and computational literature. A large number of computer programmes exist to perform multiple regression so that experimentalists with access to one of these programmes will not be concerned with methods (attention is drawn, however, to the warning in Section 12.10.2). Experimentalists not having access to such a programme are advised to seek the advice of a numerical analyst to determine the method of solution best suited to these equations.

It is convenient at this stage to write the equations (12.27) above in matrix notation. The reader with no knowledge of algebra will find no difficulty in understanding this section if he makes reference where necessary to the brief appendix on matrix algebra which appears at the end of this chapter. The equations are

$$y = X\hat{a}, \tag{12.28}$$

where y is the column vector

$$\begin{bmatrix} \Sigma(y_i - \bar{y})(x_{1i} - \bar{x}_1) \\ \Sigma(y_i - \bar{y})(x_{2i} - \bar{x}_2) \\ \cdot \quad \cdot \quad \cdot \quad \cdot \quad \cdot \\ \Sigma(y_i - \bar{y})(x_{ki} - \bar{x}_k) \end{bmatrix},$$

$\hat{a}$ is the column vector

$$\begin{bmatrix} \hat{a}_1 \\ \hat{a}_2 \\ \cdot \cdot \\ \hat{a}_k \end{bmatrix},$$

and X is the symmetric matrix

$$\begin{bmatrix} \Sigma(x_{1i} - \bar{x}_1)^2 & \Sigma(x_{1i} - \bar{x}_1)(x_{2i} - \bar{x}_2) & \Sigma(x_{1i} - \bar{x}_1)(x_{ki} - \bar{x}_k) \\ \Sigma(x_{1i} - \bar{x}_1)(x_{2i} - \bar{x}_2) & \Sigma(x_{2i} - \bar{x}_2)^2 & \Sigma(x_{2i} - \bar{x}_2)(x_{ki} - \bar{x}_k) \\ \cdot \quad \cdot \quad \cdot \quad \cdot & \cdot \quad \cdot \quad \cdot \quad \cdot & \cdot \quad \cdot \quad \cdot \quad \cdot \\ \Sigma(x_{1i} - \bar{x}_1)(x_{ki} - \bar{x}_k) & \Sigma(x_{2i} - \bar{x}_2)(x_{ki} - \bar{x}_k) & \Sigma(x_{ki} - \bar{x}_k)^2 \end{bmatrix}$$

and where all sums are for $i = 1, 2, ..., n$.

The total sum of squares $\sum\limits_{i=1}^{n}(y_i - \bar{y})^2$ may be split into two sums of squares. The first of these provides an estimate of the residual variability σ^2 and the second provides a measure of the amount of variation removed from the total variation by the fitted function. The sum of squares due to fitting the function is:

$$\hat{a}_1\Sigma(y_i - \bar{y})(x_{1i} - \bar{x}_1) + \hat{a}_2\Sigma(y_i - \bar{y})(x_{2i} - \bar{x}_2) + \cdots + \hat{a}_k\Sigma(y_i - \bar{y})(x_{ki} - \bar{x}_k)$$

or in matrix notation $y'\hat{a}$—the scalar product between vectors y' and $\hat{a}$. (Note: the product between a *row* vector y' and a *column* vector $\hat{a}$, following the normal matrix multiplication rules, is a scalar, or, in other words, one number.) The complete analysis of variance is given in Table 12.19.

TABLE 12.19. ANALYSIS OF VARIANCE FOR MULTIPLE REGRESSION

Source	S.S.	D.F.
Due to function	$\hat{a}_1 \Sigma(y_i - \bar{y})(x_{1i} - \bar{x}_1) + \hat{a}_2 \Sigma(\acute{y}_i - y)(x_{2i} - \bar{x}_2) +$ $\cdots + \hat{a}_k \Sigma(y_i - \bar{y})(x_k{}^i - x_k)$	k
About function	$\Sigma(y_i - \bar{y})^2$ $- \hat{a}_1 \Sigma(y_i - \bar{y})(x_{1i} - \bar{x}_1) - \cdots - \hat{a}_k \Sigma(y_i - \bar{y})(x_{ki} - \bar{x}_k)$	$n - k - 1$
Total	$\Sigma(y_i - \bar{y})^2$	$n - 1$

The variance σ^2 of the error term z_i in the model (12.26) is estimated by $\hat{\sigma}^2$ the „about function" mean square.

The inverse X^{-1}, of the matrix X, multiplied by the variance σ^2 may be shown to contain the variances and covariances of the estimated coefficients $\hat{a}_1, \hat{a}_2, \ldots, \hat{a}_k$. The variances appear in the diagonal of the matrix $\sigma^2 X^{-1}$ and the off-diagonal elements are the covariances between pairs of estimated coefficients. This matrix is the variance–covariance matrix of estimates:

$$\sigma^2 X^{-1} = V = \begin{bmatrix} \text{var}(\hat{a}_1) & \text{covar}(\hat{a}_1\hat{a}_2) & \text{covar}(\hat{a}_1\hat{a}_k) \\ \text{covar}(\hat{a}_1\hat{a}_2) & \text{var}(\hat{a}_2) & \text{covar}(\hat{a}_2\hat{a}_k) \\ \cdot \cdot \cdot & \cdot \cdot \cdot & \cdot \cdot \cdot \\ \text{covar}(\hat{a}_1\hat{a}_k) & \text{covar}(\hat{a}_2\hat{a}_k) & \text{var}(\hat{a}_k) \end{bmatrix}$$

The variability of individual y-values about the fitted function for given values of variables $x_1, x_2, \ldots, x_k$ is given by the variance of y for the given values $x_{1T}, x_{2T}, \ldots, x_{kT}$ of the variables $x_1, x_2, \ldots, x_k$. In matrix notation this is:

$$\text{var}(y) = \frac{(n + 1)}{n} \sigma^2 + x'Vx, \tag{12.29}$$

where $V (= \sigma^2 X^{-1})$ is the variance-covariance matrix, x' is the row vector $((x_{1T} - \bar{x}_1)(x_{2T} - \bar{x}_2) \cdots (x_{kT} - \bar{x}_k))$ and x is the corresponding column vector. The expanded form of this variance is

$$\text{var}(y) = \frac{(n + 1)\sigma^2}{n} + \sum_{j=1}^{k} (x_{jT} - \bar{x}_j)^2 \text{var}(\hat{a}_j)$$

$$+ 2 \sum_{j=1}^{k-1} \sum_{l=j+1}^{k} (x_{jT} - \bar{x}_j)(x_{lT} - \bar{x}_l) \text{covar}(\hat{a}_j\hat{a}_l) \tag{12.30}$$

or an alternative way of writing this is:

$$\text{var}(y) = \frac{(n + 1)\sigma^2}{n} + \sum_{j=1}^{k} (x_{jT} - \bar{x}_j)^2 \text{var}(\hat{a}_j)$$

$$+ \sum_{\substack{j=1 \\ j \ne l}}^{k} \sum_{l=1}^{k} (x_{jT} - \bar{x}_j)(x_{lT} - \bar{x}_l) \text{covar}(\hat{a}_j\hat{a}_l). \tag{12.31}$$

The computational procedure is illustrated by an example in Section 12.10.1.

12.9.1. The Selection of Independent Variables

The need to describe a variable y by a function of other variables may arise for a variety of reasons such as when the measurement of y results in the destruction of the article on which the measurement is made (for example, the life of an electric light bulb) or when y is a measurement which can not be measured at the present time (for example, the adult height of a teenage boy), or possibly when the measurement of y is expensive in time or technique. The need in the first case is to predict the value of y without destroying the article, in the second we attempt to predict the value that y will take in the future (we may wish to predict the boy's eventual height to determine whether or not he should start now on an apprenticeship for a job which has a height restriction in adult life), in the third we attempt to predict y as a function of cheaper measurements. In most situations of all three types we try to predict y as economically as possible; that is, using as few independent variables as possible (or in the third case that set of variables which costs the least to measure). To do this we fit a number of different functions where the independent variables are different selections from the set of possible variables.

It is a wise precaution to fit the function using all available variables to first determine whether or not a satisfactory description of y can be obtained. In problems in which each variable is assumed to cost the same to measure the ideal approach would be to take each of the possible independent variables separately in turn and determine which of these gives the best prediction of y. If the best of these gives a satisfactory description of y we accept this description. Otherwise we continue by taking each possible pair of independent variables and determine which pair gives the best prediction. If the best pair gives a satisfactory description of y we accept this description otherwise we proceed to analyse each possible triple and so on until an acceptable description is obtained. If a large number of variables are available this procedure may involve the fitting of a prohibitively large number of functions. Even with the aid of a computer the amount of computation can be excessive. For this reason a number of procedures are being explored at present. One of these is known as "stepwise regression" in which variables are added to, or deleted from the prediction equation according to rules which attempt to decide on the information content of each variable and the adequacy of the fit.

In problems where a cost of measurement is associated with each variable the procedure would be slightly different. The possible selections of independent variables would be ordered according to cost with due consideration of the fact that the cost of measuring two variables may be less than the sum of the individual costs. The prediction of y would be determined for each selection of variables in the cost order and the first selection that met the satisfactory description criterion would be accepted.

The above discussion uses the words "satisfactory description" without comment. The experimentalist will normally require that y be estimated to a certain precision and the statistician will translate this to define an acceptable confidence interval about y. To do this the confidence probability must be agreed. The computation of confidence intervals involves the variance of y which varies according to the values of the independent variables. However, the first term

$$\frac{(n + 1)\,\sigma^2}{n}$$

is usually the dominating term in the variance of y so that from the acceptable confidence interval we may determine an acceptable value of $\hat{\sigma}^2$. This acceptable value will become our acceptance criterion. The analysis, at each decision point, which first produces an "about function" mean square less than the acceptable value of $\hat{\sigma}^2$ will be accepted.

The following analysis obtained from two descriptions of the dependent variable y, may contain useful information for the experimentalist. If we describe y in terms of the variables x_1 and x_2 we obtain the analysis of variance given in Table 12.20.

TABLE 12.20. ANALYSIS OF VARIANCE FOR A MULTIPLE REGRESSION USING x_1 AND x_2

Source	S.S.	D.F.
Fitting x_1 and x_2	F_2	2
About function	R_2	$n - 3$
Total	$\Sigma(y_i - \bar{y})^2$	$n - 1$

If variables x_1, x_2 and x_3 are now used the analysis of variance in Table 12.21 is obtained.

TABLE 12.21. ANALYSIS OF VARIANCE FOR A MULTIPLE REGRESSION USING x_1, x_2 AND x_3

Source	S.S.	D.F.
Fitting x_1, x_2, x_3	F_3	3
About function	R_3	$n - 4$
Total	$\Sigma(y - \bar{y})^2$	$n - 1$

From these two analyses we may compute the useful combined analysis of variance table given in Table 12.22.

TABLE 12.22. ANALYSIS OF VARIANCE COMBINING THE INFORMATION FROM TWO FITS

Source	S.S.	D.F.
Fitting x_1 and x_2	F_2	2
Extra information in x_3	$F_3 - F_2$	1
About function	R_3	$n - 4$
Total	$\Sigma(y_i - \bar{y})^2$	$n - 1$

The second sum of squares in this table is a measure of the extra information obtained by adding x_3 to the regression equation containing x_1 and x_2.

12.10. POLYNOMIAL REGRESSION

One class of functions frequently used to describe experimental data is the class of polynomial:

$$y_i = a'_0 + a_1 x_i + a_2 x_i^2 + a_3 x_i^3 + \cdots + a_k x_i^k + z_i. \qquad (12.32)$$

Unlike the multiple regression case it is not possible to rearrange this basic model to reduce the number of parameters. In fact the point with coordinates $(\bar{x}, \bar{y})$ does not necessarily lie on the fitted function if the polynomial is of second degree or more. However, it is useful to rearrange the model as follows:

$$y_i - \bar{y} = a_0 + a_1 x_i + a_2 x_i^2 + \cdots + a_k x^k + z_i. \qquad (12.33)$$

The coefficients $a_0, a_1, a_2, \ldots, a_k$ are estimated by minimizing

$$S = \sum_{i=1}^{n} z_i^2 = \sum_{i=1}^{n} (y_i - \bar{y} - a_0 - a_1 x_i - a_2 x_i^2 - \cdots - a_k x_i^k)^2$$

with respect to $a_0, a_1, a_2, \ldots, a_k$. This gives equations (12.34) which may be solved for the parameter estimates.

$$\Sigma(y_i - \bar{y}) = \hat{a}_0 n + \hat{a}_1 \Sigma x_i + \hat{a}_2 \Sigma x_i^2 + \cdots + \hat{a}_k \Sigma x_i^k$$

$$\Sigma(y_i - \bar{y}) x_i = \hat{a}_0 \Sigma x_i + \hat{a}_1 \Sigma x_i^2 + \hat{a}_2 \Sigma x_i^3 + \cdots + \hat{a}_k \Sigma x_i^{k+1}$$

$$\Sigma(y_i - \bar{y}) x_i^2 = \hat{a}_0 \Sigma x_i^2 + \hat{a}_1 \Sigma x_i^3 + \hat{a}_2 \Sigma x_i^4 + \cdots + \hat{a}_k \Sigma x_i^{k+2} \qquad (12.34)$$

$$\cdot \quad \cdot \quad \cdot \quad \cdot \quad \cdot \quad \cdot \quad \cdot \quad \cdot$$

$$\Sigma(y_i - \bar{y}) x_i^k = \hat{a}_0 \Sigma x_i^k + \hat{a}_1 \Sigma x_i^{k+1} + \hat{a}_2 \Sigma x_i^{k+2} + \cdots + \hat{a}_k \Sigma x_i^{2k},$$

where all summations are for $i = 1, 2, ..., n$. These equations written in matrix notation are:

$$y = X\hat{a}, \qquad (12.35)$$

where y is the column vector

$$\begin{bmatrix} 0 \\ \Sigma(y_i - \bar{y})\, x_i \\ \Sigma(y_i - \bar{y})\, x_i^2 \\ \cdot \quad \cdot \quad \cdot \\ \Sigma(y_i - \bar{y})\, x_i^k \end{bmatrix},$$

a is the column vector

$$\begin{bmatrix} \hat{a}_0 \\ \hat{a}_1 \\ \hat{a}_2 \\ \cdot \quad \cdot \\ \hat{a}_k \end{bmatrix}$$

and where X is the symmetric matrix:

$$\begin{bmatrix} k & \Sigma x_i & \Sigma x_i^2 & \Sigma x_i^3 & \cdots & \Sigma x_i^k \\ \Sigma x_i & \Sigma x_i^2 & \Sigma x_i^3 & \Sigma x_i^4 & \cdots & \Sigma x_i^{k+1} \\ \Sigma x_i^2 & \Sigma x_i^3 & \Sigma x_i^4 & \Sigma x_i^5 & \cdots & \Sigma x_i^{k+2} \\ \cdot & \cdot & \cdot & \cdot & \cdot & \cdot \\ \Sigma x_i^k & \Sigma x_i^{k+1} & \Sigma x_i^{k+2} & \Sigma x_i^{k+3} & \cdots & \Sigma x_i^{2k} \end{bmatrix}.$$

The total sum of squares

$$\sum_{i=1}^{n} (y_i - \bar{y})^2$$

may be split up into two sums of squares the first of which provides an estimate of the residual variability σ^2 and the second provides a measure of the amount of variation removed from the total by the fitted function. The sum of squares due to the fitted function in matrix notation and its expanded form is:

$$y'\hat{a} = \hat{a}_1 \Sigma(y_i - \bar{y})\, x_i + \hat{a}_2 \Sigma(y_i - \bar{y})\, x_i^2 + \cdots + \hat{a}_k \sum (y_i - \bar{y})\, x_i^k. \quad (12.36)$$

The analysis of variance table in Table 12.23 shows the total sum of squares split according to two separate sources of variation.

TABLE 12.23. ANALYSIS OF VARIANCE FOR POLYNOMIAL REGRESSION

Source	S.S.	D.F.
Due to function	$\hat{a}_1\Sigma(y_i - \bar{y})\,x_i + \hat{a}_2\Sigma(y_i - \bar{y})\,x_i^2 + \cdots + \hat{a}_k\Sigma(y_i - \bar{y})\,x_i^k$	k
About function	$\Sigma(y_i - \bar{y})^2 - \hat{a}_1\Sigma(y_i - \bar{y})\,x_i + \cdots + \hat{a}_k\Sigma(y_i - \bar{y})\,x_i^k$	$n - k - 1$
Total	$\Sigma(y_i - \bar{y})^2$	$n - 1$

The "about function" mean square is taken as an estimate of the residual variance σ^2 with $(n - k - 1)$ degreees of freedom. This estimate will, of course, contain all sources of variation remaining after the fitting of the particular degree polynomial chosen to describe the dependent variable. If the "true" relationship between x and y is of degree 3 and we fit a quadratic in x the "about function" mean square will be inflated by the contribution due to the necessary term in x^3 not included in the model. In taking the "about function" mean square as an estimate of σ^2 there is an implicit assumption that we have chosen the correct model. For this reason a plot of the data together with the fitted polynomial should always be inspected critically. If some, or all, of the y-values are replicated (that is we have situation 5 described above) the analysis of variance table may be expanded to include a residual sum of squares. In this situation the residual mean square is taken as the estimate of σ^2 and the adequacy of fit of the chosen polynomial may be tested by comparing the "about function" mean square with the residual mean square. Having estimated σ^2 we may multiply the inverse X^{-1} of the matrix X by this estimate $\hat{\sigma}^2$ to estimate the variance–covariance matrix containing the variances and covariances of the estimated coefficients $\hat{a}_0, \hat{a}_1, \hat{a}_2, ..., \hat{a}_k$. Denoting the variance–covariance matrix by $V(= \sigma^2 X^{-1})$ we have the following expression for the variance of individual y-values about the fitted function.

$$\text{Var}(y) = \frac{(n + 1)\,\sigma^2}{n} + x'Vx, \qquad (12.37)$$

where $V(= \sigma^2 X^{-1})$ is the variance–covariance matrix, and estimated by $\sigma^2 X^{-1}$
$\quad$ x' is the row vector $(1, x_T, x_T^2, ..., x_T^k)$, and
$\quad$ x is the corresponding column vector.
The expanded form of this variance may be written in either of the following two equivalent ways:

$$\text{var}(y) = \frac{(n + 1)\,\sigma^2}{n} + \sum_{j=0}^{k} x_T^{2j}\,\text{var}(\hat{a}_j) + 2\sum_{j=0}^{k-1}\sum_{l=j+1}^{k} x_T^{j+l}\,\text{covar}(\hat{a}_j\hat{a}_l)$$

$$(12.38)$$

$$\text{var}(y) = \frac{(n + 1)\,\sigma^2}{n} + \sum_{j=0}^{k} x_T^{2j}\,\text{var}(\hat{a}_j) + \sum_{\substack{j=0 \\ j \neq l}}^{k}\sum_{l=0}^{k} x_T^{j+l}\,\text{covar}(\hat{a}_j\hat{a}_l).$$

12.10.1. Example

We illustrate the computing procedure by fitting the quadratic

$$y_i - \bar{y} = a_0 + a_1 x_i + a_2 x_i^2 + z_i \tag{12.39}$$

to the data given in Table 12.24.

TABLE 12.24. EXAMPLE DATA FOR POLYNOMIAL REGRESSION

x	0	1	2	3	4	5	6
y	0·43	2·37	8·14	24·01	34·21	57·32	76·73

Firstly we compute the y-mean

$$\bar{y} = 29·03$$

followed by the differences between the y-values and the y-mean:

x_i	0	1	2	3	4	5	6
$y_i - \bar{y}$	$-28·60$	$-26·66$	$-20·89$	$-5·02$	$5·18$	$28·29$	$47·70$

From this we compute the basic sums:

$$\Sigma(y_i - \bar{y}) x_i = 364·87 \qquad \Sigma x_i = 21,$$
$$\Sigma(y_i - \bar{y}) x_i^2 = 2351·93 \qquad \Sigma x_i^2 = 91,$$
$$\Sigma(y_i - \bar{y})^2 = 5092·7546 \qquad \Sigma x_i^3 = 441,$$
$$\Sigma x_i^4 = 2275$$

and hence obtain equations:

$$0 = 7\hat{a}_0 + 21\hat{a}_1 + 91\hat{a}_2,$$
$$364·87 = 21\hat{a}_0 + 91\hat{a}_1 + 441\hat{a}_2,$$
$$2351·93 = 91\hat{a}_0 + 441\hat{a}_1 + 2275\hat{a}_2.$$

The solution of these equations gives

$$\hat{a}_0 = -29·4083, \qquad \hat{a}_1 = 1·4091, \qquad \hat{a}_2 = 1·9370,$$

the sum of squares "due to function" is

$$1·4091 \times 364·87 + 1·9370 \times 2351·93 = 5069·8267$$

and the complete analysis of variance table is given in Table 12.25.

We now illustrate the calculation of confidence limits for the point $x_T = 6$. The matrix X is

$$\begin{bmatrix} 7 & 21 & 91 \\ 21 & 91 & 441 \\ 91 & 441 & 2275 \end{bmatrix}.$$

TABLE 12.25. EXAMPLE ANALYSIS OF VARIANCE FOR FITTING A QUADRATIC

Source	S.S.	D.F.	M.S.	F.	Prob.
Due to function	5069·8267	2	2534·9134	442	<0·001
About function	22·9279	4	5·7320		
Total	5092·7546	6			

The inverse of this matrix is:

$$X^{-1} = \begin{bmatrix} 0·761920 & -0·464295 & 0·059525 \\ -0·464295 & 0·464295 & -0·071430 \\ 0·059525 & -0·071430 & 0·011905 \end{bmatrix}.$$

The variance–covariance matrix is now:

$$\hat{V} = \hat{\sigma}^2 X^{-1} = \begin{bmatrix} 4·36733 & -2·66134 & 0·34120 \\ -2·66134 & 2·66134 & -0·40944 \\ 0·34120 & -0·40944 & 0·06824 \end{bmatrix}.$$

Computing the term $x'Vx$ we have

$$x'\hat{V} = \begin{bmatrix} 1 & 6 & 36 \end{bmatrix} \begin{bmatrix} 4·36733 & -2·66134 & 0·34120 \\ -2·66134 & 2·66134 & -0·40944 \\ 0·34120 & -0·40944 & 0·06824 \end{bmatrix}$$

$$= [0·68249 \quad -1·43314 \quad 0·34120],$$

$$x'\hat{V}x = [0·68249 \quad -1·43314 \quad 0·34120] \begin{bmatrix} 1 \\ 6 \\ 36 \end{bmatrix}$$

$$= 4·36685,$$

$$\text{vâr}(y \mid x_T = 6) = \frac{8}{7} \times 5·7320 + 4·3669 = 10·9177.$$

The 95 per cent confidence limits for y are given by

$$29·03 - 29·4083 + 6 \times 1·4091 + 36 \times 1·9370 \pm 2·776 \times (10·9177)^{\frac{1}{2}},$$

that is

$$77·808 \pm 2·776 \times 3·304,$$

or

$$68·636 \text{ to } 86·980.$$

The summary and plot of the fitted quadratic are given in Table 12.26 and Fig. 12.6.

TABLE 12.26. SUMMARY OF THE FITTED QUADRATIC

x_i	Observed y_i	Fitted y_i	Vâr (y_i)	$(\text{vâr}(y_i))^{1/2}$	95 per cent confidence limits
0	0·43	−0·38	6·5509	2·579	−6·78 to 7·54
1	2·37	2·97	8·1886	2·862	−4·97 to 10·91
2	8·14	10·29	8·1886	2·862	2·24 to 18·13
3	24·01	21·28	8·4615	2·909	13·21 to 29·35
4	34·21	36·25	8·1884	2·862	28·31 to 44·19
5	57·32	55·09	8·1883	2·862	47·15 to 63·04
6	76·73	77·81	10·9177	3·304	68·64 to 86·98

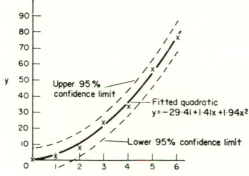

FIG. 12.6. Plot of the fitted quadratic.

12.10.2. Computational Accuracy

The least squares equations (12.34) in the case of polynomial fitting are described by numerical analysts as "ill-conditioned" and special computational techniques are necessary to guarantee accurate solutions in all cases. The methods outlined above are correct mathematically but may prove difficult from a computational point of view. Special techniques may be avoided, however, by using the methods described in Section 12.10.3 below. This method to be described is particularly easy if the values of the independent variable x are equally spaced.

12.10.3. Polynomial Regression Using Orthogonal Polynomials

It is clear that the equations (12.34) or (12.35) would be easy to solve if the off-diagonal elements of the matrix X were all zero. This would also avoid much of the ill-condition in the equations. To achieve this we reformulate model (12.33) in terms of orthogonal polynomials. The special property of orthogonality will be defined later. At this stage we simply define $P_r(x)$, a polynomial of degree r in x, as:

$$P_r(x) = C_{r,r}x^r + C_{r,r-1}x^{r-1} + C_{r,r-2}x^{r-2} + \cdots + C_{r,1}x + C_{r,0}, \qquad (12.40)$$

where the coefficients $C_{r,j}$ will be determined later. Model (12.33) rewritten in terms of orthogonal polynomials is:

$$y_i - \bar{y} = b_0 P_0(x_i) + b_1 P_1(x_i) + b_2 P_2(x_i) + \cdots + b_k P_k(x_i) + z_i. \quad (12.41)$$

To estimate the parameters $b_0, b_1, b_2, \ldots, b_k$ in this model we minimize

$$S = \sum_{i=1}^{n} z_i^2 = \sum_{i=1}^{n} (y_i - \bar{y} - b_0 P_0(x_i) - b_1 P_1(x_i) - \cdots - b_k P_k(x_i))^2$$

with respect to $b_0, b_1, b_2, \ldots, b_k$ to obtain equations:

$$y = X\hat{b}, \qquad (12.42)$$

where y is the column vector

$$\begin{bmatrix} \Sigma(y_i - \bar{y}) \, P_0(x_i) \\ \Sigma(y_i - \bar{y}) \, P_1(x_i) \\ \cdot \quad \cdot \quad \cdot \quad \cdot \quad \cdot \\ \Sigma(y_i - \bar{y}) \, P_k(x_i) \end{bmatrix},$$

b is the column vector

$$\begin{bmatrix} \hat{b}_0 \\ \hat{b}_1 \\ \cdot \\ \hat{b}_k \end{bmatrix}$$

and where X is the symmetric matrix

$$\begin{bmatrix} \Sigma[P_0(x_i)]^2 & \Sigma[P_1(x_i) \, P_0(x_i)] & \cdots & \Sigma[P_0(x_i) \, P_k(x_i)] \\ \Sigma[P_0(x_i) \, P_1(x_i)] & \Sigma[P_1(x_i)]^2 & \cdots & \Sigma[P_1(x_i) \, P_k(x_i)] \\ \cdot \quad \cdot \quad \cdot \quad \cdot & \cdot \quad \cdot \quad \cdot \quad \cdot & \cdots & \cdot \quad \cdot \quad \cdot \quad \cdot \\ \Sigma(P_0(x_i) \, P_k(x_i)] & \Sigma[P_1(x_i) \, P_k(x_i)] & \cdots & \Sigma[P_k(x_i)]^2 \end{bmatrix}$$

and where all summations are for $i = 1, 2, \ldots, n$.

The advantage that we seek in using orthogonal polynomials is that the off-diagonal elements of matrix X be zero. This is the orthogonality property possessed by orthogonal polynomials and it is

$$\sum_{i=1}^{n} [P_r(x_i) \, P_s(x_i)] = 0 \qquad (12.43)$$

for $r, s = 1, 2, ..., k$ and $r \neq s$. Equations (12.42) may now be written

$$\sum_{i=1}^{n} (y_i - \bar{y}) P_r(x_i) = \hat{b}_r \sum_{i=1}^{n} [P_r(x_i)^2] \qquad (12.44)$$

for $r = 1, 2, ..., k$. These equations are directly solvable and we see that:

$$\hat{b}_r = \frac{\sum_{i=1}^{n} (y_i - \bar{y}) P_r(x_i)}{\sum_{i=1}^{n} [P_r(x_i)]^2}. \qquad (12.45)$$

The polynomials, in addition to being orthogonal, are each standardized if:

$$\sum_{i=1}^{n} [P_r(x_i)]^2 = 1 \qquad (12.46)$$

so that if $P_r(x_i)$ is standardized the expression (12.45) becomes:

$$\hat{b}_r = \sum_{i=1}^{n} (y_i - \bar{y}) P_r(x_i). \qquad (12.47)$$

The matrix $V(= \hat{\sigma}^2 X^{-1})$ is once again the variance–covariance matrix of the estimates $\hat{b}_r$. Since matrix X is diagonal the inverse of X is also diagonal with the ith diagonal element of the inverse equal to the reciprocal of the ith diagonal element of X. Thus the covariances between the estimated coefficients are all zero and the variances are given by

$$\text{var} (\hat{b}_r) = \sigma^2 \left\{ \sum_{i=1}^{n} [P_r(x_i)]^2 \right\}^{-1}. \qquad (12.48)$$

The variances are equal to σ^2 if the polynomials are standardized. Thus we see that if values of the coefficients C can be chosen so that the orthogonality property (12.43) is satisfied the computation of the estimates of the parameters b in the model (12.41) readily follows. It will be noticed that the orthogonality property (12.43), and hence the orthogonal polynomial coefficients C, depends on the actual x-values. A change in one of the x-values or the inclusion of an extra x-value would produce a change in the coefficients C. The polynomials are only orthogonal for the particular set of x-values. The expression (12.45) for the parameters in the model to be fitted involve the values of the relevant polynomial at each data point. Forsythe (1957) gives the following relation between polynomials:

$$\lambda_r P_r(x_i) = -d_{r,r-2} P_{r-2}(x_i) - d_{r,r-1} P_{r-1}(x_i) + P_1(x_i) P_{r-1}(x_i), \qquad (12.49)$$

where

$$d_{r,r-2} = \sum_{i=1}^{n} P_1(x_i) P_{r-1}(x_i) P_{r-2}(x_i)$$

$$d_{r,r-1} = \sum_{i=1}^{n} P_1(x_i) [P_{r-1}(x_i)]^2,$$

where λ_r is a constant determined by standardization and where the polynomials P_r, P_{r-1}, P_{r-2} are standardized.

The first two orthogonal polynomials in unstandardized form are

$$P'_0(x_i) = 1$$

$$P'_1(x_i) = x_i - \bar{x} \tag{12.50}$$

and in standardized form are:

$$P_0(x_i) = S_0$$

$$P_1(x_i) = S_1(x_i - \bar{x}), \tag{12.51}$$

where S_0 and S_1 are standardization values determined as follows:

$$S_0 = \left\{ \sum_{i=1}^{n} [P'_0(x_i)]^2 \right\}^{-\frac{1}{2}} = 1/\sqrt{n},$$

$$S_1 = \left\{ \sum_{i=1}^{n} [P'_1(x_i)]^2 \right\}^{-\frac{1}{2}} = \left\{ \sum_{i=1}^{n} (x_i - \bar{x})^2 \right\}^{-\frac{1}{2}} = \frac{1}{\sigma_x}.$$

Since we may easily obtain the values of polynomials $P_0(x_i)$ and $P_1(x_i)$ we may generate the values of $P_2(x_i)$, $P_3(x_i)$, ... by successive use of relation (12.49). For example, we may calculate $P_2(x_i)$ as follows:

$$d_{2,0} = \sum_{i=1}^{n} S_1(x_i - \bar{x}) S_1(x_i - \bar{x}) S_0 = \frac{1}{\sigma_x^2 \sqrt{n}} \sum_{i=1}^{n} (x_i - \bar{x})^2 = \frac{1}{\sqrt{n}},$$

$$d_{1,0} = \sum_{i=1}^{n} S_1(x_i - \bar{x}) [S_1(x_i - \bar{x})]^2 = \frac{1}{\sigma_x^3} \sum_{i=1}^{n} (x_i - \bar{x})^3,$$

$$\lambda_2 P_2(x_i) = -\frac{1}{n} - \left[\frac{1}{\sigma_x^3} \sum_{i=1}^{n} (x_i - \bar{x})^3 \right] (x_i - \bar{x}) + \frac{1}{\sigma_x^2} (x_i - \bar{x})^2. \tag{12.52}$$

The value of λ_2 is now obtained by standardization. That is $P_2(x_i)$ is the standardized version of $\lambda_2 P_2(x_i)$. The use of relation (12.49) will be further clarified by the example which follows in Section 12.10.4.

The steps in the estimation of the parameters b may be listed:

1. Write down polynomials P_0 and P_1 (expressions (12.51)) and list the values of these polynomials for each data point x_i.
2. Using expression (12.49) as many times as necessary generate the required polynomials P_2, P_3, ..., and list the values of the polynomials for each data point x_i.
3. Compute sums $\sum_{i=1}^{n} (y_i - \bar{y}) P_r(x_i)$ and hence estimate parameters $b_1, b_2, b_3, ...$
4. The variances of these estimates are σ^2 since the polynomials are standardized.

5. Having estimated the parameters b, we may expand the model (12.41) in powers of $(x_i - \bar{x})$ and hence obtain the fitted function as an ordinary polynomial in x_i.

The analysis of variance procedure associated with polynomial fitting using orthogonal polynomials is simplified by the fact that each term in the statistical model (12.41) is estimated independently and has an independent entry in the analysis of variance table. The sum of squares due to the rth orthogonal polynomial is

$$\hat{b}_r^2 \sum_{i=1}^{n} [P_r(x_i)]^2$$

or simply $\hat{b}_r^2$ if P_r is standardized. The total sum of squares

$$\sum_{i=1}^{n} (y_i - \bar{y})^2$$

is split up into k sums of squares representing k† individual orthogonal polynomials. An example analysis of variance up to degree three is given in Table 12.27.

TABLE 12.27. EXAMPLE ANALYSIS OF VARIANCE FOR FIT TO DEGREE 3

Source	S.S.	D.F.
Linear	$\hat{b}_1^2 \sum_{i=1}^{n} [P_1(x_i)]^2$	1
Quadratic	$\hat{b}_2^2 \sum_{i=1}^{n} [P_2(x_i)]^2$	1
Cubic	$\hat{b}_3^2 \sum_{i=1}^{n} [P_3(x_i)]^2$	1
About function	$\sum_{i=1}^{n} (y_i - \bar{y})^2 - \sum_{j=1}^{3} \hat{b}_j^2 \sum_{i=1}^{n} [P_j(x_i)]^2$	$n - 4$
Total	$\sum_{i=1}^{n} (y_i - \bar{y})^2$	$n - 1$

12.10.4. Example

As an example of the method of fitting polynomials using orthogonal polynomials we will fit a cubic to the data in Table 12.24. The x-values in

† The polynomial P_0 is accounted for by taking the total sum of squares about the mean $\bar{y}$ and not about zero.

this data are in fact equally spaced and a simple procedure, described in Section 12.10.5, is available in this case. We will analyse this data, however, as though the x-values were unequally spaced as an example of the general method.

Step 1.

$$P_0(x_i) = S_0 = 1/\sqrt{7}$$

$$P_1(x_i) = S_1(x_i - \bar{x})$$

$$S_1 = \left\{ \sum_{i=1}^{n} (x_i - \bar{x}) \right\}^{-\frac{1}{2}} = \{(0 - 3)^2 + (1 - 3)^2 + \cdots + (6 - 3)^2\}^{-\frac{1}{2}}$$

$$= 1/\sqrt{28}.$$

The list of values of these polynomials is given in columns 3 and 4 of Table 12.28.

Step 2. Compute $P_2(x_i)$ using expression (12.49)

$$d_{2,0} = (-3/\sqrt{28}) (-3/\sqrt{28}) (1/\sqrt{7}) + (-2/\sqrt{28}) (-2/\sqrt{28}) (1/\sqrt{7})$$
$$+ \cdots + (3/\sqrt{28}) (3/\sqrt{28}) (1/\sqrt{7}) = 1/\sqrt{7},$$

$$d_{2,1} = (-3/\sqrt{28}) (-3/\sqrt{28})^2 + (-2/\sqrt{28}) (-2/\sqrt{28})^2$$
$$+ \cdots + (3/\sqrt{28}) (3/\sqrt{28})^2 = 0,$$

$$\lambda_2 P_2(x_i) = -(1/\sqrt{7}) (1/\sqrt{7}) + ((x_i - \bar{x})/\sqrt{28}) ((x_i - \bar{x})/\sqrt{28})$$
$$= (x_i - \bar{x})^2/28 - 1/7.$$

List the values of $\lambda_2 P_2(x_i)$ and standardize to obtain the values of $P_2(x_i)$ listed as column 5 of Table 12.28. From these values we compute:

$$\lambda_2 = ((5/28)^2 + (0)^2 + \cdots + (5/28)^2)^{\frac{1}{2}} = \sqrt{(3/28)}$$

and hence the standardized second-degree polynomial is

$$P_2(x_i) = (x_i - \bar{x})^2/\sqrt{84} - 4/\sqrt{84}.$$

Compute $P_3(x_i)$ again using expression (12.49)

$$d_{3,1} = (-3/\sqrt{28}) (5/\sqrt{84}) (-3/\sqrt{28}) + (-2/\sqrt{28}) (0) (-2/\sqrt{28})$$
$$+ \cdots + (3/\sqrt{28}) (5/\sqrt{84}) (3/\sqrt{28}) = 84/(28 \sqrt{84}) = \sqrt{(3/28)},$$

$$d_{3,2} = (-3/\sqrt{28}) (5/\sqrt{84})^2 + (-2/\sqrt{28}) (0)^2 + \cdots + (3/\sqrt{28}) (5/\sqrt{84})^2$$
$$= 0,$$

$$\lambda_3 P_3(x_i) = -\sqrt{(3/28)}\,(x_i - \bar{x})/\sqrt{28} + ((x_i - \bar{x})/\sqrt{28})((x_i - \bar{x})^2/\sqrt{84} - 4/\sqrt{84})$$

$$= (x_i - \bar{x})^3/(28\sqrt{3}) - 7(x_i - \bar{x})/(28\sqrt{3})$$

$$= [(x_i - \bar{x})^3 - 7(x_i - \bar{x})]/(28\sqrt{3}.)$$

TABLE 12.28. LISTS OF VALUES OF POLYNOMIALS UP TO DEGREE 3

$(y_i - \bar{y})$	x_i	$P_0(x_i)$	$P_1(x_i)$	$P_2(x_i)$	$\lambda_3 P_3(x_i)$	$P_3(x_i)$
$-28{\cdot}60$	0	$1/\sqrt{7}$	$-3/\sqrt{28}$	$5/\sqrt{84}$	$-6/28\sqrt{3}$	$-1/\sqrt{6}$
$-26{\cdot}66$	1	$1/\sqrt{7}$	$-2/\sqrt{28}$	0	$6/28\sqrt{3}$	$1/\sqrt{6}$
$-20{\cdot}89$	2	$1/\sqrt{7}$	$-1/\sqrt{28}$	$-3/\sqrt{84}$	$6/28\sqrt{3}$	$1/\sqrt{6}$
$-\ 5{\cdot}02$	3	$1/\sqrt{7}$	0	$-4/\sqrt{84}$	0	0
$5{\cdot}18$	4	$1/\sqrt{7}$	$1/\sqrt{28}$	$-3/\sqrt{84}$	$-6/28\sqrt{3}$	$-1/\sqrt{6}$
$28{\cdot}29$	5	$1/\sqrt{7}$	$2/\sqrt{28}$	0	$-6/28\sqrt{3}$	$-1/\sqrt{6}$
$47{\cdot}70$	6	$1/\sqrt{7}$	$3/\sqrt{28}$	$5/\sqrt{84}$	$6/28\sqrt{3}$	$1/\sqrt{6}$

From the values of $\lambda_3 P_3(x_i)$ shown in column 7 of Table 12.28 we may obtain the standardized form of the third-degree polynomial as:

$$P_3(x_i) = [(x_i - \bar{x})^3 - 7(x_i - \bar{x})]/(6\sqrt{6}).$$

Step 3.

$$\hat{b}_1 = \sum_{i=1}^{7} (y_i - \bar{y})\, P_1(x_i) = (-28{\cdot}60)\,(-3/\sqrt{28}) + (-26{\cdot}66)\,(-2/\sqrt{28})$$

$$+ \cdots + (47{\cdot}70)\,(3/\sqrt{28}) = 364{\cdot}87/\sqrt{28},$$

$$\hat{b}_2 = \sum_{i=1}^{7} (y_i - \bar{y})\, P_2(x_i) = (-28{\cdot}60)\,(5/\sqrt{84}) + (-26{\cdot}66)\,(0)$$

$$+ \cdots + (47{\cdot}70)\,(5/\sqrt{84}) = 162{\cdot}71/\sqrt{84},$$

$$\hat{b}_3 = \sum_{i=1}^{7} (y_i - \bar{y})\, P_3(x_i) = (-28{\cdot}60)\,(-1/\sqrt{6}) + \cdots + (47{\cdot}70)\,(1/\sqrt{6})$$

$$= -\ 4{\cdot}72/\sqrt{6}.$$

The estimates $\hat{b}_1, \hat{b}_2, \hat{b}_3$ are recorded in Table 12.29 together with the three corresponding orthogonal polynomials.

TABLE 12.29. ESTIMATES AND ORTHOGONAL POLYNOMIALS UP TO DEGREE 3

Degree	Estimate	Orthogonal polynomial
1	$364{\cdot}87/\sqrt{28}$	$(x_i - \bar{x})/\sqrt{28}$
2	$162{\cdot}71/\sqrt{84}$	$[(x_i - \bar{x})^2 - 4]/\sqrt{84}$
3	$-4{\cdot}72/\sqrt{6}$	$[(x_i - \bar{x})^3 - 7(x_i - \bar{x})]/(6\sqrt{6})$

The fitted polynomial is:

$$y = 29\cdot03 - 4\cdot72[(x_i - \bar{x})^3 - 7(x_i - \bar{x})]/36 + 162\cdot71[(x_i - \bar{x})^2 - 4]/84$$
$$+ 364\cdot87(x_i - \bar{x})/28$$

or

$$y = -0\cdot1311(x_i - \bar{x})^3 + 1\cdot9370(x_i - \bar{x})^2 + 13\cdot9489(x_i - \bar{x}) + 21\cdot2819,$$

that is:

$$y = 0\cdot4082 - 1\cdot2131x_i + 3\cdot1170x_i^2 - 0\cdot1311x_i^3.$$

The sum of squares due to fitting the rth polynomial is

$$\hat{b}_r^2 \sum_{i=1}^{n} [P_r(x_i)]^2$$

or simply $\hat{b}_r^2$ if the rth polynomial is standardized. The values for the first three polynomials are 4754·6470, 315·1731 and 3·7130. The analysis of variance table is shown in Table 12.30.

TABLE 12.30. ANALYSIS OF VARIANCE FITTING THREE ORTHOGONAL POLYNOMIALS TO DATA IN TABLE 12.24

Source	S.S.	D.F.	M.S.	F.	Prob.
1st degree	4754·6470	1	4754·6470	742·08	≪0·001
2nd degree	315·1731	1	315·1731	49·19	<0·01
3rd degree	3·7130	1	3·7130	0·58	N.S.
Residual	19·2215	3	6·4072		
Total	5092·7546	6			

It should be noted that the sum of the sums of squares due to fitting the first- and second-degree polynomials is equal to the sum of squares „due to function" in Table 12.25. In fact if the fit is confined to the first- and second-degree polynomials the fitted function and analysis is the same as was obtained by the normal least squares method in Section 12.10.1.

12.10.5. Polynomial Regression Using Orthogonal Polynomials (Equally-spaced x-values)

The fitting method described in Section 12.10.3 applies to both equally-spaced and unequally spaced x-values. However, if the x-values are equally-spaced, the values of the first six orthogonal polynomials are given in Table 47 of Pearson and Hartley for data containing up to 50 points. These are tabulated in an unstandardized form in which all values are integers.

The sum of squares of the values is also included in the table so that the standardized values, as shown, for example, in Table 12.29 above, may be obtained by dividing each value by the square root of the sum of squares.

12.11. MULTIPLE AND POLYNOMIAL REGRESSION IN SITUATIONS 2 TO 5

The five situations discussed in this chapter can arise in the multiple-regression case and in polynomial regression. The solution for the second, third and fourth situations is obtainable from the procedures described above by including weights w_i in all summations and replacing n by Σw_i. Situation 5 may be analysed by including weights $w_i = r_i$, by using $\bar{y}_i$. instead of y_i, by using $\bar{y}..$ instead of $\bar{y}.$ and by computing the extra residual sum of squares:

$$\sum_{i=1}^{n} \sum_{j=1}^{r_i} (y_{ij} - \bar{y}_{i.})^2.$$

12.12. FITTING OTHER FUNCTIONS

The procedures described above apply to all functions in which the coefficients to be estimated enter linearly into the equation. Functions such as

$$y_i = a + b \log x_i + z_i$$

and

$$y_i = a + b \cos x_i + c \sin t_i + z_i$$

may be analysed by presenting $\log x_i$, $\cos x_i$ and $\sin t_i$ as independent variables.

Functions such as

$$y_i = a + \log (x_i - b) + z_i$$

cannot be fitted into this procedure and require a different approach. It would appear to be possible to minimize:

$$S = \sum_{i=1}^{n} z_i^2 = \sum_{i=1}^{n} (y_i - a - \log (x_i - b))^2$$

with respect to a and b, but the equations obtained for non-linear functions of this type are frequently difficult to solve. Estimates of a and b may be obtained by calculating S for all possible combinations of values of a and b and choosing the values of a and b which minimize S. Complete scans of this type, particularly with many variables, are normally too lengthy to be possible and techniques for directing the scan more efficiently towards the minimum are being currently developed.

Appendix. Matrix Algebra

For the purposes of Chapter 12 we may describe a matrix as a rectangular arrangement of numbers. The arrangement may have any number of rows and columns from one upwards. Thus a single number may be regarded as a matrix of size one by one and such a matrix is normally called a scalar. A matrix with one row is called a row vector and a matrix with one column is called a column vector.

The algebraic equation

$$ax = b$$

has so far specified a relation between three quantities (scalars) denoted by a, x and b. Matrix equations look the same but specify relations between matrices. The rules of addition, subtraction, multiplication and division in the algebra of scalars are well known but their counterparts in Matrix Algebra do not display the same properties. For example, we are used to being able to write ax or xa to denote the product of x and a. The multiplication rules in matrix algebra are such that ax is not necessarily the same as xa, in fact the product may not even exist. Thus we must be more careful in writing down matrix equations and in solving them than with equations involving only scalars.

The Equality of Two Matrices

The equality of two matrices implies that each element in one is equal to the element in the other occupying the same row and column and is only defined if they have the same number of rows and the same number of columns. Thus if A and B are two such matrices with r rows and c columns the relation

$$A = B$$

specifies that the element a_{ij} in the ith row and jth column in A equals the element b_{ij} in the ith row and jth column of B for all values of i and j. For example:

$$\begin{bmatrix} 2 & 4 \\ 3 & 9 \\ 6 & 1 \end{bmatrix} = \begin{bmatrix} 2 & 4 \\ 3 & 9 \\ 6 & 1 \end{bmatrix}.$$

Multiplication by Constant

Multiplication of a matrix by a constant multiplies each element of the matrix by that constant. Thus

$$2 \cdot \begin{bmatrix} 1 & 8 & 9 \\ 4 & 7 & 2 \end{bmatrix} = \begin{bmatrix} 2 & 16 & 18 \\ 8 & 14 & 4 \end{bmatrix}.$$

Addition and Subtraction

Matrices may be added or subtracted only if they have the same number of rows and the same number of columns. Addition, or subtraction, is performed by adding, or subtracting, corresponding elements in the two matrices to obtain a new matrix with the same dimensions as the original matrices. For example, if

$$A = \begin{bmatrix} 2 & 4 \\ 1 & 8 \\ 9 & 3 \end{bmatrix}; \quad B = \begin{bmatrix} 3 & 2 \\ 1 & 8 \\ 9 & 4 \end{bmatrix}; \quad C = \begin{bmatrix} 1 & 3 \\ 2 & 6 \\ 7 & 5 \end{bmatrix},$$

then

$$D = A + B - C = \begin{bmatrix} 4 & 3 \\ 0 & 10 \\ 11 & 2 \end{bmatrix}.$$

The order in which additions and subtractions are performed is not important and we may quickly verify that

$$A + B - C = B + A - C = -C + B + A.$$

Transposition

The transpose of a matrix A, written A' or A^T, is obtained by interchanging rows with columns. That is if a_{ij} is the element of A in the ith row and jth column this becomes the element in the jth row and ith column of the transpose. For example, if

$$A = \begin{bmatrix} 2 & 4 \\ 1 & 8 \\ 9 & 3 \end{bmatrix} \text{ then } A' = \begin{bmatrix} 2 & 1 & 9 \\ 4 & 8 & 3 \end{bmatrix}.$$

Thus we see that the shape of the matrix will be changed by transposition unless the matrix is square.

Symmetric Matrices

A matrix A is symmetric if it is equal to its transpose. That is, if $A = A'$ or alternatively $a_{ij} = a_{ji}$ for all i and j. It is clear that this is only possible for square matrices. A great many matrices encountered in statistics are symmetric.

Multiplication of Matrices

The order in which multiplication of two matrices is performed is important and the product only exists if the number of columns of the first is equal to the number of rows of the second. The number of rows in the product is equal to the number of rows in the first matrix, and the number of columns in the product is equal to the number of columns in the second matrix.

Elements in the product are produced in the following way. Consider the ith row in the first matrix and the jth column of the second. Sum the products of corresponding elements in these to obtain the elements in the ith row and jth column of the product. For example, if

$$A = \begin{bmatrix} 4 & 3 \\ 1 & 2 \\ 6 & 5 \end{bmatrix} \quad \text{and} \quad B = \begin{bmatrix} 1 & 2 & 6 & 2 \\ 1 & 3 & 1 & 4 \end{bmatrix},$$

then

$$AB = \begin{bmatrix} 7 & 17 & 27 & 20 \\ 3 & 8 & 8 & 10 \\ 11 & 27 & 41 & 32 \end{bmatrix},$$

whereas BA does not exist since there are four columns in B and three rows in A.

Identity Matrix (Unit Matrix)

The Identity Matrix, or Unit Matrix, is a generalization of the scalar unity. It is a square matrix, denoted by I with unity in the top-left–bottom-right diagonal, and zeros elsewhere. That is the unit matrix of size 4 is:

$$\begin{bmatrix} 1 & 0 & 0 & 0 \\ 0 & 1 & 0 & 0 \\ 0 & 0 & 1 & 0 \\ 0 & 0 & 0 & 1 \end{bmatrix}.$$

It may be easily verified that multiplication of a matrix by the identity matrix of the appropriate size does not change the matrix. That is: $AI = A$ and $IA = A$.

Determinant of a Square Matrix

The determinant of a square matrix is a single numerical value computed according to specific rules from the elements of the matrix. The rules need not concern us in this description since it is sufficient to know of its existence.

Matrix Inverse

The inverse of a matrix A is a generalization of the reciprocal of a scalar. It is usually written A^{-1} and only exists for square matrices. It is well known that if a is a scalar its reciprocal a^{-1} does not exist (is infinite) if a is zero. Similarly A^{-1} will not exist if A is "singular", that is if the determinant of A is zero. The rules for the calculation of an inverse may not be written down in explicit terms and its computation is tedious unless electronic help is available. The inverse A^{-1} of square matrix A has the same number of rows (and columns) as A. It is defined as the matrix such that:

$$AA^{-1} = A^{-1}A = I.$$

Thus it may be looked on as the reciprocal of A. If A is symmetric it is clear that A^{-1} must be symmetric also. It is sufficient for the purposes of Chapter 12 to know that the inverse of a matrix exists if A is square and "non-singular" and that it may be computed from A itself. The computation is best left to a well-written matrix subroutine for a computer. A well-written subroutine is one which takes all possible care to ensure an accurate inverse and which warns the user of singularity when appropriate to do so.

Vector Operations

The same rules of addition, subtraction, transposition, multiplication apply to vectors but it is appropriate to observe that the transpose of a column vector is a row vector and vice versa. The product of a column vector and a row vector is a matrix whereas the product of a row vector and a column vector is a scalar. For example:

$$\begin{bmatrix} 1 \\ 3 \end{bmatrix} \begin{bmatrix} 2 & 5 \end{bmatrix} = \begin{bmatrix} 2 & 5 \\ 6 & 15 \end{bmatrix} \quad \text{but} \quad \begin{bmatrix} 2 & 5 \end{bmatrix} \begin{bmatrix} 1 \\ 3 \end{bmatrix} = 17.$$

The Solution of Equations

Let column vectors y and a and square matrix X be defined as follows:

$$y = \begin{bmatrix} y_1 \\ y_2 \\ y_3 \\ y_4 \end{bmatrix}; \quad a = \begin{bmatrix} a_1 \\ a_2 \\ a_3 \\ a_4 \end{bmatrix}; \quad X = \begin{bmatrix} x_{11} & x_{12} & x_{13} & x_{14} \\ x_{21} & x_{22} & x_{23} & x_{24} \\ x_{31} & x_{32} & x_{33} & x_{34} \\ x_{41} & x_{42} & x_{43} & x_{44} \end{bmatrix}.$$

In regression problems we need to solve a set of equations of the type:

$$x_{11}a_1 + x_{12}a_2 + x_{13}a_3 + x_{14}a_4 = y_1$$

$$x_{21}a_1 + x_{22}a_2 + x_{23}a_3 + x_{24}a_4 = y_2$$

$$x_{31}a_1 + x_{32}a_2 + x_{33}a_3 + x_{34}a_4 = y_3$$

$$x_{41}a_1 + x_{42}a_2 + x_{43}a_3 + x_{44}a_4 = y_4$$

for the values of a_1, a_2, a_3 and a_4 given all the x's and y's. These equations may be written in matrix form as:

$$y = Xa.$$

Pre-multiplying both sides of this equation by the inverse of X, taking care to place X^{-1} in front of the expression on each side, we obtain:

$$X^{-1}y = X^{-1}Xa = Ia = a.$$

Thus the required solution of these equations, $X^{-1}y$, is a column vector. In fact it is possible to obtain the solution of these equations without computing the inverse X^{-1} and this would normally be done but the inverse is required if we wish to compute confidence limits as described in Sections 12.9 and 12.10 and exampled in Section 12.10.1.

QUESTIONS

1. Consider the data given in Table 11.7 and ignore the fact that it has been obtained using different treatments and arranged in balanced blocks.

(a) Is there evidence for believing that there is an association between temperature and yield?

(b) Fit a straight line describing yield in terms of temperature and compare the slope of this line with the estimate ($\hat{C} = 0.7007$) obtained in Section 11.4.1.

(c) Why does this estimate differ from that obtained here?

(d) Compute the residuals from this fitted straight line, rearrange them into the form of the original two-way factorial design given in Table 11.7 and perform a two-way analysis of variance on these residuals. Compare the residual mean square obtained with those in the analysis of variance and covariance of the original data given in Tables 11.11 and 11.10 respectively.

2 (a). Fit a straight line to the data used in question 1 which passes through the point (Yield $= 0.0$, Temp. $= 110.0$).

(b) Compute a graduation of the fitted function at the given values of temperature including the fitted value and 95 per cent confidence limits.

(c) Is it reasonable to believe that the straight line

$$\text{Yield} = 0.4 \text{ Temp.} + 110.0$$

describes the data?

(d) Repeat question (c) for all combinations of slope and constant specified by the ranges slope $= 0.0 (0.25) 1.0$ and constant $= 80.0 (10.0) 150$.

Plot a Y for an acceptance and N for a refusal on a simple plot of yield against temperature. Draw in the boundary of the region containing all the acceptances. Repeat the above test with further slopes and constants if necessary to define the boundary more precisely. This region is a confidence region with confidence probability equal to the significance level used in the test.

3. Eight pairs of values for two variables were obtained during an experiment and it is believed that the variability in variable y is proportional to the value of x and that the constant of proportionality is 1.0. That is the variance of y_t is equal to x_t^2. The values obtained are given below:

x	1.0	2.0	3.0	4.0	5.0	6.0	7.0	8.0
y	6.5	21.8	27.6	44.0	51.3	58.3	73.5	79.2

(a) Fit a straight line to this data, describing y in terms of x under these conditions.

(b) Does the fitted straight line describe the data adequately?

(c) How would you modify the fitting procedure if the constant of proportionality had been unknown? How would you estimate this constant?

4 (a). Fit a quadratic function to the data given in question 3 above assuming that an assumption of equal variance in variable y is justified.

(b) How would you modify the fitting process to take account of the unequal variances as described in question 3?

(c) Fit the quadratic allowing for the unequal variances.

5. The matrix given below contains the sums of squares and cross-products for three variables each measured on thirty-two individuals. The means for these variables are also given:

	X	Y	T
X	9·88913	1·53223	1·90244
Y	1·53223	14·86653	5·73238
T	1·90244	5·73238	16·01014
Means	18·832	20·428	20·005

(a) Describe Y by a linear function of X and T.

(b) Describe Y by a linear function of X alone and test the hypothesis that the addition of variable T to this description makes no improvement to the description of Y.

CHAPTER 13

THE POISSON PROCESS AND COUNTING PROBLEMS

THIS chapter describes the Poisson Process and its application as a description of the random process underlying many counting problems.

13.1. THE POISSON PROCESS

The Poisson process is a random process in which events are occurring at random on a time scale (or perhaps a distance scale). A common application of the Poisson process is the description of nuclear disintegrations occurring at random in time. Other applications include the occurrence of accidents at a particular highway intersection, and the appearance (in area) of yeast cells. Figure 13.1 illustrates the random occurrence of events on a time scale. Events are shown by an x and the times zero, t, and $t + \Delta t$ are also marked.

FIG. 13.1. Random process example.

The Poisson process makes the following assumptions:

1. The probability of an event in any small interval of time Δt approaches $\lambda \Delta t$ as Δt approaches zero, where λ is a positive quantity.
2. The probability of more than one event occurring in Δt is of order smaller than Δt. (We write this as $o(\Delta t)$.)
3. Events occurring in non-overlapping intervals are independent.

From these assumptions we seek the probability that there are exactly n events in an interval of time of length t.

Let $P_n(t)$ be the required probability that exactly n events are observed in the time interval from time zero to time t (written $(0, t)$). We consider two adjacent and non-overlapping intervals $(0, t)$ and $(t, t + \Delta t)$ and we first consider the case when $n = 0$.

No events occur in interval $(0, t + \Delta t)$ if none occur in interval $(0, t)$ and none occur in interval $(t, t + \Delta t)$ so that

$$P_0(t + \Delta t) = P_0(t) [1 - \lambda \Delta t - o (\Delta t)], \tag{13.1}$$

$[1 - \lambda \Delta t - 0(\Delta t)]$ is the probability that no events occur in interval $(t, t + \Delta t)$. Rearranging expression (13.1) gives:

$$\frac{P_0(t + \Delta t) - P_0(t)}{\Delta t} = -\lambda P_0(t) + o(\Delta t).$$

As $\Delta t \to 0$ we obtain

$$P'_0(t) = -\lambda P_0(t)$$

$$\frac{P'_0(t)}{P_0(t)} = -\lambda \tag{13.2}$$

and integrating (13.2) we have

$$\log_e(P_0(t)) = -\lambda t \quad (\text{since } P_0(0) = 1)$$

and hence

$$P_0(t) = e^{-\lambda t}. \tag{13.3}$$

Considering now the case when $n > 0$, n events can occur within time interval $(0, t + \Delta t)$ in any of the ways listed below.

  n events occur in $(0, t)$ and 0 events occur in $(t, t + \Delta t)$,
  $n - 1$ events occur in $(0, t)$ and 1 event occurs in $(t, t + \Delta t)$,
  $n - 2$ events occur in $(0, t)$ and 2 events occur in $(t, t + \Delta t)$,
  and so on.
Thus

$$P_n(t + \Delta t) = P_n(t) [1 - \lambda \Delta t - o(\Delta t)] + P_{n-1}(t) \lambda \Delta t + o(\Delta t). \tag{13.4}$$

Rearranging expression (13.4) gives:

$$\frac{P_n(t + \Delta t) - P_n(t)}{\Delta t} = -\lambda P_n(t) + \lambda P_{n-1}(t) + o(\Delta t)$$

and as $\Delta t \to 0$ we obtain:

$$P'_n(t) = -\lambda P_n(t) + \lambda P_{n-1}(t). \tag{13.5}$$

The solution of this equation may be obtained by direct means but it is sufficient for present purposes to show by substitution that

$$P_n(t) = \frac{e^{-\lambda t}(\lambda t)^n}{n!} \tag{13.6}$$

is the required solution.

$$\text{L.H.S. of } (13.5) = P_n'(t) = \frac{dP_n(t)}{dt}$$

$$= -\lambda e^{-\lambda t} \frac{(\lambda t)^n}{n!} + e^{-\lambda t} n\lambda \frac{(\lambda t)^{n-1}}{n!}$$

$$= -\lambda e^{-\lambda t} \frac{(\lambda t)^n}{n!} + \lambda e^{-\lambda t} \frac{(\lambda t)^{n-1}}{(n-1)!}$$

$$= -\lambda P_n(t) + \lambda P_{n-1}(t)$$

$$= \text{R.H.S. of } (13.5).$$

The probability of observing n events in time t is thus seen to be:

$$P_n(t) = e^{-\lambda t} \frac{(\lambda t)^n}{n!}. \tag{13.7}$$

A complete set of these probabilities for $n = 0, 1, 2, \ldots$ forms a Poisson distribution with parameter λt. That is, writing k for λt,

$$e^{-k}, \quad ke^{-k}, \quad \frac{k^2}{2}e^{-k}, \quad \frac{k^3}{6}e^{-k}\ldots$$

forms a Poisson distribution with parameter k. Table 39 of Pearson and Hartley (1958) gives the values of these probabilities for values of k up to 15.

13.1.1. Mode, Mean and Variance of the Poisson Distribution

1. The modal value (or values) is the value (or values) of r satisfying the inequality:

$$k - 1 \leqq r \leqq k.$$

2. The mean value is given by

$$\mu_1' = \sum_{r=0}^{\infty} rP_r(t) = \sum_{r=1}^{\infty} re^{-k} \frac{k^r}{r!}$$

$$= ke^{-k} \sum_{r=1}^{\infty} \frac{k^{r-1}}{(r-1)!} = ke^{-k}e^k = k.$$

The mean value is k.

3. The variance is given by:

$$\mu_2 = \sum_{r=0}^{\infty} (r - k)^2 \, P_r(t) = \sum_{r=0}^{\infty} (r^2 - 2rk + k^2) \, e^{-k} \frac{k_r}{r!}$$

$$= \sum_{r=1}^{\infty} r^2 e^{-k} \frac{k^r}{r!} - 2k \sum_{r=1}^{\infty} re^{-k} \frac{k^r}{r!} + k^2 \sum_{r=0}^{\infty} e^{-k} \frac{k^r}{r!}$$

$$= \sum_{r=1}^{\infty} [r(r - 1) + r] \, e^{-k} \frac{k^r}{r!} - 2k^2 + k^2$$

$$= k^2 \sum_{r=2}^{\infty} e^{-k} \frac{k^{r-2}}{(r - 2)!} + k - k^2$$

$$= k^2 + k - k^2$$

$$\mu_2 = k.$$

The variance is also equal to k.

13.1.2. Further Moments of the Poisson Distribution

The third and fourth central moments and the moment ratios are:

$$\mu_3 = k; \qquad \mu_4 = k + 3k^2,$$

$$\beta_1 = \frac{k^2}{k^3} = \frac{1}{k} \to 0 \quad \text{as} \quad k \to \infty,$$

$$\beta_2 = \frac{k + 3k^2}{k^2} = 3 + \frac{1}{k} \to 3 \quad \text{as} \quad k \to \infty.$$

13.1.3. Properties of the Poisson Distribution

1. The mean and variance are both equal to k.
2. The distribution tends to normality as k increases.
3. If two independent variables (counts) c_1 and c_2 have Poisson distributions with means k_1 and k_2 then the sum $c_1 + c_2$ has a Poisson distribution with mean $k_1 + k_2$.

The size of the interval we choose to observe a random process has no effect on the process itself. If we observe a process for 2 minutes we would expect the same form of distribution to apply with only a change in parameter as if we had observed the process for 1 minute. If the interval is 1 minute and the Poisson parameter is k, the third property states, as we would naturally expect, that the parameter for a 2-minute interval will be $2k$.

The parameter may be changed simply by a change of interval so that the actual interval used is important information in the description of the process. Such a description should include the interval as well as the value of the parameter.

13.1.4. Fitting a Poisson Distribution

An active source was counted for successive periods of 2 minutes and the numbers of counts for each 2-minute period were noted. The observed distribution given in Table 13.1 was obtained from 400 such 2-minute periods.

TABLE 13.1. OBSERVED DISTRIBUTION OF INTERVAL COUNT FREQUENCIES

No. of counts	0	1	2	3	4	5	6	7	8	9	10	11	12	Total
No. of intervals	0	20	43	53	86	70	54	37	18	10	5	2	2	400

The mean and variance estimates obtained from this distribution are

$$\bar{x} = \hat{k} = 4 \cdot 68 \qquad \hat{\sigma}^2 = 4 \cdot 46.$$

These two values agree well and are consistent with the first property of the Poisson distribution.

The fitting process consists of computing a set of expected frequencies obtained by multiplying the terms

$$e^{-k}, \quad ke^{-k}, \quad \frac{k}{2} e^{-k} \ldots$$

by the sample size 400. To do this we use the mean of 4·68 as the value of k. The observed and expected distributions are given in Table 13.2.

TABLE 13.2. COMPARISON OF OBSERVED DISTRIBUTION WITH EXPECTED POISSON DISTRIBUTION

No of counts (2 minutes)	0 and 1	2	3	4	5	6	7	8	9	≥ 10	Total
Observed no. of intervals	20	43	53	86	70	54	37	18	10	9	400
Expected no. of intervals	21·1	40·6	63·4	74·2	69·4	54·2	36·2	21·2	11·0	8·7	400·0

The agreement between the observed the expected frequencies may be tested by computing the value of the statistic ϕ^2 described in Section 5.4.2 as

$$\phi^2 = \sum_{i=1}^{k} \frac{(O_i - E_i)^2}{E_i},$$

where O_i is the ith observed frequency, E_i is the ith expected frequency, and where there are k frequencies in each set.

$$\phi^2 = \frac{(20 - 21 \cdot 1)^2}{21 \cdot 1} + \frac{(43 - 40 \cdot 6)^2}{40 \cdot 6} + \cdots + \frac{(9 - 8 \cdot 7)^2}{8 \cdot 7}$$

$$= 4 \cdot 38.$$

It should be noted that certain groups have been combined so that there are now only ten contributions to ϕ^2. The reason for this combination is that the expected frequencies for the 0, 11, 12 and greater groups were lower than five and therefore too low to be considered as separate groups.

$$\Pr \{\chi^2 \geqq 4 \cdot 38 \mid \nu = 8\} = 0 \cdot 82.$$

The conclusion is that the agreement between the observed and expected frequencies is acceptable and we conclude that it is reasonable to believe that the counts are occurring randomly in time in accordance with a Poisson process.

13.2. A SINGLE COUNT

If we observe a random process for a time t and obtain c counts we may:

1. Estimate the population mean count k in time t and hence the count rate.
2. Compute confidence limits for the population mean count k, and hence for the count rate, if we are prepared to assume that the random process is a Poisson process.

The estimate of the population mean count k in time t is simply the observed count c and the population count rate is estimated as c/t. Confidence limits for the population count k may be computed by the following procedure. If the population mean count in time t is k then the probability of observing a count of c, or greater than c, is

$$P_l = \sum_{n=c}^{\infty} e^{-k} \frac{k^n}{n!}.$$

If this value is less than the acceptable value of $0 \cdot 05$ (say) we conclude that it is not reasonable to believe that the true count is k. If we repeat this testing procedure for a range of values of k we will find a *lower* limit for the value of k. Values of k below this limit would be rejected at the 5 per cent level by the above procedure. This limit will be the lower confidence limit for k. The upper confidence limit may be found by applying the same procedure using the probability of observing c counts or less:

$$P_u = \sum_{n=0}^{c} e^{-k} \frac{k^n}{n!}.$$

If a 5 per cent level is used in both cases we obtain 90 per cent confidence limits for k. Confidence limits obtained by this process are tabulated for confidence probabilities 0·90, 0·95, 0·98, 0·99 and 0·998 for values of c less than or equal to 50 in Table 40 of Pearson and Hartley (1958). Confidence limits for the count rate may be found by dividing each of the above limits by the time interval t.

13.2.1. Example

We have observed a count of 10 in 2 minutes.

1. The population mean count in 2 minutes is estimated as 10 and hence the count rate is estimated as 5 counts per minute.
2. Calculation of the lower 5 per cent confidence limit for the population count gives:

$$k = 4 \qquad P_L = 0\cdot005 \qquad \text{rejected}$$

$$k = 5 \qquad P_L = 0\cdot031 \qquad \text{rejected}$$

$$k = 6 \qquad P_L = 0\cdot082 \qquad \text{accepted}$$

$$k = 5\cdot2 \quad P_L = 0\cdot0390 \quad \text{rejected}$$

$$k = 5\cdot4 \quad P_L = 0\cdot0488 \quad \text{rejected}$$

$$k = 5\cdot5 \quad P_L = 0\cdot0537 \quad \text{accepted.}$$

The lower limit is found, by continuing the process, to be 5·43.

The upper 5 per cent confidence limit, calculated by a similar process using P_u, is found to be 16·96. Thus 90 per cent confidence limits for the population count (in 2 minutes) are 5·43 to 16·96 and hence 90 per cent confidence limits for the count rate are 2·71 to 8·48 counts per minute.

These limits are not symmetrical about the mean count rate and the short-cut method for other sets of confidence limits described in Section 5.2.7 is not possible in this case because the shape of the Poisson distribution changes as k changes. If k is large, however (say greater than 50), the change is sufficiently small to allow the following approximation to confidence limits for count k:

$$c \pm x_\alpha \sqrt{c},$$

where c is the observed count and x_α is the Normal distribution value corresponding to confidence probability α. Confidence limits for the population count rate may be obtained from these limits by dividing both limits by the time interval t.

The approximate method for $c = 50$ gives the following 90 per cent confidence limits:

$$50 \pm 1\cdot6449 \sqrt{50}.$$

That is the approximate limits are 38·37 to 61·63 and the exact limits are 38·96 to 63·29.

13.2.2. Two Warnings

1. If the results of counting are stated as follows: "A count of 10 was observed in 2 minutes," it is possible to estimate the count rate and to calculate confidence limits. If only the count rate is recorded it is not possible to compute confidence limits.

2. The above confidence limits were computed on the *total count* and not on the count rate. Ninety per cent confidence limits for a count of 5 are 1·97 to 10·51 and form a wider interval than those calculated by considering the total count of 10 in 2 minutes. These were 2·71 to 8·48.

13.3. A Number of Counts of the Same Source

There is more information contained in r counts each for a time t than one count for time rt. The overall estimate of the population count rate is the same in both cases but the first allows us to test the validity of the assumptions made by the Poisson process.

We have r counts c_i of the same source each made over the same size interval t. From this information we may estimate the mean count and the variance of the counts:

$$\hat{k} = \bar{c} = \frac{1}{r} \sum_{i=1}^{r} c_i \quad \text{and} \quad \hat{\sigma}_c^2 = \frac{1}{r-1} \sum_{i=1}^{r} (c_i - \bar{c})^2.$$

If the random process is a Poisson process the mean and variance should agree reasonably. If the variance is significantly greater than the mean, there is more variability than is normally present in a Poisson process. The events are therefore occurring in batches and not at random. If the variance is significantly less than the mean, the counts may be tending to occur at equal intervals.

The test of significance uses the statistic

$$\chi^2_{r-1} = \frac{(r-1)\,\hat{\sigma}_r^2}{\bar{c}} = \frac{1}{\bar{c}} \sum_{i=1}^{r} (c_i - \bar{c})^2. \tag{13.8}$$

13.3.1. Example

The following five counts were obtained for five separate periods of 5 minutes each;

$$204 \quad 182 \quad 186 \quad 212 \quad 216.$$

From these we estimate:

$$\hat{k} = \bar{c} = 200 \quad \text{and} \quad \hat{\sigma}_c^2 = 234,$$

$$\chi_{r-1}^2 = \frac{4 \times 234}{200} = 4 \cdot 68,$$

$$\Pr\{\chi^2 \geq 4 \cdot 68 \mid \nu = 4\} = 0 \cdot 322.$$

The agreement between the mean and the variance estimate is acceptable, and we have no grounds for believing that the process is not a Poisson process.

13.3.2. Confidence Limits for the Count Rate

The total count was 1000 in 25 minutes so that the count rate is estimated as 40 counts per minute. Approximate 90 per cent confidence limits for the total count are given by:

$$1000 \pm 1 \cdot 6449 \sqrt{1000},$$

that is, 94·798 to 1052·02.

Ninety per cent confidence limits for the count rate are therefore 37·92 to 42·08 (90 per cent confidence limits for 40 counts would have been 30·20 to 52·07).

13.4. Function Fitting Involving Counts

The fitting of functions to data consisting of counts c_i made over intervals of time t_i for each of a number of values x_i is an example of situation 4 in function fitting described in Section 12.6.

The function is fitted between y_i ($= c_i/t_i$) and x_i and the variance of y_i depends on the fitted value. The statistical model is

$$y_i = f(x_i) + z_i, \tag{13.9}$$

where z_i is a random error with zero mean and variance $f(x_i)/t_i$. That is the variable $t_i y_i$ has a Poisson distribution with mean, and hence variance, equal to $t_i f(x_i)$. The variance of y_i may be derived as follows. If c is a variable with a Poisson distribution with mean count k in t minutes then the variance of c is also k. The variance of the count rate c/t is therefore c/t^2. Thus if $y = c/t$ the variance of y is c/t^2 or y/t.

The formulae given in Section 12.6 apply to counts with the weights w_i equal to $t_i/f(x_i)$. The iteration process begins by using the observed value y_i in the weights:

$$w_{1i} = t_i/y_i.$$

The fitting process using these weights estimates the parameters in the function $f(x_i)$ and new weights for the next iteration may be calculated using this fitted function:

$$w_{2i} = t_i/f_1(x_i).$$

This process continues until the estimated parameters agree, to an acceptable accuracy, from one iteration to the next. The value of the minimized weighted sum of squares S of the residuals z_i is given by:

$$S = \sum_{i=1}^{n} w_i z_i^2 = \sum_{i=1}^{n} w_i(y_i - f(x_i))^2.$$

The value of S for the last iteration is distributed as χ^2 with $n - p$ degrees of freedom and this provides a test for the goodness of fit of the fitted function.

13.4.1. Example

We will illustrate this fitting process by fitting a straight line to the data given in Table 13.3.

TABLE 13.3. EXAMPLE DATA

Value	x_i	1	2	3	4	5	6
Count	c_i	147	127	224	170	219	135
Time	t_i	1	1	2	2	3	3
Value	$y_i = c_i/t_i$	147	127	112	85	73	45
First weight	$w_{1i} = t_i/y_i$	1/147	1/127	2/112	2/85	3/73	3/45

The computation for this fit is:

$$\sum_{i=1}^{6} w_i = 0.163826 \qquad \sum_{i=1}^{6} w_i x_i = 0.775719 \qquad \sum_{i=1}^{6} w_i y_i = 12,$$

$$\sum_{i=1}^{6} w_i x_i^2 = 4.002881 \qquad \sum_{i=1}^{6} w_i x_i y_i = 50 \qquad \sum_{i=1}^{6} w_i y_i^2 = 1022,$$

$$\sum_{i=1}^{6} w_i(x_i - \bar{x})^2 = 4.002881 - 0.775719^2/0.163826 = 0.329838,$$

$$\sum_{i=1}^{6} w_i(x_i - \bar{x})(y_i - \bar{y}) = 50 - 0.775719 \times 12/0.163826 = -6.820212,$$

$$\sum_{i=1}^{6} w_i(y_i - \bar{y})^2 = 1022 - 12^2/0.163826 = 143.018640.$$

$$\hat{b} = \frac{-6\cdot820212}{0\cdot329838} = -20\cdot6775,$$

$$\hat{a} = \frac{12 + 20\cdot6775 \times 0\cdot775719}{0\cdot163826} = 171\cdot1568.$$

The first fitted line is $y = 171\cdot1568 - 20\cdot6775\,x$. The data for the next iteration is given in Table 13.4.

TABLE 13.4. EXAMPLE DATA FOR SECOND ITERATION

Value x_i	1	2	3	4	5	6
Value y_i	147	127	112	85	73	45
Time t_i	1	1	2	2	3	3
Function value $f(x_i)$	150·48	129·80	109·12	88·45	67·77	47·09
Weight $w_{2i} = t_i/f(x_i)$	1/150·48	1/129·80	2/109·12	2/88·45	3/67·77	3/47·09

(For example, $w_{21} = 1/f(x_i) = 1/(171\cdot1568 - 20\cdot6775x_i)$.)

Following the same fitting process for this data we find the second fitted line to be:

$$y = 171\cdot2634 - 20\cdot6634x.$$

This process is continued until there is acceptably small change in the slope and constant of the fitted line from one iteration to the next. For purposes of illustration we will accept this fit as the final fit.

The sum of squares due to fit is

$$\hat{b}^2 \sum_{i=1}^{n} w_i(x_i - \bar{x})^2 = (-20\cdot6634)^2 \times 0\cdot322759 = 137\cdot8098$$

and the complete analysis of variance table is given in Table 13.5.

TABLE 13.5. ANALYSIS OF VARIANCE FOR POISSON REGRESSION EXAMPLE

Source	S.S.	D.F.	M.S.	F.	Prob.
Due to function	137·8098	1	137·8098	268·48	≪0·005
About function	2·0530	4	0·5133		
Total	139·8628	5			

The "about function" *sum of squares* is used as the goodness of fit criterion

$$\text{Pr}\,\{\chi^2 \geq 2\cdot0530 \mid \nu = 4\} = 0\cdot73$$

and we see that the fit is acceptable. The variance of the slope is given by

$$\text{var}(\hat{b}) = \frac{1}{\sum\limits_{i=1}^{n} w_i(x_i - \bar{x})^2} = 3\cdot0318$$

(since $\sigma^2 = 1$ is part of the null hypothesis).

13.4.2. Interpretation of a Significantly Bad Fit

The following interpretations of a significant result in the goodness of fit test are:

(a) that the function does not fit,
(b) that the variance assumptions are incorrect,
(c) that both (a) and (b) apply.

It is a common practice to accept the second possibility, when it is "genuinely" believed that the function does fit, and to scale the variances of the slope, and of the y-values, by multiplying by the about line mean square. This is equivalent to saying that the variance of z_i in the original model

$$y_i = f(x_i) + z_i$$

is equal to $kf(x_i)/t_i$ where k is unknown.

The about line mean square is an estimate of the constant k.

It must be appreciated that this assumption is being made and the fit should be accepted only if this assumption can be justified.

QUESTIONS

1. An active source is counted for successive periods of 1 minute. The observed distribution obtained from 250 such periods is given below:

No. of counts	0	1	2	3	4	5	6	7	8	9 and over	Total
No. of intervals	6	20	35	40	53	39	30	17	6	4	250

(a) Estimate the mean number of counts in 1 minute and the variance of the number of counts. Test the significance of the difference between these.

(b) Fit a Poisson distribution with mean equal to the estimated mean and assess the adequacy of the fit.

2. Two hundred counts are observed in a period of 10 minutes.

(a) Estimate the population mean count rate per minute.

(b) Compute 95 per cent confidence limits for the count rate.

3. Repeat question 2 for a count of 20 in 1 minute and compare the results with those of question 2.

4. Perform one more iteration in the example in Section 13.4.1 starting from the straight line $y = 171 \cdot 2634 - 20 \cdot 6634x$ and repeat the goodness-of-fit test.

CHAPTER 14

DISTRIBUTION-FREE TESTS

THE majority of tests of significance described in previous chapters assume a known functional form for the population distribution. Whilst it is possible to develop similar tests assuming any particular form of population distribution, the derivation of the distribution of the chosen test statistic is frequently complex. In addition, situations arise in which the form of the population distribution is not known. For these reasons, much attention has recently been given to the development of tests which are free from this assumption of a particular form for the population distribution. This chapter briefly describes the method of randomization, and, in more detail, the steadily increasing range of rank tests which are currently being developed. The description is mainly confined to consideration of two sample tests and tests for correlation although other situations are briefly considered. The aim of this chapter is to indicate the latest developments in this area rather than to present a detailed exposition.

14.1. THE RANDOMIZATION METHOD

The randomization method replaces the assumption of a particular form for the population distribution by the assumption that each possible arrangement of the observed data within the experimental design is equally likely. If, for example, we have two samples of size n and m and wish to test the hypothesis that the population means are equal the randomization method assumes that each of the $^{n+m}C_n$ possible arrangements of $n + m$ values into the two samples is equally likely. The distribution of the chosen statistic (say t as defined in Student's test for this hypothesis) is made up of the $^{n+m}C_n$ values of t that may be obtained by arrangement of the $n + m$ observed values into two samples. One property of this method is that the distribution of the statistic is dependent on the observed values and cannot therefore be computed before the experiment is performed. In fact, the distribution of the statistic differs from one set of observations to another. Whereas, with other methods, it is possible to produce tables of the chosen statistic, this is clearly impossible with the randomization method. The amount of work necessary in the performance of a randomization test is a serious drawback to the method but the availability of computers to compute the required distribution makes the method more feasible.

14.1.1. Example

We will illustrate the randomization method testing the hypothesis that two populations have the same variance. The data consists of two samples of six observations given in Table 14.1.

TABLE 14.1. EXAMPLE TWO-SAMPLE DATA

Sample 1	26·4	27·9	31·2	24·9	28·1	30·1
Sample 2	27·1	20·1	23·4	26·6	22·1	25·9

The statistic to be used is the ratio F of the two sample estimates of variance $\hat{\sigma}_1^2$ and $\hat{\sigma}_2^2$ or, since the samples are of equal size, F is also the ratio of the sample sums of squares about their means. That is, F is defined as

$$F = \frac{\sum_{i=1}^{6} (x_i - \bar{x})^2}{\sum_{j=1}^{6} (y_j - \bar{y})^2} = \frac{\hat{\sigma}_1^2}{\hat{\sigma}_2^2}. \tag{14.1}$$

The value of F for the example data given in Table 14.1 is 0·688. There are $^{12}C_6 = 924$ different arrangements of twelve observations into two samples of size six and there are 602 arrangements which produce values of F greater than that observed. That is, the observed arrangement of the data is significant at the $602/924 = 0·65$ level. The actual distribution made up of the 924 values of F is shown in Fig. 14.1.

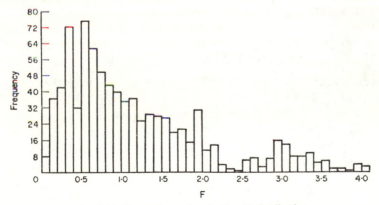

FIG. 14.1. Example randomization F-distribution.

14.2. The Ranking Method

The ranking method is very similar to the randomization method but avoids the dependence of the distribution of the chosen statistic on the observed data. The ranking method replaces the actual observations by their ranks in a single ordering of all of the data values and the statistic is computed on the ranks. The data in Table 14.1 would become the two samples of ranks given in Table 14.2.

TABLE 14.2. RANKS CORRESPONDING TO EXAMPLE TWO-SAMPLE DATA

| Sample 1 | 6 | 9 | 12 | 4 | 10 | 11 |
| Sample 2 | 8 | 1 | 3 | 7 | 2 | 5 |

The lowest observation 20·1 is replaced by rank 1, the second lowest observation 22·1 is replaced by the rank 2, and so on until finally the highest observation 31·2 is replaced by the last rank 12. The statistic is now computed on these ranks and its distribution is made up of the $^{12}C_6$ possible arrangements of the ranks 1 to 12 into two samples of six. The use of ranks in place of the observations means that the distribution is no longer dependent on the observations so that we may compute the distribution once and for all and produce tables of its distribution, or of its important significance levels

The randomization method uses the actually observed values and these, of course, vary from application to application. The ranking method, however, computes the chosen statistic from the ranks and these remain constant from application to application. The ranks are always the first twelve integers and we are concerned with the observed arrangement of the ranks into the two samples. It is clear that if we know the ranks in one sample those in the other sample may be determined. For this reason a number of rank tests use the ranks from one sample only. For example, one form of the Wilcoxon test to be described in Section 14.3.1 uses the sum of the ranks of one sample as the statistic. Examples of rank tests testing hypotheses concerning two populations will now be given in three sections:

1. Testing the hypothesis that the two populations have the same mean (Location Tests).
2. Testing the hypothesis that the two populations have the same variance (Dispersion Tests).
3. Testing the hypothesis that the two populations have the same distribution (Distribution Tests).

Distribution tests are, of course, sensitive to both location differences and dispersion differences.

14.2.1. *The Problem of Ties*

Unlike tests using the observed data values, rank tests assume that the data values can be placed in a unique order, that is, all observations are different. Equal observations are, however, not uncommon and if rank tests are to be useful they must be able to cope with equal observations (or ties). In many tests, it is found that if the equal observations are in the same sample they have no effect on the value of the statistic. This is usually only affected by ties involving observations from different samples. Two possible methods of dealing with ties have been suggested. The first assigns an order to the tied values at random. The second assigns the average of the ranks involved in the tie to all observations in the tie. That is, if four observations are all equal and if unequal would be assigned ranks 4, 5, 6 and 7 then all four observations are assigned the value $5\frac{1}{2}$. The first method introduces a random element that affects the significance of the statistic. It is possible for the statistic to be significant with one possible order and not with another. Significance in this case is determined, therefore, by the random selection. The second method has the appeal of fairness but the distribution of the statistic differs from the distribution with no ties. Usually, the effect of ties is not very great provided not more than half the observations are tied. Results on the effects of ties are included wherever these exist.

14.3. Two-sample Location Rank Tests

Student's *t*-test as described in Section 6.1.3 tests the null hypothesis that two populations have the same mean and it assumes that the population distributions are normal with common but unknown variance. Tests described in this section test this same hypothesis without assuming normality or a common population variance. They may be used whether the population distribution is normal or not. When the distributions are known to be normal these tests become competitors of the *t*-test. Power comparisons with the *t*-test are given in Section 14.6. In the descriptions that follow the two samples will be denoted by:

$$\{x_1, x_2, ..., x_i, ... x_n\} \quad \text{and} \quad \{y_1, y_2, ..., y_j, ..., y_m\}.$$

14.3.1. *Wilcoxon's Test*

The location test introduced by Wilcoxon (1945, 1947) uses the statistic U which is the sum of the ranks in one sample:

$$U_1 = \sum_{i=1}^{n} r_i, \qquad (14.2)$$

where n is the number of observations in the chosen sample, and where r_i is the rank of the ith observation in the chosen sample. This statistic is simple to compute and may be expected to be a faithful measure of location differences. Wilcoxon considered the case of two equal samples of size n and gave tables of the lower 1 per cent, 2 per cent and 5 per cent significance levels. The distributions of U_1 has been shown by Mann and Whitney (1947) to tend to normality as the sample size increases. The mean and variance of U_1 for equal sample sizes were given by Wilcoxon as

$$\mathscr{E}(U_1) = \frac{n}{2}(2n + 1),$$

$$\sigma^2(U_1) = \frac{n^2}{12}(2n + 1). \tag{14.3}$$

Mann and Whitney also suggested a slightly different, but equivalent, form of the test. Each member of the x-sample is compared with each member with the y-sample making nm comparisons in all, and Mann and Whitney's statistic U is the number of comparisons in which the x-value was more than the y-value. That is:

$$\text{if } x_i < y_j \text{ score } d_{ij} = 0,$$

$$\text{if } x_i > y_j \text{ score } d_{ij} = 1, \tag{14.4}$$

$$\text{then } U = \sum_{i=1}^{n} \sum_{j=1}^{m} d_{ij}.$$

Another way of computing U is to take each member of the y-sample and count the number of values in the x-sample greater than it. U is then the sum of these counts. The Mann–Whitney form of the Wilcoxon test is very easy to compute since it does not even involve ranking the observations. The distribution of U is the same as the distribution of Wilcoxon's U_1 except for a change in mean value. The two statistics, and their means, are related:

$$U_1 = mn + \frac{m}{2}(m + 1) - U. \tag{14.5}$$

Thus

$$\mathscr{E}(U_1) = mn + \frac{m}{2}(m + 1) - \mathscr{E}(U).$$

The moments of the Mann–Whitney statistic are:

$$\mathscr{E}(U) = \frac{nm}{2},$$

$$\sigma^2(U) = \frac{nm}{12}(n + m + 1). \tag{14.6}$$

Mann and Whitney tabulated the distribution for sample sizes n, $m \leqq 8$ and suggested that for sample sizes outside this range the distribution was sufficiently close to normal with moments given by expressions (14.6). Allowance should be made for the usual continuity correction (Section 7.2.3) when a continuous distribution is used to approximate to a discrete distribution. Table 9 in the Table Section includes significance levels of the distribution of U in the Mann–Whitney form.

14.3.2. Example

Applying Wilcoxon's test to the two samples of ranks in Table 14.2 we have

$$U_1 = 6 + 9 + 12 + 4 + 10 + 11 = 52.$$

Applying the test in its Mann–Whitney form to the data in Table 14.1 we have

$$U = 4 + 6 + 6 + 4 + 6 + 5 = 31.$$

Mann and Whitney's tables show that

$$\Pr \{U \geq 31 \mid n = m = 6\} = 0 \cdot 021.$$

Using a 5 per cent significance level, we would conclude that the population means are significantly different.

14.3.3. The Effect of Ties on the Wilcoxon Test

The rule concerning ties that will be followed is to assign the average of the ranks involved in the tie to each of the equal observations. The equivalent rule in the Mann–Whitney form of the test is to score half if x_i and y_j are equal. The effect of ties is to change the second and subsequent moments of the statistic. The mean is unaffected but the variance becomes:

$$\sigma^2(U_1) = \sigma^2(U) = \frac{nm}{12}(n + m + 1)$$

$$+ \frac{nm}{12(n + m)(n + m - 1)} \sum_{i=1}^{T} t_i(t_i^2 - 1), \quad (14.7)$$

where t_i is the number of observations in the ith tie and where T is the number of ties.

The effect on subsequent moments depends on which ranks are involved in the tie. The testing procedure in the case of ties differs from that without ties only to the extent that the variance is reduced by the extra term given in expression (14.7). The effect of ties on the normality assumption is not serious unless more than half the observations are involved in ties.

14.3.4. Example

TABLE 14.3. EXAMPLE TWO-SAMPLE DATA WITH TIES

| Sample 1 | 12·4 | 13·8 | 16·1 | 14·1 | 17·1 | 12·4 | 15·9 | 12·9 |
| Sample 2 | 10·9 | 15·2 | 12·4 | 13·2 | 15·5 | 14·1 | 11·3 | 14·7 |

The example two-sample data given in Table 14.3 contains two ties, one involving three values (12.4) and the other involving two values (14.1). Using the Mann–Whitney form of the test we have

$$U = 8 + 3 + 7 + 5 + 3 + 3\tfrac{1}{2} + 8 + 3 = 40\tfrac{1}{2}.$$

The Mann and Whitney tables give

$$\Pr\{U \geq 40 \mid n = m = 8\} = 0{\cdot}221,$$

$$\Pr\{U \geq 41 \mid n = m = 8\} = 0{\cdot}191.$$

The conclusion is, therefore, that the population means do not differ significantly. Using this data as an example of the normal approximation we compute first the mean which is unaffected by the ties:

$$\mathscr{E}(U) = \frac{8 \times 8}{2} = 32$$

followed by the variance. There are two ties $(T = 2)$ of size $t_1 = 3$ and $t_2 = 2$ so that the variance is:

$$\sigma^2(U) = \frac{8 \times 8 \times 17}{12} - \frac{8 \times 8}{12 \times 16 \times 15}[3 \times 8 + 2 \times 3]$$

$$= 90\tfrac{2}{3} - \tfrac{2}{3} = 90.$$

The normal approximation is applied by standardizing U:

$$U_s = \frac{40\tfrac{1}{2} - 32 - \tfrac{1}{2}}{\sqrt{90}} = \frac{8}{9{\cdot}49} = 0{\cdot}85$$

$$\Pr\{U_s \geq 0{\cdot}85 \mid U_s \sim N(0, 1)\} = 0{\cdot}20.$$

The conclusion is, therefore, as before that the population means do not differ significantly.

14.3.5. Wilcoxon's Test with Correlated Samples

The Wilcoxon test may also be applied in the situation discussed in Section 6.1.7 where the samples are equal in size and the observations in the two samples are correlated in pairs. The correlated form of Student's t-test was described in Section 6.1.7. The Wilcoxon test applied to this situation involves computing the differences

$$d_i = x_i - y_i$$

between corresponding observations and ranking the differences ignoring sign. The statistic is U_2 the sum of the ranks corresponding to positive differences and the sample sizes n and m are taken to be the number of positive differences and the number of negative differences respectively. The Mann–Whitney form for this situation would require the differences to be divided into two samples according to the sign and the test performed in the usual way taking all differences as positive. The sample sizes in this form of the test are in fact random variables although this fact is not considered. The sample sizes are treated as though they were fixed before the experiment was performed.

14.3.6. Example

TABLE 14.4. EXAMPLE TWO-SAMPLE CORRELATED DATA

Sample 1 (x_i)	12·4	13·8	16·1	14·1	17·1	12·4	15·9	12·9	13·1	15·3
Sample 2 (y_i)	10·9	15·2	12·4	13·2	15·5	14·1	11·3	14·7	15·2	14·8
Differences (d_i)	1·5	−1·4	3·7	0·9	1·6	−1·7	4·6	−1·8	−2·1	0·5
Ranks	4	3	9	2	5	6	10	7	8	1

Table 14.4 shows example two-sample correlated data, the differences, and the ranks obtained by ranking the differences without regard to sign. The number of positive differences is $n = 6$, and the number of negative differences is $m = 4$. The sum of the ranks corresponding to positive differences is

$$U_2 = 4 + 9 + 2 + 5 + 10 + 1 = 31.$$

The conclusion is that the two population means are not significantly different.

14.3.7. Mood's Median Test

This test has been described already in Section 7.2 under the heading 2×2 tables. The procedure consists of finding the joint median of the $n + m$ observations and counting the numbers of observations in each sample less than, and greater than, the joint median. These four counts are the four entries in the 2 × 2 table, an example of which is shown in Table 14.5.

TABLE 14.5. 2 × 2 TABLE

	Less than median	More than median	Totals
Sample 1	a	c	n
Sample 2	b	d	m
Totals	r	s	N

The analysis of Table 14.5 is given in Section 7.2. Although this test is introduced in Chapter 7 as a special case of a test for distributional differences it may be shown that it is only sensitive to location differences.

The calculation of the median value differs according to whether the total number of observations is odd or even. If odd, the median is the middle value; if even, it is the average of the two middle values. If the total number of observations is odd we have difficulty in determining how to score the median value itself. The usual procedure is, in fact, to ignore the median value and not score it at all.

14.3.8. Example

As an example of the Mood's median test, we will apply this test to the data recorded in Table 14.1. The two middle values are 26·4 and 26·6 giving a median of 26·5. The 2 × 2 table is made up of counts of the number of observations less than and greater than 26·5 in each sample. This table is shown in Table 14.6.

TABLE 14.6. EXAMPLE 2 × 2 TABLE FORMED FROM DATA IN TABLE 14.1

	Less than median	More than median	Totals
Sample 1	2	4	6
Sample 2	4	2	6
Totals	6	6	12

Table 38 in Pearson and Hartley as described in Section 7.2 shows that the value of $b = 4$ is not significant at the 5 per cent significance level and we conclude that the populations have the same means.

14.4. TWO-SAMPLE DISPERSION RANK TESTS

The F ratio test as described in Section 6.2.1 tests the null hypothesis that two populations have the same variance and it assumes that the population distributions are normal. Tests described in this section test this same hypothesis without assuming normality so that they are available in situations where the distributions are not normal or when the form of distribution is not known. Investigations of these tests are being actively pursued at the present time. So far some statistics are without tables of their distribution or of their important significance levels and there has been little work done on the power of these tests.

14.4.1. Mood's Test and David's Test for Dispersion Differences

Mood (1954) suggested, as a measure of dispersion differences, the statistic

$$W = \sum_{i=1}^{n} \left(r_i - \frac{N+1}{2} \right)^2, \qquad (14.8)$$

where r_i is the rank of the ith observation in the x-sample, and where $N(= n + m)$ is the total number of observations. If the ranks in the x-sample are spread out, the value of W will be large, but if the ranks are close to the centre rank $(N + 1)/2$, the value of W will be small. Thus, the critical region may consist of both low and high values of W. It is assumed, however, that the population means are the same. If the means are different there is a tendency for one sample to contain the low ranks and the other sample the high ranks. A high value of W will be obtained because of the location difference rather than a dispersion difference. To reduce, but unfortunately not eliminate, this sensitivity to location differences David (1956) suggested that the variance of the ranks of one sample be used instead. That is

$$v = \frac{1}{n-1} \sum_{i=1}^{n} (r_i - r)^2, \qquad (14.9)$$

where r_i is the rank of the ith observation in the x-sample, and where $\bar{r}$ is the average of the ranks in the x-sample. This statistic takes the sum of squares of ranks of one sample about the average rank in that sample and is, therefore, less sensitive to location differences. However, like all dispersion tests described in this section, the statistic is not completely independent of location differences. This dependence is discussed in Section 14.4.7.

So far neither the distributions nor the important significance levels of these statistics have been tabulated. The distributions have been shown to approach normality as the sample sizes increase and the means and variances of W and v with ties not present are:

$$\mathscr{E}(W) = \frac{n}{12}(N^2 - 1),$$

$$\sigma^2(W) = \frac{nm}{180}(N + 1)(N^2 - 4), \qquad (14.10)$$

$$\mathscr{E}(v) = \frac{N}{12}(N+1),$$

$$\sigma^2(v) = \frac{mN(N+1)}{360n(n-1)}[3(N+1)(n+1) - Nn].$$

(14.11)

The method of obtaining moments of certain rank statistics suggested by David (1956) may also be used to obtain the moments when ties are present and the mean and variances of W and v with ties present are given by Cooper (1957). The expressions are somewhat lengthy, and it is usually found that the corrections necessary because of the presence of ties, are small.

14.4.2. Example

Applying Mood's test to the ranks in Table 14.2 we have

$$W = (6 - 6\tfrac{1}{2})^2 + (9 - 6\tfrac{1}{2})^2 + \cdots + (11 - 6\tfrac{1}{2})^2 = 75\tfrac{1}{2},$$

$$\mathscr{E}(W) = \frac{6}{12} \times 143 = 71\tfrac{1}{2},$$

$$\sigma^2(W) = \frac{6 \times 6 \times 13 \times 140}{180} = 364,$$

$$X = \frac{W - \mathscr{E}(W)}{\sigma(W)} = \frac{75\tfrac{1}{2} - 71\tfrac{1}{2}}{\sqrt{364}} = 0 \cdot 21.$$

Using the normal approximation to the distribution of W provides

$$\Pr\{X \geqq 0 \cdot 21\} = 0 \cdot 42$$

and the conclusion is that the population variances are not significantly different.

Applying David's test to the same ranks we have:

$$\bar{r} = \frac{52}{6},$$

$$v = \frac{1}{5}\left[\left(6 - \frac{52}{6}\right)^2 + \left(9 - \frac{52}{6}\right)^2 + \cdots + \left(11 - \frac{52}{6}\right)^2\right] = 9 \cdot 4667,$$

$$\mathscr{E}(v) = \frac{12 \times 13}{12} = 13,$$

$$\sigma^2(v) = \frac{6 \times 156 \times 201}{360 \times 30} = 17 \cdot 42,$$

$$X = \frac{v - \mathscr{E}(v)}{\sigma(v)} = \frac{9 \cdot 4667 - 13}{\sqrt{17 \cdot 42}} = -0 \cdot 20.$$

Using the normal approximation to the distribution of v provides

$$\Pr\{X \leqq -0{\cdot}20\} = 0{\cdot}42$$

from which we again conclude that the population variances are not significantly different. The change of sign X from Mood's to David's test is worth noting. There is a suggestion that the mean of the first population is higher than that of the second.

14.4.3. Kamat's Dispersion Test

The test suggested by Kamat (1956) is based on the ranges of the ranks in the two samples. If the samples are of different sizes the x-sample is chosen to be the smaller so that we may assume $m \geqq n$. If R_n and R_m are the ranges of the ranks (the difference between the highest and lowest ranks) in the x- and y-samples, respectively, Kamat defines the statistic $D_{n,m}$ as

$$D_{n,m} = R_n - R_m + m, \tag{14.12}$$

$D_{n,m}$ can take values 0, 1, 2, ..., $n + m$ and the critical region will consist of low and high values of $D_{n,m}$. Kamat gives tables of the upper and lower 5, 2·5, 1 and 0·5 per cent points of the distribution of $D_{n,m}$ for $n, m \leqq 10$. The limiting distribution (as n and m increases) is irregular in shape and is far from normal. However, the shape of the distribution for finite n and m is also irregular and similar in shape to the limiting distribution. Barton in an addendum to Kamat's paper describes an approximate procedure based on this similarity for obtaining significance probabilities.

14.4.4. Example

Applying Kamat's test to the ranks in Table 14.2 we have

$$R_n = 12 - 6 = 6, \qquad R_m = 8 - 1 = 7,$$

$$D_{n,m} = 6 - 7 + 6 = 5.$$

The lower 5 per cent significance level for $n = m = 6$ is 1 so that the observed value of 5 is not significant at the 5 per cent level.

14.4.5. Wilks' Dispersion Test

The statistic r chosen by Wilks (1942) is the number of observations in the y-sample outside the range of the x-sample. Large values of r form the critical region. Rosenbaum (1953) tabulated the upper 5 per cent and 1 per cent significance points of r and obtained the limiting distribution. If n and m are large and not very different then the probability that there are r_0 y-values outside the range of the x-sample is given by

$$\Pr\{r - r_0\} = \frac{r_0 + 1}{2^{r_0 + 2}}. \tag{14.13}$$

It is interesting to note that this probability is independent of the sample size so that if the sample sizes are large ($n, m \geq 30$) and nearly equal the upper 5 per cent significance level is 7 or more and the upper 1 per cent significance level is 10 or more. Rosenbaum (1954) pointed out that considering only one side of the samples, for example large values, provided a test for location differences.

14.4.6. Example

Applying Wilks' test to the data in Table 14.1 (or the ranks in Table 14.2) we see that the extreme values of the y-sample are 20·1 and 27·1 (extreme ranks are 1 and 8). The number of x-values outside this range is 4. The upper 5 per cent significance point given in Rosenbaum's tables for $n = m = 6$ is 5 so that the observed value of 4 is not significant. It should be noted that this value of 4 is made up of no x-values less than the lowest y-value and 4 x-values above the highest y-value. This would seem to suggest a location difference rather than a dispersion difference. This sensitivity of dispersion tests to location differences is discussed further in Section 14.4.7.

14.4.7. A Problem with Dispersion Tests

The rank tests for dispersion described in the preceding sections are sensitive to location differences as well as dispersion differences. This has been discussed briefly in some sections but the sensitivity to location differences may best be seen by considering the effect of a very large location difference. If the mean of the x-population is considerably less than the mean of the y-population, the ranks for the x-sample will be $1, 2, \dots, n$ and the ranks corresponding to the y-sample will be $n + 1, n + 2, \dots, n + m$. This arrangement of the ranks will occur when the two populations hardly overlap and no statistic based on these ranks alone will measure the difference between the population variances. Dispersion tests as described above depend on the overlap between the samples to provide the measure of the difference between the variances. It has been suggested that the medians of the two samples be used to ensure that the two samples are centred at the same point. If the median of the x-sample is subtracted from each x-value and the median of the y-sample is subtracted from each y-value, the tests described above may be applied to these differences without serious effect on the test. From the data given in Table 14.1 we compute the differences recorded in Table 14.7 by subtracting the medians 28·0 and 24·65 from observations in the x- and y-samples respectively.

TABLE 14.7. DATA IN TABLE 14.1 CORRECTED FOR MEDIANS

| Sample 1 | −1·6 | −0·1 | 3·2 | −3·1 | 0·1 | 2·1 |
| Sample 2 | 2·45 | −4·55 | −1·25 | 1·95 | −2·55 | 1·25 |

The testing procedure would now consist of ranking these differences and applying the test to the ranks so obtained.

14.5. TWO-SAMPLE DISTRIBUTION RANK TESTS

The tests to be described in this section test the hypothesis that the two populations have the same distributions. They are sensitive to any difference between the two population distributions including location and dispersion differences as well as general distributional differences.

14.5.1. The Kolmogorov–Smirnov Distribution Test

The Kolmogorov–Smirnov test is based on a comparison between the two-sample cumulative distributions. The two-sample cumulative distributions for the data given in Table 14.1 are plotted in Fig. 14.2. The upward steps in a sample cumulative distribution are of length $1/n$ where n is the size of the sample. In the example data plotted these upward steps are $1/6$ in both the x- and the y-sample. One step upward is taken at each point corresponding to an observation so that the final value reached by the cumulative distribution is unity. The first step in the x-distribution occurs

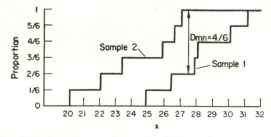

FIG. 14.2. Sample cumulative distributions for data in Table 14.1.

at 24·9, the second at 26·4 and so on. The Kolmogorov–Smirnov statistic D_{mn} is the maximum difference between these two distributions. In the example plotted in Fig. 14.1, the maximum difference of 4/6 occurs between values 27·1 and 27·9. Tables of the distribution of D_{mn} are given by Massey (1951). An approximation to the upper 5 per cent significance level of D_{mn} is:

$$1·36 \left(\frac{n + m}{nm} \right)^{\frac{1}{2}}. \tag{14.14}$$

This approximation is reasonable for $n, m \geqq 12$. If the two sample sizes are unequal, it should be realized that the upward steps taken by the two cumulative distributions will be unequal. This is illustrated in the example which follows:

14.5.2. Example

TABLE 14.8. EXAMPLE DATA WITH UNEQUAL SAMPLE SIZES

Sample 1	69·2	70·4	70·8	72·2	72·4	72·8	74·2	74·3
	74·7	74·9	76·8	77·4	78·1			
Sample 2	72·2	73·8	74·0	74·1	74·4	75·2	75·8	76·3
	76·8	77·1	77·7	78·2	18·4	79·3	82·8	

The example data recorded in Table 14.9 has been ordered for ease of plotting. The cumulative frequency distributions are plotted in Fig. 14.3 and it will be seen that the maximum difference between the two distributions occurs near the value 75 and that its value is:

$$D_{mn} = 10/13 = 5/15 = 0·436.$$

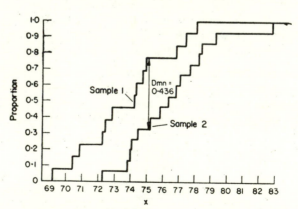

FIG. 14.3. Sample cumulative distributions for unequal sample size data.

The upper 5 per cent significance point for $n = 13$ and $m = 15$ is given approximately as

$$1 \cdot 36 \left(\frac{13 + 15}{13 \times 15} \right)^{\frac{1}{2}} = 0 \cdot 515$$

so that the observed value of $0 \cdot 436$ is not significant at the 5 per cent significance level.

14.5.3. *The Wald and Wolfowitz Distribution Test*

Wald and Wolfowitz originally suggested this test as a test for randomness in a sequence of zero–one observations but it may be used as a test of distributional differences also. Its original use will become obvious in the description of the two-sample application which follows. The observations in both samples are jointly ranked and each observation in the sequence is replaced by the letter x or the letter y depending on the sample of origin. The data is thereby replaced by a sequence of the letters x and y such as *xxyxxyyyyxyxx*. Wald and Wolfowitz statistic R is the number of runs in this sequence where a run is defined as an unbroken sequence of x's or of y's. There are seven runs in the sequence given two sentences above. The distribution of R is made up of probabilities:

$$\Pr\{R = 2k\} = 2 \, {}^{m-1}C_{k-1} {}^{n-1}C_{k-1} / {}^{n+m}C_n \qquad \text{(i.e. } R \text{ even)}, \tag{14.15}$$

$$\Pr\{R = 2K + 1\} = ({}^{m-1}C_k {}^{n-1}C_{k-1} + {}^{m-1}C_{k-1} {}^{n-1}C_k)/{}^{n+m}C_n \quad \text{(i.e. } R \text{ odd)}. \tag{14.16}$$

The calculation of the exact distribution from these probabilities is made easier by the following relation between consecutive probabilities

$$\Pr\{R = 2k + 1\} = \Pr\{R = 2k\} \left(\frac{n + m - 2k}{2k} \right). \tag{14.17}$$

The distribution of R tends rapidly to normality, allowing the usual discrete distribution continuity correction of a half, as the sample sizes n,m increase. The approximation is good for n,m as small as five. The mean and variance of R necessary in the application of the normal approximation are:

$$\mathscr{E}(R) = 1 + \frac{2nm}{n + m},$$

$$\sigma^2(R) = \frac{2nm(2nm - n - m)}{(n + m)^2 (n + m - 1)}. \tag{14.18}$$

14.5.4. Example

Replacing the observations in Table 14.1 (or equivalently the ranks in Table 14.2) by a sequence of the letters x and y we have:

$$xxxyxyxxyyyy.$$

There are six $(R = 6)$ runs in this sequence. The expected number is seven $(\mathscr{E}(R) = 7)$ so that we do not expect $R = 6$ to be significant. We proceed, however, with the normal approximation by way of example. We have

$$R = 6 \qquad \mathscr{E}(R) = 7$$
$$\sigma^2(R) = 30/11$$
$$X = \frac{6 - 7 + \frac{1}{2}}{\sqrt{(30/11)}} = -0.30$$
$$\Pr\{X \leq -0.30\} = 0.382.$$

The exact probability is 0.392 which we see to be in good agreement with the approximate value. In either case, the observed value of R is not significant.

14.6. POWER COMPARISONS FOR LOCATION TESTS

The work of a number of authors on the power of rank tests for location differences may be briefly summarized as follows. When the two samples have, in fact, been drawn from normal populations, the power of the Wilcoxon test for location differences is only marginally less than that for Student's t-test. Since Wilcoxon's test in its original form or in its Mann–Whitney form is simple computationally it is a serious competitor to Student's test. Wilcoxon's test, of course, has the advantage over Student's that it may be used whether or not the population distributions are normal. Comparisons of the power of Wilcoxon's test with Mood's test, the Kolmogorov–Smirnov test, and the Wald and Wolfowitz test, with normal samples and for location differences, have indicated that Wilcoxon's test is better than Mood's which in turn is better than the other two which scarcely differ in power.

14.7. LOCATION RANK TESTS FOR MANY SAMPLES

If the several populations represented by the samples have normal distributions, the simple hierarchical analysis of variance described in Sections 7.5 and 8.1 is available to test the null hypothesis that the populations have equal means. It is assumed by this test, in addition to assuming normality, that the populations have equal variance. This section now describes two tests which use the ranking method to test the same hypothesis. The tests do not explicitly assume that the population variances are equal. The distributions of the statistics when the null hypothesis is true are unaffected by variance differences.

14.7.1. Median Test for Many Samples

It will be immediately apparent on re-inspection of Mood's median test described in Section 14.3.7, that this test may be readily applied in the same manner to several samples. The procedure consists simply of computing the median of all the observations and counting the numbers of observations in each sample less than, and the numbers of observations greater than, the median and forming a frequency table of the form of Table 14.9.

TABLE 14.9. FREQUENCY TABLE FOR MANY-SAMPLE MEDIAN TESTS

	Less than median	More than median	Total
Sample 1	n_{11}	n_{12}	$n_{1.}$
Sample 2	n_{21}	n_{22}	$n_{2.}$
. . . . .	. . . . .	. . . . .	. . . . .
Sample k	n_{k1}	n_{k2}	$n_{k.}$
Total	$N/2$	$N/2$	N

The method of analysing this type of table has been given in Section 7.1.3. The statistic measuring the agreement with the null hypothesis is ϕ^2 and the restriction that an expected value must not be less than (say) 5 must be borne in mind when this is applied. Since the expected values are all equal to half the relevant sample size, this restriction may be translated to mean that sample sizes of less than 10 cannot be accommodated.

14.7.2. Example

TABLE 14.10. EXAMPLE DATA FROM FOUR SAMPLES

Sample 1	12·1	14·3	17·9	12·3	10·9	15·3	14·8	16·5	13·9	15·1
	17·2	13·8								
Sample 2	15·0	14·0	18·9	19·2	15·7	16·6	13·0	10·4	13·6	12·0
Sample 3	10·1	14·2	13·5	12·2	9·7	8·3	11·8	10·0	14·7	15·2
	12·9									
Sample 4	9·2	14·4	7·9	8·2	10·8	13·2	12·5	11·7	10·2	8·8
	10·7	14·1	13·3	12·8	10·6					

Applying this test to the data in Table 14.10 we first of all compute the median of the forty-eight observations. This data has been ranked and the ranks appear in Table 14.12 in the next section. The median is half-way between the 24th and 25th observations and is equal to $\frac{1}{2}(13\cdot0 + 13\cdot2)$ = 13·1. The frequency table compiled from this data is given in Table 14.11 and the expected frequencies are shown in brackets.

TABLE 14.11. FREQUENCY TABLE OF OBSERVATIONS LESS THAN AND MORE THAN THE JOINT MEDIAN

	Less than median	More than median	Totals
Sample 1	3 (6·0)	9 (6·0)	12
Sample 2	3 (5·0)	7 (5·0)	10
Sample 3	7 (5·5)	4 (5·5)	11
Sample 4	11 (7·5)	4 (7·5)	15
Totals	24	24	48

The value of ϕ_1^2 computed from Table 14.12 for the "less than median" entries only is:

$$\phi_1^2 = \frac{(3 - 6\cdot0)^2}{6\cdot0} + \frac{(3 - 5\cdot0)^2}{5\cdot0} + \frac{(7 - 5\cdot5)^2}{5\cdot5} + \frac{(11 - 7\cdot5)^2}{7\cdot5}$$

$$\phi_1^2 = 5\cdot07.$$

The value of ϕ^2 for the whole table is twice this value:

$$\phi^2 = 2 \times \phi_1^2 = 10\cdot14,$$

$$\Pr\{\chi^2 \geqq 10\cdot14 \mid \nu = 3\} = 0\cdot018.$$

The conclusion from this test would normally be that there is a significant difference between the population means. Inspection of Table 14.11 suggests that the means of the first two populations are greater than those of the last two populations.

14.7.3. The Kruskal–Wallis Many-sample Rank Test for Location Differences

The Kruskal–Wallis (1952) test may be regarded as an extension of the Wilcoxon test. The procedure consists of replacing all observations with their ranks and computing a statistic H based on the sums of the ranks in each sample. The samples do not have to be of equal size. We define

k = the number of samples,

n_j = the number of observations in the jth sample,

N = the total number of observations ($= \sum_{j=1}^{k} n_j$),

R_j = the sum of the ranks in the jth sample,

and the Kruskal–Wallis statistic is defined as

$$H = \frac{12}{N(N+1)} \sum_{j=1}^{k} \frac{R_j^2}{n_j} - 3(N+1). \qquad (14.19)$$

It may be shown, provided the sample sizes are not too small (say not less than five), that

$$H \doteq \chi_{k-1}^2.$$

The critical region contains high values of χ^2.

14.7.4. Example

TABLE 14.12. RANKS CORRESPONDING TO FOUR-SAMPLE DATA IN TABLE 14.10

Sample	Ranks										Totals	Sample size
Sample 1	18	34	46	20	14	41	37	43	30	38		
	45	29									395	12
Sample 2	40	31	47	48	42	44	24	10	28	17	331	10
Sample 3	8	33	27	19	6	3	16	7	36	39		
	23										217	11
Sample 4	5	35	1	2	13	25	21	15	9	4		
	12	32	26	22	11						233	15
								Totals			1176	48

The ranks corresponding to the example data given in Table 14.10 are given in Table 14.12. The totals of the ranks in each sample are also shown. Applying the Kruskal–Wallis test to this data we have

$$H = \frac{12}{48 \times 49} [395^2/12 + 331^2/10 + 217^2/11 + 233^2/15] - 3 \times 49$$

$$= \frac{31858 \cdot 27}{196} - 147$$

$$= 15 \cdot 54$$

$\Pr \{x^2 \geq 15 \cdot 54 \mid v = 3\} = 0 \cdot 0014$.

The conclusion from this test would be the same as that drawn from the median test, namely that the population means differ significantly.

14.8. Two-way Factorial Analysis by Ranks

The analysis of variance of a two-way factorial experiment without replication is discussed in Section 9.1. This section describes a test using ranks for the two-way factorial experiment thereby avoiding the assumption of normality and the assumption of equal variance.

14.8.1. Friedman's Test

The data consists of one observation for each combination of levels of two factors which we call rows and columns. The Friedman (1937) test inspects one factor at a time and we will illustrate the method by testing for differences between columns. To do this, we first of all rank the observations in each row separately from 1 to c, where c is the number of columns. We then add up the ranks in each column to give one total for each column and compute the statistic F_c, defined below, based on the column totals. We define:

$r = $ the number of rows,
$c = $ the number of columns,
$C_j = $ the sum of the ranks in the jth column

and F_c is defined as

$$F_c = \frac{12}{rc(c + 1)} \sum_{j=1}^{c} (C_j)^2 - 3r(c + 1). \qquad (14.20)$$

Friedman has shown that if neither the number of rows nor the number of columns is too small (not less than five), then

$$F_c \approx \chi^2_{c-1}.$$

To test for the difference between rows, we rank each column separately from 1 to r, compute the sums of ranks for each row R_i and compute

$$F_r = \frac{12}{cr(r + 1)} \sum_{i=1}^{r} (R_i)^2 - 3c(r + 1). \qquad (14.21)$$

Similarly, we have that if neither the number of rows nor the number of columns is too small (not less than five), then

$$F_r \approx \chi^2_{r-1}.$$

14.8.2. Example

Example two-way data is given in Table 14.13. To test for column differences we rank the observations within each row separately from 1 to 7. These

TABLE 14.13. EXAMPLE TWO-WAY FACTORIAL DATA

Rows	Columns						
	1	2	3	4	5	6	7
1	14·1	15·8	12·9	13·4	14·9	16·2	15·6
2	12·1	14·3	15·1	14·1	12·7	13·2	14·7
3	10·1	11·2	10·8	12·1	14·3	14·2	14·0
4	12·8	13·8	13·9	15·1	14·8	15·2	14·9
5	13·4	11·8	15·1	12·7	13·8	12·9	14·3
6	12·8	13·4	12·2	14·3	14·1	15·8	15·1

TABLE 14.14. RANKS IN ROWS FOR EXAMPLE DATA IN TABLE 14.13

Rows	Columns						
	1	2	3	4	5	6	7
1	3	6	1	2	4	7	5
2	1	5	7	4	2	3	6
3	1	3	2	4	7	6	5
4	1	2	3	6	4	7	5
5	4	1	7	2	5	3	6
6	2	3	1	5	4	7	6
Totals	12	20	21	23	26	33	33	168

are given in Table 14.14. Having obtained these totals, we compute F_c according to expression (14.20).

$$F_c = \frac{12}{6 \times 7 \times 8} [12^2 + 20^2 + \cdots + 33^2] - 3 \times 6 \times 8$$

$$= \frac{1}{28} \times 4368 - 144$$

$$= 12$$

$$\Pr \{\chi^2 \geqq 12 \mid \nu = 5] = 0·061.$$

Thus we have a suggestion that the columns are significantly different.

14.9. RANK CORRELATION

Section 12.1 describes the measurement of the degree of association (or correlation) between two variables by the product moment correlation coefficient r defined as

$$r = \frac{\sum\limits_{i=1}^{n} (x_i - \bar{x})(y_i - \bar{y})}{\left[\sum\limits_{i=1}^{n} (x_i - \bar{x})^2 \sum\limits_{i=1}^{n} (y_i - \bar{y})^2\right]^{\frac{1}{2}}}, \tag{14.22}$$

where $\bar{x}$ is the mean of the x-values $\{x_1, x_2, ..., x_n\}$ and $\bar{y}$ is the mean of the y-values $\{y_1, y_2, ..., y_n\}$.

The correlation coefficient r may take values between -1 and $+1$. Values of r near -1 indicate high negative correlation, values near $+1$ indicate high positive correlation and values near zero indicate little or no correlation. A testing procedure based on r for testing the null hypothesis that the population correlation coefficient is zero was introduced in Section 12.1.3. The rank tests now described in this section have been derived for the same purpose and to avoid the assumption of bivariate normality that was necessary in Section 12.1.3. The rank correlation methods require the x-values to be ranked from 1 to n and the y-values to be separately ranked from 1 to n also. The measures of rank correlation are based on the agreement, or disagreement, in the two series of ranks. Both tests given below have been described by Kendall (1948).

14.9.1. Spearman's Rank Correlation Coefficient

The product moment correlation coefficient defined in expression (14.22) may be rearranged to give:

$$r = \frac{\sum\limits_{i=1}^{n} x_i y_i - \frac{1}{n} \sum\limits_{i=1}^{n} x_i \sum\limits_{i=1}^{n} y_i}{\left[\left(\sum\limits_{i=1}^{n} x_i^2 - \frac{1}{n}\left(\sum\limits_{i=1}^{n} x_i\right)^2\right)\left(\sum\limits_{i=1}^{n} y_i^2 - \frac{1}{n}\left(\sum\limits_{i=1}^{n} y_i\right)^2\right)\right]^{\frac{1}{2}}}. \tag{14.23}$$

Spearman's rank correlation coefficient r_s is in fact the product moment correlation coefficient calculated on the ranks instead of the observations. Spearman's r_s may be calculated using expression (14.23) but since the ranks involved are the numbers 1 to n, many of the sums in expression (14.23) take values depending on n only. We may readily show that

$$\sum_{i=1}^{n} r_i = \frac{n}{2}(n + 1)$$

$$\sum_{i=1}^{n} r_i^2 = \frac{n}{6}(n + 1)(2n + 1),$$

so that

$$\sum_{i=1}^{n} (r_i - \bar{r})^2 = \frac{n}{12}(n^2 - 1).$$

Substituting for these sums in expression (14.23) and with a little rearrangement we obtain the more familiar form of Spearman's rank correlation coefficient

$$r_s = 1 - \frac{6}{n(n^2 - 1)} \sum_{i=1}^{n} d_i^2, \tag{14.24}$$

where d_i is the difference between the x-rank and the y-rank for the ith observation. If ties are present, the following expression for r_s incorporates corrections for ties.

$$r_s = \frac{\sum_{i=1}^{n} x_i^2 + \sum_{i=1}^{n} y_i^2 - \sum_{i=1}^{n} d_i^2}{2\sqrt{\left[\sum_{i=1}^{n} x_i^2 \sum_{i=1}^{n} y_i^2\right]}}, \tag{14.25}$$

where

$$\sum_{i=1}^{n} x_i^2 = \frac{n}{12}(n^2 - 1) - \sum_{j=1}^{T_x} \frac{t_j}{12}(t_j^2 - 1)$$

$$\sum_{i=1}^{n} y_i^2 = \frac{n}{12}(n^2 - 1) - \sum_{l=1}^{T_y} \frac{t_l'}{12}(t_l'^2 - 1)$$

where t_j is the number of x-values involved in the jth tie, t_l' is the number of y-values involved in the lth tie, T_x is the number of ties in the x-values, T_y is the number of ties in the y-values.

To test the null hypothesis that the population correlation coefficient is zero, we may use the approximate procedure based on the t-distribution

$$t = r_s \sqrt{\left(\frac{n - 2}{1 - r_s^2}\right)}. \tag{14.26}$$

For large sample sizes (larger than $n = 10$) t as defined in (14.26) has a t-distribution with $n - 2$ degrees of freedom.

14.9.2. Example

TABLE 14.15. EXAMPLE RANK CORRELATION DATA

Variable x_i	1	8	10	4	6	2·5	5	2·5	8	12	11	8
Variable y_i	3	6	9	2	5	1	7·5	4	7·5	11	10	12
Differences d_i	−2	2	1	2	1	1·5	−2·5	−1·5	0·5	1	1	−4

The example rank data in Table 14.15 contains three ties, two in the x-values and one in the y-values. Computing Spearman's rank correlation coefficient for this data we have:

$$Tx = 2 \qquad Ty = 1$$
$$t_1 = 2 \qquad t_1' = 2$$
$$t_2 = 3$$

$$\sum_{i=1}^{12} x_i^2 = \frac{12}{12} \times 143 - \frac{2 \times 3}{12} - \frac{3 \times 8}{12} = 140 \cdot 5,$$

$$\sum_{i=1}^{12} y_i^2 = \frac{12 \times 143}{12} - \frac{2 \times 3}{12} = 142 \cdot 5,$$

$$r_s = \frac{140 \cdot 5 + 142 \cdot 5 - ((-2)^2 + (2)^2 + \cdots + (-4)^2)}{2 \times \sqrt{(140 \cdot 5 \times 142 \cdot 5)}},$$

$$= \frac{240}{282 \cdot 99} = 0 \cdot 8481.$$

Using the t approximation we have:

$$t = 0 \cdot 8481 \sqrt{\left[\frac{10}{(1 - 0 \cdot 8481^2)} \right]} = 6 \cdot 88,$$

$$\Pr \{ t \geqq 6 \cdot 88 \mid \nu = 10 \} = 0 \cdot 046.$$

From this probability we would assume, using a 5 per cent significance level, that there was significant correlation between variables x and y.

14.9.3. Kendall's Rank Correlation Coefficient

Kendall's rank correlation coefficient (Kendall, 1948) is based on comparisons between the x- and y-values of one point with those of another. We compare (x_i, y_i) with (x_j, y_j) and

score $+1$ if $x_i > x_j$ and $y_i > y_j$,

score $+1$ if $x_i < x_j$ and $y_i < y_j$,

score -1 if $x_i > x_j$ and $y_i < y_j$,

score -1 if $x_i < x_j$ and $y_i > y_j$.

There are $\frac{1}{2}n(n-1)$ such comparisons and if we define S as the sum of all such scores then Kendall's rank correlation coefficient τ is defined as

$$\tau = \frac{2n}{n(n-1)} \qquad (14.27)$$

If both sequences of ranks are precisely the same, all comparisons will result in a score of $+1$. Thus, the highest value that τ may take is $+1$ and the lowest value is -1 and we see that τ has the same range as the product moment correlation coefficient.

If ties occur, the scoring system and the denominator of τ are modified. If $x_i = x_j$ or $y_i = y_j$, we score zero and the modified expression for τ is

$$\tau = \frac{S}{\left[\left(\frac{n}{2}(n-1) - \sum_{j=1}^{T_x} \frac{t_j}{2}(t_j - 1) \right) \left(\frac{n}{2}(n-1) - \sum_{l=1}^{T_y} \frac{t'_l}{2}(t'_l - 1) \right) \right]^{\frac{1}{2}}},$$

where t_j is the number of x-values involved in the ith tie, t'_l is the number of y-values involved in the lth tie, T_x is the number of ties in the x-values, T_y is the number of ties in the y-values.

The distribution of τ, assuming no correlation in the population, approaches the normal distribution as the sample size increases. For sample sizes of ten or greater, we may assume that the distribution of τ is normal with mean and variance:

$$\mathscr{E}(\tau) = 0,$$

$$\sigma^2(\tau) = \frac{2(2n+5)}{9n(n-1)}.$$

14.9.4. Example

Calculating Kendall's τ for the ranks given in Table 14.15. Taking each x-rank (except the last) and its corresponding y-rank in turn we compare these with all the x- and y-rank pairs following it. The scores for all these comparisons are given in Table 14.16.

TABLE 14.16. COMPARISON SCORES FOR KENDALL'S RANK CORRELATION

j	2	3	4	5	6	7	8	9	10	11	12	Total
$i=1$	1	1	−1	1	−1	1	1	1	1	1	1	7
2		1	1	1	1	−1	1	0	1	1	0	6
3			1	1	1	1	1	1	1	1	−1	7
4				1	1	1	−1	1	1	1	1	6
5					1	−1	1	1	1	1	1	5
6						1	0	1	1	1	1	5
7							1	0	1	1	1	4
8								1	1	1	1	4
9									1	1	0	2
10										1	−1	0
11											−1	−1
									Total (S) =			45

$$\tau = \frac{45}{\left[\left(66 - \frac{3 \times 2}{2} - \frac{2 \times 1}{2}\right)\left(66 - \frac{2 \times 1}{2}\right)\right]^{\frac{1}{2}}} = 0.7089.$$

Using the normal approximation to the distribution of τ we compute:

$$\sigma^2(\tau) = \frac{2 \times 29}{9 \times 12 \times 11} = 0.048822,$$

$$X = \frac{0.7089}{\sqrt{0.048822}} = 3.21,$$

$$\Pr\{X \geqq 3.21\} = 0.0006.$$

From this probability we would normally reject the null hypothesis of zero correlation in the population.

Questions

1. Consider the data given in question 1 after Chapter 6 and avoid the assumption of normality in the analyses that you perform on this data.

   (a) Compute the medians for both samples.

   (b) Is there any difference in variability between the two samples? Perform all the variance tests described in this chapter on the data as it stands and repeat each test after subtracting the sample medians from the individual observations.

   (c) Is there any evidence that the average length differs in the two types of nest? Perform all location tests described in this chapter.

   (d) Perform distributional test also.

2. Add the extra sample given in question 2 after Chapter 6 to the two considered in question 1 above. Again avoid the assumption of normality.

   (a) Is there any evidence that the average length differs in the three types of nest? Perform both location tests described in this chapter.

3. Consider the two-way data given in Section 9.1 (Table 9.1) and avoid the assumption of normality in the analyses that you perform on this data.

   (a) Test the significance of the differences between doses. Relax the χ^2 condition of 5 for the minimum number of levels of either factor.

   (b) Test the significance of the differences between heat treatments again relaxing the χ^2 condition.

4. Compare the correlation coefficients obtained by Spearman's and by Kendall's methods for the data given in Table 12.2 with that obtained in Section 12.1.4. Test the significance of the coefficients you obtain.

HINTS AND ANSWERS TO PROBLEMS

Chapter 1

1. This is a reference to Section 1.2.2 which the reader should re-read if he is uncertain of the answer to this question. The first set of confidence limits include variability in the experimentalist's technique as used in his method.

The use of different methods introduces an additional source of variation which may or may not have a significant effect on the estimation process.

2. It is frequently instructive to reconsider conclusions drawn from data with an extreme value omitted. This may bring out the fact that a significant result is largely due to one value. The experimentalist may therefore decide to continue experimenting for a further period to obtain a more reliable conclusion.

3. (a) Arrangements 2, 3 and 5 have each treatment in each column once and once only so that any cooling effect producing differences between columns will not affect the differences between treatments.

(b) Arrangement 3 is the only arrangement which has each treatment in each column and each row once and once only and is therefore balanced in both diections.

4. (a) The design is a Latin square which is discussed in Section 1.2.7.

(b) The fault would have had no effect on the comparison between the tyres if the additional wear due to the fault was constant in each period since each tyre appears once in the faulty position. General experimental variability may, however be increased.

(c) The fault would now cause additional wear on one tyre only and would therefore affect the tyre comparisons. This possibility is not balanced in this design although a plot of residuals as outlined in Section 9.1.2 may reveal the fault.

5. The new design is an example of a Youden square as discussed in Section 1.2.8. There are five treatments to compare with only four positions available at any one time.

(b) See answer to question 4 (b).

(c) The duplicate measurement would provide information on the variability due to measuring the wear but would not increase the information on the variability normally experienced by two tyres used under identical conditions. These duplicate measurements are repeats rather than replicates with respect to tyre comparisons.

(d) To obtain replicate measurements we must double the number of time periods so that each tyre appears in each position twice. We may reduce the length of each time period to avoid a long experiment but this would increase the variability in wear measurements a little.

Chapter 2

1. (a) 0·15.   (b) 0·50.   (c) 0·35.

(d) $(0·15)^2 = 0·0225$.   (e) $(0·50)^2 = 0·2500$.

(f) $(0·35)^2 = 0·1225$.   (g) $0·0225 + 0·2500 + 0·1225 = 0·3950$.

(h) $1 - 0·3950 = 0·6050$.

Answers for the finite group become:

(d) $\dfrac{3}{20} \times \dfrac{2}{19} = \dfrac{6}{380} = 0{\cdot}0158.$　(e) $\dfrac{10}{20} \times \dfrac{9}{19} = \dfrac{90}{380} = 0{\cdot}2368.$

(f) $\dfrac{7}{20} \times \dfrac{6}{19} = \dfrac{42}{380} = 0{\cdot}1105.$　(g) $\dfrac{6}{380} + \dfrac{90}{380} + \dfrac{42}{380} = \dfrac{138}{380} = 0{\cdot}3632.$

(h) $1 - \dfrac{138}{380} = 0{\cdot}6368.$

2. (a) $\dfrac{1}{4},\ \dfrac{1}{2},\ \dfrac{1}{4}.$　(b) $1,\ 0,\ 0.$　(c) $\dfrac{1}{2},\ \dfrac{1}{2},\ 0.$

(d) $\Pr\{AA\} = \dfrac{1}{4} \times \dfrac{1}{4} + 1 \times \dfrac{1}{4} + 2 \times \dfrac{1}{2} \times \dfrac{1}{4} = \dfrac{9}{16}.$

(e) $\Pr\{Aa\} = \dfrac{1}{2} \times \dfrac{1}{4} + 0 \times \dfrac{1}{4} + 2 \times \dfrac{1}{2} \times \dfrac{1}{4} = \dfrac{6}{16}.$

(f) $\Pr\{aa\} = \dfrac{1}{4} \times \dfrac{1}{4} + 0 \times \dfrac{1}{4} + 2 \times 0 \times \dfrac{1}{4} = \dfrac{1}{16}.$

$$\text{Total} = \dfrac{16}{16}.$$

5. (a) $\left[\dfrac{2}{3}\right]^8 = \dfrac{256}{6561}.$　(b) $8\left[\dfrac{1}{3}\right]\left[\dfrac{2}{3}\right]^7 = \dfrac{1024}{6561}.$

(c) $\left[\dfrac{2}{3}\right]^8 + 8\left[\dfrac{1}{3}\right]\left[\dfrac{2}{3}\right]^7 + \dfrac{8 \times 7}{2}\left[\dfrac{1}{3}\right]^2\left[\dfrac{2}{3}\right]^6 = \dfrac{3072}{6561}.$

(d) 2 or 3.

Chapter 3

1. (a) The population of pellets that would be produced by this process if it were continued indefinitely under the same conditions.

(b) That the same conditions will prevail in the future.

(c) $X = \dfrac{0{\cdot}400 - 0{\cdot}342}{0{\cdot}022} = 2{\cdot}64;$　Table 1 shows $P = 0{\cdot}9959.$

(d) $X = \dfrac{0{\cdot}300 - 0{\cdot}342}{0{\cdot}022} = -1{\cdot}91;$　Table 1 for $X = +1{\cdot}91$ shows $P = 0{\cdot}9719.$

Thus there are $0{\cdot}0041$ above 400 and $0{\cdot}0281$ below 300, and $0{\cdot}9678$ pellets are predicted to have lengths within the range 300 to 400 cm.

(e) $N(5{\cdot}472,\ 0{\cdot}007744).$

(f) $X_1 = \dfrac{5{\cdot}460 - 5{\cdot}472}{0{\cdot}088} = -0{\cdot}136;\ \Pr(-\infty \text{ to } X_1) = 1 - {\cdot}5541 = 0{\cdot}4459.$

$X_2 = \dfrac{5{\cdot}480 - 5{\cdot}472}{0{\cdot}088} =\ \ 0{\cdot}091;\ \Pr(X_2 \text{ to } +\infty) = 1 - 0{\cdot}5363 = 0{\cdot}4637.$

Thus, $1 - 0.4459 - 0.4637 = 0.0904$ columns are predicted to have lengths between 5·460 and 5·480 cm.

3.(a) 0·0049, 0·0166, 0·0440, 0·0919, 0·1498, 0·1915, 0·1915, 0·1498,

  0·0919, 0·0440, 0·0166, 0·0049.

(b) Proportion less than 59 is 0·0013.
  Proportion greater than 77 is 0·0013.
  Proportion not catered for is $0.0013 + 0.0013 = 0.0026$.

4. 2·08 and 9·92.

5.
$$\mu(x) = \frac{1}{2}, \quad \sigma^2(x) = \frac{1}{12}, \quad \beta_1 = 0.$$

$$\mu_3(x) = 0, \quad \mu_4(x) = \frac{1}{80}, \quad \beta_2 = 1.8.$$

6. Probability of finding a skull of length equal to, or greater than, 13 inches is 0·0007. We feel that it is highly unlikely that this skull belonged to an individual of the same country and we therefore reject this hypothesis.

Chapter 4

1. (a)
$$\mathscr{E}(t_1) = \frac{1}{4}\mathscr{E}(x_1) + \frac{1}{2}\mathscr{E}(x_2) + \frac{1}{4}\mathscr{E}(x_3) = \mu,$$

$$\mathscr{E}(t_2) = \frac{1}{5}\mathscr{E}(x_1) + \frac{2}{5}\mathscr{E}(x_2) + \frac{2}{5}\mathscr{E}(x_2) = \mu,$$

$$\mathscr{E}(t_3) = \frac{1}{6}\mathscr{E}(x_1) + \frac{2}{6}\mathscr{E}(x_2) + \frac{3}{6}\mathscr{E}(x_3) = \mu,$$

since
$$\mathscr{E}(x_1) = \mathscr{E}(x_2) = \mathscr{E}(x_3) = \mu.$$

(b) Using expression (4.6) we have:

$$\sigma^2(t_1) = \left(\frac{1}{16} + \frac{1}{4} + \frac{1}{16}\right)\sigma^2 = \frac{3}{8}\sigma^2 = 0.375\sigma^2,$$

$$\sigma^2(t_2) = \left(\frac{1}{25} + \frac{4}{25} + \frac{4}{25}\right)\sigma^2 = \frac{9}{25}\sigma^2 = 0.360\sigma^2,$$

$$\sigma^2(t_3) = \left(\frac{1}{36} + \frac{4}{36} + \frac{9}{36}\right)\sigma^2 = \frac{7}{18}\sigma^2 = 0.367\sigma^2.$$

(c) t_2 has the smallest variance.

(d) The sample mean $\bar{x} = \frac{1}{3}x_1 + \frac{1}{3}x_2 + \frac{1}{3}x_3$ has an even smaller variance of $0.333\,\sigma^2$.

2. (a) As in 1 (a).

(b) $\sigma^2(t_1) =$

$$\left(\frac{1}{16} + \frac{1}{4} + \frac{1}{16}\right)\sigma^2 + 2\left(\frac{1}{4}\cdot\frac{1}{2}\cdot\frac{1}{2} + \frac{1}{4}\cdot\frac{1}{4}\cdot\frac{1}{3} + \frac{1}{2}\cdot\frac{1}{4}\cdot\frac{1}{4}\right)\sigma^2 = 0{\cdot}604\sigma^2,$$

$\sigma^2(t_2) =$

$$\left(\frac{1}{25} + \frac{4}{25} + \frac{4}{25}\right)\sigma^2 + 2\left(\frac{1}{5}\cdot\frac{2}{5}\cdot\frac{1}{2} + \frac{1}{5}\cdot\frac{2}{5}\cdot\frac{1}{3} + \frac{2}{5}\cdot\frac{2}{5}\cdot\frac{1}{4}\right)\sigma^2 = 0{\cdot}573\sigma^2,$$

$\sigma^2(t_3) =$

$$\left(\frac{1}{36} + \frac{4}{36} + \frac{9}{36}\right)\sigma^2 + 2\left(\frac{1}{6}\cdot\frac{2}{6}\cdot\frac{1}{2} + \frac{1}{6}\cdot\frac{3}{6}\cdot\frac{1}{3} + \frac{2}{6}\cdot\frac{3}{6}\cdot\frac{1}{4}\right)\sigma^2 = 0{\cdot}583\sigma^2.$$

(c) t_2 has the smallest variance.

(d) The sample mean is not now the estimator with the smallest variance. The variance of the mean is $0{\cdot}574\,\sigma^2$.

3. (a) As in 1 (a).

(b) Similar to 2 (b).

(c) t_1 now has the smallest variance.

(d) The sample mean has a variance smaller than that of t_1 but there could be a linear function of x_1, x_2, and x_3 which has an even smaller variance.

5. (a) $\hat{\mu} = \bar{x} = 10{\cdot}113$.

(b) $\hat{\sigma}^2 = \dfrac{1}{11}(1227{\cdot}316814 - 1227{\cdot}273228) = \dfrac{0{\cdot}043586}{11} = 0{\cdot}003962,$

$\hat{\sigma}^2 = \dfrac{1}{11}((10{\cdot}023 - 10{\cdot}113)^2 + \cdots + (10{\cdot}087 - 10{\cdot}113)^2) = 0{\cdot}003962.$

(c) $\hat{\sigma}^2 = \dfrac{1}{11}(1227{\cdot}32 - 1227{\cdot}27) = 0{\cdot}0045$

$\hat{\sigma}^2 = \dfrac{1}{11}((10{\cdot}023 - 10{\cdot}113)^2 + \cdots + (10{\cdot}087 - 10{\cdot}113)^2) = 0{\cdot}003962.$

6. (a) $L = \dfrac{n!}{r!\,(n-r)!}\,p^r(1-p)^{n-r}.$

(b) $\text{Log } L = \log\left(\dfrac{n!}{r!(n-r)!}\right) + r\log p + (n-r)\log(1-p),$

$$\frac{\partial \log L}{\partial p} = \frac{r}{p} - \frac{n-r}{1-p},$$

equating to zero gives $\hat{p} = r/n$.

7.

$$S = \sum_{i=1}^{n}(y_i - bx_i)^2$$

$$\frac{\partial S}{\partial b} = -2\sum_{i=1}^{n}(y_i - bx_i)\,x_i,$$

$$\text{equating to zero gives } \hat{b} = \frac{\sum\limits_{i=1}^{n} y_i x_i}{\sum\limits_{i=1}^{n} x_i^2} \, .$$

Chapter 5

1. (a) 1 10 45 120 210 252 210 120 45 10 1 all divided by 1024.

(b) We would test the null hypothesis that the probability of a correct decision is 0·5, that is, that she is guessing.

(c) The critical region of size (say) about 0·05 would contain values 10, 9 and 8. The exact size would be 0·0547.

(d) The value 9 falls in our critical region so that we would reject the *null* hypothesis and accept her claim.

(e) The probability that we are wrong is 11/1024 = 0·0107.

(f) $(0.6)^{10} + 10(0.6)^9(0.4) + 45(0.6)^8(0.4)^2 = 0.1673$.

(g) One point on the power function corresponding to the significance level of 0·0547 is given in (c), another point is given in (f). For others repeat (f) with probabilities 0·7 0·8, 0·9 and 0·95 and plot the probability of rejection against the probability of a correct decision.

2. (a) The mean is estimated as 10·113 with 95 per cent confidence limits 10·076, 10·150· (Section 5.2.7 using the *Normal* distribution.)

(b) The 95 per cent confidence limits are now 10·073 to 10·153.

(c) It has been assumed that the population of results represented by this sample is normally distributed.

3. (a) $t = 2.38$; double-tailed probability $= 2 \Pr\{t \geq 2.38 \,|\, v = 9\} = 0.0412$. If we use a 5 per cent significance level we would conclude that the values obtained were not consistent with the belief that the mean of the population of all such values was 523 lbs.

(b) $X = 2.96$; $2 \Pr\{X \geq 2.96\} = 0.0030$. Conclusion as in 3(a).

(c) $C = 13.85$; $\Pr\{C \geq 13.85\} = 0.128$.

4. (b) The mean is estimated as 6·392, the variance is estimated as 2·7443 (with continuity correction of 1/2).

(c) Fitting a Normal distribution by the methods of Section 5.4.1 and testing by the methods of Section 5.4.2 we obtain $\phi^2 = 5.11$ and $v = 5$ (last two groups are combined). Since $\Pr\{\chi^2 \geq 5.11 \,|\, v = 5\} = 0.40$ we accept the null hypothesis that the population distribution is Normal.

5. All expected frequencies are 100; $\phi^2 = 18.8$; $\Pr\{\chi^2 \geq 18.8 \,|\, v = 9\} = 0.026$. Thus we feel confident in rejecting the null hypothesis that all digits appear equally often.

Chapter 6

1. (a) $F = \dfrac{0.5689}{0.4225} = 1.34$; $v_1 = 14$; $v_2 = 8$.

Thus there is no evidence for believing that the variances are different.

(b) $t = \dfrac{1.08}{0.3028} - 3.57$; $v = 22$; $p = 0.001$.

Thus we would conclude that there was a difference in lengths.

2. (a) Bartlett's test $M' = 2\cdot57$ and $Pr\{\chi^2 \geq 2\cdot57 \mid \nu = 2\} = 0\cdot27$.
From this we would conclude that the variances do not differ.

(b) The probability plot is very interesting and indicates, bearing in mind that points in the centre are more reliable than the extremes, that the additional sample contains some extreme values which are not present in the other two. This data is discussed further in questions after Chapter 7.

3. (a) The probability of a significant result for one test when the null hypothesis is true is $1/20$. The probability of no significant results from three independent tests is therefore $(19/20)^3 = 0\cdot8573$. The probability of at least one significant result is therefore $0\cdot1427$. Although we have falsely assumed independence this result illustrates the point that the more tests we perform the more likely we are to obtain a significant result when the null hypothesis is, in fact, true.

(b) We would expect one significant result even though the null hypothesis was true

4. (a) The null hypothesis to be tested is that the population mean difference is zero. Let D denote the population mean difference and d the sample mean difference. From Section 6.1.9 we see that the distribution of d is Normal with mean D and variance $2\sigma^2/n$, where both samples are of size n. If $X = (d - D)/\sqrt{(2\sigma^2/n)}$ then X is a standardized Normal variable. If we use a $0\cdot05$ significance level in the significance test we will reject the null hypothesis that $D = 0$ if $X \geq 1\cdot6445$. That is if $d \geq 1\cdot6445 \times \sqrt{(2\sigma^2/n)}$. The probability of detecting the alternative hypothesis that D is some positive value D_1 is the probability that $d - D_1$ is greater than, or equal to, the critical value given above, that is $1\cdot6445 \times \sqrt{(2\sigma^2/n)}$. This is equal to the probability that X is greater than, or equal to, X_1, where $X_1 = (1\cdot6445\sqrt{(2\sigma^2/n)} - D_1)/\sqrt{(2\sigma^2/n)}$ and where X is a standardized Normal variable.

Applying this to the values given we have:
The critical value $= 1\cdot6445 \times \sqrt{(200/10)} = 7\cdot35$,
if $D_1 = 5$; $X_1 = (7\cdot35 - 5\cdot00)/\sqrt{(200/10)} = 0\cdot525$,
$Pr\{X \geq 0\cdot525\} = 0\cdot30$.

(b) Repeat (a) with different values.

(c) The results in question (b) show that the required sample size (that is both samples) is about 70.

(e) The plot in (d) is not the power function. The reader should re-read Section 6.1.9 if wrong on this question.

5. (a) Testing variances: S.S. $= 29\cdot54$; $C = 29\cdot54/1\cdot9^2 = 8\cdot18$.
$Pr\{\chi^2 \geq 7\cdot94 \mid \nu = 7\} = 0\cdot32$; variances consistent.
Testing means: $\bar{x} = 14\cdot20$; $X = 1\cdot04$
$2 \times Pr\{X \geq 1\cdot04\} = 2 \times 0\cdot15 = 0\cdot30$; means consistent.

(b) Variances consistent; Pr $= 0\cdot34$.
Means consistent but close to $0\cdot05$ significance level; Pr $= 0\cdot06$.

(c) Correlated t-test $t = 1\cdot83$ with $\nu = 7$; $Pr\{t \geq 1\cdot83 \mid \nu = 7\} = 0\cdot055$. Double-tailed probability $= 0\cdot11$: dyes not inconsistent but close to significance.

(d) The differences have a Normal distribution.

Chapter 7

1. (a)

Source	S.S.	D.F.	M.S.	F.	Prob.
Between nests	28·8947	2	14·4473	19·70	0·001
Within nests	25·6611	35	0·7332		
Total	54·5558	37			

We would conclude that the lengths of cuckoo's eggs in the three nests are different.

2. (a) $\phi^2 = 9\cdot36$; $\nu = 6$; $\Pr\{\chi^2 \geq 9\cdot36 \ \nu = 6\} = 0\cdot154$.

(b) Compute the expected values in each group as the number in the group times $0\cdot25$, $0\cdot50$ and $0\cdot25$, respectively. We obtain $\phi^2 = 9\cdot63$ and $\nu = 8 = (3-1) \times 4$ since the totals for the types do not agree.

(c) The dominant and hybrid types would be combined and the total could be analysed using the short cut described in Section 7.1.2.

(d)
$$L = \{(1-p)\}^{262}\{2p(1-p)\}^{537}\{p^2\}^{257} \ 262!537!257!/1056!$$

$$\log L = 262 \times 2 \times \log(1-p) + 537 \times \{\log(2) + \log(p) + \log(1-p)\}$$
$$+ 257 \times 2 \times \log(p) + \text{const},$$

$$\frac{\log L}{p} = 0 \quad \text{gives:} \quad \frac{-524}{1-p} + \frac{537}{p} - \frac{537}{1-p} + \frac{514}{p} = 0.$$

Hence $\hat{p} = \dfrac{537 + 514}{2 \times 1056} = 0\cdot498$.

The general expression is $\hat{p} = \dfrac{(h + 2r)}{2n}$ where h, r and n are the numbers of hybrids, recessives, and the total respectively.

(e) The expected values are now given by multiplying the proportions $(1-p)^2, 2p(1-p)$ and p^2 by the number of individuals in the group. The degrees of freedom are now $\nu = (3-1) \times 4 - 1 = 7$ since this case differs from that in 2 (b) by the replacement of the given value of $p = \frac{1}{2}$ by an estimate.

3.
$$\mu(a) = 17; \quad \sigma^2(a) = 7\cdot0905,$$
$$X = 2\cdot44; \quad \Pr\{X \geq 2\cdot44\} = 0\cdot007.$$

There is evidence for the belief that a smoker is more likely to contract lung cancer than a non-smoker.

Chapter 8

1. (d) No significant variance differences.

Source	S.S.	D.F.	M.S.	F.	Prob.
Between diets	126·005	2	63·002	6·61	0·005
Within diets	257·349	27	9·531		
Total	383·354	29			

Means significantly different.

(e) Normal with equal variance.

(f) 95 per cent confidence limits for first diet are $79\cdot14 \pm 2\cdot052 \times 0\cdot976$. That is, $77\cdot14$ to $81\cdot14$.

(g) We have assumed equal weights at the beginning of the experiment so that the weights at the end will reflect the effect of the diets.

3. (a) The residual means squares in the two experiments are extremely close.

(b) Variances are consistent in experiment 2.

Source	S.S.	D.F.	M.S.	F.	Prob.
Between diets	120·691	2	60·345	6·54	0·005
Within diets	249·044	27	9·224		
Total	369·735	29			

Means are significantly different and experiment 2 gives the same results for comparisons between diets. It remains to be established whether, or not, the two experiments differ in the absolute values of the responses to the diets.

(c)

Source	S.S.	D.F.	M.S.	F.	Prob.
Between exps.	3·601	1	3·601	0·38	N.S.
Between diets/exps.	246·696	4	61·674	6·58	0·001
Within diets	506·393	54	9·377		
Total	756·690	59			

Difference between experiments is not significant.

(d) Variances within all six diets are the same.

(e) Diets would now be a random effect and experiments would be tested against diets instead of against "within diets." This is possible using the mean squares given above since there are the same numbers of observations per diet.

Chapter 9

1.

Source	S.S.	D.F.	M.S.	F.	Prob.
Dyes	1·2656	1	1·2656	3·33	N.S.
Rolls	40·3394	7	5·7628	15·17	0·001
Interaction	2·6594	7	0·3799		
Total	44·2644	15			

(a) Dyes not significantly different although probability is about 0·1.

(b) Rolls differ significantly.

(d) Although dyes do not differ significantly rolls do, so that confidence limits must be computed either separately for each roll using the residual mean square or include variability between rolls. This latter may be done by combining the rolls and interaction terms to give an estimate of variability with fourteen degrees of freedom.

2.

Source	S.S.	D.F.	M.S.	F.	Prob.
Dyes	2·2500	1	2·2500	1·94	N.S.
Rolls	55·2875	7	7·8982	6·83	0·05
Interaction	8·1000	7	1·1571		
Total	65·6375	15			

(a) Dyes do not differ significantly.

(b) Rolls differ significantly.

(c) Higher residual in second experiment but not significantly so. This accounts for lower significance of the rolls effect.

(d)

Source	S.S.	D.F.	M.S.	F.	Prob.
Experiments	0·0078	1	0·0078	0·01	N.S.
Dyes	3·4453	1	3·4453	4·48	N.S. (0·05)
Rolls/exp.	95·6269	14	6·8304	8·89	0·001
Dyes × exp.	0·0703	1	0·0703	0·09	0·001
Residual	10·7594	14	0·7685		
Total	109·9097	31			

No significant difference between experiments, rolls differ significantly, dyes do not differ significantly although this effect is now very close to the 5 per cent level.

3.

Source	S.S.	D.F.	M.S.	F.	Prob.
Experiments	0·0078	1	0·0078	0·03	N.S.
Dyes	3·4453	1	3·4453	13·60	0·01
Rolls	93·6422	7	13·3775	52·77	0·001
Dyes × exp.	0·0703	1	0·0703	0·28	N.S.
Exp. × rolls	1·9847	7	0·2835	1·12	N.S.
Dyes × rolls	8·9852	7	1·2836	5·06	0·025
Interaction	1·7742	7	0·2535		
Total	109·9097	31			

Rolls now becomes a symmetric factor.

(a) The comparison reveals the interesting situation in which the significant dyes × rolls interaction, which was part of the residual in the first form of analysis, masked the significance of the dye effect. This is now significant at the 1 per cent level and the residual variance has been reduced to 0·2535.

(b) The two dye means agree in the two experiments; this is borne out by the non-significant exp. × dyes interaction.

(c) The two-way table of means for dyes and rolls.

4.

Source	S.S.	D.F.	M.S.	F.	Prob.
Dyes	3·4473	1	3·4473	14·38	0·025
Rolls	93·6422	7	13·3775	55·81	0·001
Dyes × rolls	8·9852	7	1·2836	5·36	0·005
Residual	3·8350	16	0·2397		
Total	109·9097	31			

5. Dyes is almost certainly a fixed effect. If the two experiments were conducted by different experimentalists we might regard experiments as a fixed effect also. If performed by the same person we may argue that these are two experiments drawn at random from the population of experiments performed by this person. Similarly rolls could be considered to be a fixed effect if each roll represented something specific such as different treat-

ments or machines and random if they were randomly selected from a population of such rolls.

6. (a) Testing dyes we have $\hat{\tau} = 1\cdot0984$; $\hat{v}_2 = 5$; $F = 3\cdot14$. See Section 9.8.

7. The duplicate measurements would be repeats with respect to dye and roll comparisons and the residual obtained in the analysis would not contain differences between lengths of cloth and could not be used as a basis for comparison with the other three effects. Thus dyes and rolls must be tested against the interaction term.

Chapter 10

1. (a) Obtain analysis from answer to question 1 (c).

(c)	Source	S.S.	D.F.	M.S.	F.	Prob.
	Treats (lin)	253·009	1	253·009	393·48	0·001
	Treats (rem)	1·106	3	0·369	0·57	N.S.
	Blocks	16·836	3	5·612	8·73	0·005
	Tr (lin) × Blocks	3·861	3	1·287	2·00	N.S.
	Tr (rem) × Blocks	5·788	9	0·643		
	Total	280·600	19			

2. (a)	Source	S.S.	D.F.	M.S.	F.	Prob.
	Rows	0·6369	3	0·2123	0·50	N.S.
	Columns	2·3219	3	0·7740	1·84	N S.
	Foods	64·7069	3	21·5690	51·18	0·001
	Residual	2·5287	6	0·4214		
	Total	70·1944	15			

(b)	Source	S.S.	D.F.	M.S.	F.	Prob.
	Rows	0·1725	3	0·0575	0·50	N.S.
	Columns	1·8525	3	0·6175	0·94	N.S.
	Foods	67·3725	3	22·4575	34·03	0·001
	Residual	3·9600	6	0·6600		
	Total	73·3575	15			

(c) See answer to question (d).

(d)	Source	S.S.	D.F.	M.S.	F_1	F_2	P_1	P_2
	Treatments	132·0009	3	44·0003		56·86		0·001
	Animals (rem)	9·2863	12	0·7739	4·25		0·01	
	Weeks	0·0003	1	0·0003	0·00		N.S.	
	Wks × Trs	0·0785	3	0·0262	0·14		N.S.	
	Wks × Trs (rem)	2·1862	12	0·1822				
	Total	143·5522	31					

There is significantly more variability "between animals" than "within animals" so that we must test the treatments effect against the animals (remainder) effect. The first two effects in the table contain animal differences and the last three do not.

(e) The row and column effects, and their interactions with weeks, are contained in the second and fifth effects in the table. These may be separated if desired.

Chapter 11

1. (a)

Source	S.S.	D.F.	M.S.	F.	Prob.
Trs (adj)	0·0327718	4	0·0081929	144·70	0·001
Periods	0·0009065	4	0·0002266		
Pers (Adj)	0·0003988	4	0·0000997	1·76	N.S.
Treats	0·0332795	4	0·0083199		
Positions	0·0005748	3	0·0001916	3·42	N.S.
Residual	0·0004529	8	0·0000566		
Total	0·0347060	19			

2. (a)

Source	S.S.	D.F.	M.S.	F.	Prob.
Trs (Adj)	0·0310871	4	0·0077718	66·43	0·001
Periods	0·0012553	4	0·0003138		
Pers (Adj)	0·0003856	4	0·0000964	0·82	N.S.
Treats	0·0319568	4	0·0079892		
Positions	0·0047464	3	0·0015821	13·52	0·005
Residual	0·0009360	8	0·0001170		
Total	0·0380248	19			

(d) Position effect is different for the two automobiles.

(e) Compare the two means using a *t*-test. Estimate of variance is

$$\frac{1}{16}(0·0004529 + 0·0009360) = 0·0000868 \quad \text{and} \quad \Pr\{t \geqq 2·24 \mid v = 16\} = 0·02.$$

3. Table of sums of squares and cross-products.

Source	S.S. (y)	S.C.P. (x, y)	S.S. (x)
Treatments	254·1150	−818·475	2875·70
Blocks	16·8360	20·880	79·60
Residual	9·6490	−24·205	233·90
Totals	280·6000	−822·200	3189·20

Revised analysis testing treatment differences:

Source	S.S.	D.F.	M.S.	F.	Prob.
Treatments	221·2162	4	55·3040	85·15	0·001
Residual	7·1442	11	0·6495		
Total	228·3604	15			

(c) Residual reduced from 0·8041 to 0·6496.

Chapter 12

1. (a) $r = 0·307$. Not significant.

(b) Slope estimated as 0·4124 which is rather different from the value of 0·7007 obtained during the analysis of covariance.

(c) The estimate obtained in question (b) ignores the effect of Blocks and Quantities.

(d) Analysis of variance of residuals:

Source	S.S.	D.F.	M.S.	F.	Prob.
Quantities	1152·88	2	576·44	12·24	0·01
Blocks	137·77	3	45·92	0·98	N.S.
Residual	282·61	6	47·10		
Total	1573·26	11			

The analysis of variance, the analysis of regression, and the analysis of covariance have residual mean squares of 94·250, 47·101 and 44·988, respectively.

2. (a) Slope estimated as 1·1244. Draw this line on a plot of the data.

(c) Disagreement in constant.

3. Fitted line is $y = -3·8334 + 11·0493x$.

(b) Residual sum of squares is 5·7208. Since variances are known we may test this against a χ^2-distribution with 6 degrees of freedom when, in fact, it is not significant and we conclude that the fit is adequate.

(c) Proceed as in (b) and use the residual sum of squares as a estimate of the proportionality constant.

4. (a) Fitted line is $y = -4·6429 + 12·4083x - 0·2321x^2$.

(b) The equations to be solved to estimate the parameters of the quadratic are given as expression (12.34) (or (12.35)). To include the weights we would include these in each sum and replace n by $\sum_{i=1}^{n} w_i$.

5. Fit: $y = 11·8161 + 0·088074x + 0·34758t$.

Sum of squares due to regression is 65·9499.

(b) Fit: $y = 17·5102 + 0·15494x$.

Source	S.S.	D.F.	M.S.	F.	Prob.
Fitting x	7·3596	1	7·3596	0·54	N.S.
Fitting $t\|x$	58·5903	1	58·5903	4·30	0·05
Residual	394·9124	29	13·6177		
Total	460·8623	31			

Fit with x alone is not significant but addition of t significantly improves the description. Next step would normally be to try t on its own.

Chapter 13

1. (a) Estimates: Mean $= 4$; Variance $= 3 \cdot 8714$.
(b) Fitting a Poisson distribution with mean equal to 4 yields $\phi^2 = 4 \cdot 18$.

$$\Pr \{ \chi^2 \geq 4 \cdot 18 \mid = 8 \} = 0 \cdot 84 \quad \text{Good fit.}$$

2. (a) 20 counts per minute.
(b) $17 \cdot 64$ to $22 \cdot 33$.

3. (a) 20 counts per minute.
(b) $12 \cdot 22$ to $30 \cdot 89$. That is, much wider than in question 2.

Chapter 14

1. (a) Medians are $22 \cdot 0$ and $21 \cdot 0$.
(b) Mood's statistic is $W = 431 \cdot 75$ and with medians correction $W = 351 \cdot 00$. David's statistic $v = 22 \cdot 757$ and with medians correction $v = 44 \cdot 750$.

The value $22 \cdot 757$ for v is the only one of these which is significant and this is because of the location difference rather than the dispersion difference. The changes in the values of these statistics due to medians correction illustrate their sensitivity to location differences.

(c) Wilcoxon's test in its Mann–Whitney form gives $U = 20 \cdot 5$ and this indicates a significant difference in location.

2. The median test (Section 14.7.1), ignoring the five values of $21 \cdot 0$ which are jointly the median value, give $\phi^2 = 12 \cdot 59$.
The Kruskal–Wallis test gives $H = 18 \cdot 78$.
Both of these are highly significant.

3. (a) Testing doses we have: $F_c = 8 \cdot 60$ which is significant at the 5 per cent level.
(b) Testing heat treatments we have: $F_r = 6 \cdot 18$ which is not significant.

4. Spearman's $r_s = 0 \cdot 452$ Significant at the 5 per cent level.
Kendall's $\tau = 0 \cdot 337$ Significant at the 5 per cent level.
Normal $r = 0 \cdot 487$ Significant at the 5 per cent level.
Although these differ a little in magnitude they are all close to their respective 5 per cent significance level.

REFERENCES

BARTLETT, M. S. (1937) Properties of sufficiency and statistical tests. *Proc. Roy. Soc.* Series A, **160**, p. 268.

BIRNBAUM, Z. W. (1962) *Introduction to Probability and Mathematical Statistics*. Harper & Bros., New York.

COCHRAN, W. G. and COX, G. M. (1957) *Experimental Designs*, 2nd edition. John Wiley & Sons Ltd., New York.

COOPER, B. E. (1957) The effect of ties on the moments of rank criteria. *Biometrika* **44**, p. 265.

COX, D. R. (1958) *The Planning of Experiments*. John Wiley & Sons Ltd., New York.

CRAMER, H. (1951) *Mathematical Methods of Statistics*. Harper & Bros., New York.

DAVID, F. N. (1951) *Probability Theory for Statistical Methods*. McGraw-Hill Book Co., New York and London.

DAVID, F. N. (1956) A note on Wilcoxon's and allied tests. *Biometrika* **43**, p. 485.

ELDERTON, W. P. (1935) *Frequency Curves and Correlation*, 4th edition. Cambridge University Press.

FELLER, W. (1957) *Introduction to Probability and its Applications*. John Wiley & Sons Ltd., New York.

FINNEY, D. J. (1952a) *Probit Analysis*, 2nd edition. Cambridge University Press.

FINNEY, D. J. (1952b) *Statistical Method in Biological Assay*. Charles Griffin & Co. Ltd., London.

FISHER, R. A. (1960) *The Design of Experiments*, 7th edition. Oliver & Boyd Ltd., Edinburgh.

FISHER, R. A. and YATES, F. (1953) *Statistical Tables for Biological, Agricultural, Medical Research*, 4th edition. Oliver & Boyd, Edinburgh.

FORSYTHE, G. E. (1957) Generation and use of orthogonal polynomials for data fitting with a digital computer. *J. Soc. Ind. Appl. Maths.* **5**, p. 74.

FRIEDMAN, M. (1937) The use of ranks to avoid the assumption of normality implicit in analysis of variance. *J. Amer. Statist. Ass.* **32**, p. 675.

HARTLEY, H. O. (1940) Testing the homogeneity of a set of variances. *Biometrika* **31**, p. 249.

JOHNSON, N. L. (1949) Systems of frequency curves generated by methods of translation. *Biometrika* **36**, p. 149.

KAMAT, A. R. (1956) A two-sample distribution-free test. *Biometrika* **43**, p. 377.

KEMPTHORNE, O. (1952) *Design and Analysis of Experiments*. John Wiley & Sons Ltd., New York.

KENDALL, M. G. (1949) Rank and product moment correlation. *Biometrika* **36**, p. 177.

KENDALL, M. G. and BUCKLAND, W. R. (1957) *Dictionary of Statistical Terms*. Oliver & Boyd, London and Edinburgh, and Hafner Publishing Co., New York.

KENDALL, M. G. and STUART, A. (1961) *The Advanced Theory of Statistics*, Vol. **2**. Charles Griffin & Co. Ltd., London.

KENDALL, M. G. and STUART, A. (1963) *The Advanced Theory of Statistics*, Vol. 1, 2nd edition. Charles Griffin & Co. Ltd., London.

KRUSKALL, W. H. and WALLIS, W. A. (1952) Use of ranks in one-criterion variance analysis. *J. Amer. Statist. Ass.* **47**, p. 583.

LINDLEY, D. V. and MILLER, J. C. P. (1953) *Cambridge Elementary Statistical Tables*. Cambridge University Press.

MANN, H. B. and WHITNEY, D. R. (1947) On a test whether one of two random variables is stochastically larger than the other. *Ann. Maths. Statist.* **18**, p. 50.

Massey, F. J. Jr. (1951) The distribution of the maximum deviation between two sample cumulative step functions. *Ann. Maths. Statist.* **22**, p. 125.

Mood, A. M. (1954) On the asymptotic efficiency of certain non-parametric two-sample tests. *Ann. Maths. Statist.* **25**, p. 514.

National Bureau of Standards (1950) *Tables of the Binomial Probability Distribution.* Applied Mathematics Series. Number 6.

Owen, D. B. (1962) *Handbook of Statistical Tables.* Pergamon Press, London.

Pearson, E. S. and Hartley, H. O. (1958) *Biometrika Tables for Statisticians,* Vol. 1. Cambridge University Press.

Quenouille, M. H. (1953) *Design and Analysis of Experiments.* Charles Griffin & Co. Ltd., London.

Rosenbaum, S. (1953) Tables for a non-parametric test of dispersion. *Ann. Maths. Statist.* **24**, p. 663.

Rosenbaum, S. (1954) Tables for a non-parametric test of location. *Ann. Maths. Statist.* **25**, p. 146.

Scheffe, H. (1956) *The Analysis of Variance.* John Wiley & Sons Ltd., New York.

Siegel, S. (1956) *Nonparametric Statistics for the Behavioural Sciences.* McGraw-Hill Book Co., New York and London.

Wald, A. (1947) *Sequential Analysis.* John Wiley & Sons Ltd., New York.

Welch, B. L. (1947) The generalisation of Student's problem when several different population variances are involved. *Biometrika* **34**, p. 28.

Wetherill, G. B. (1966) *Sequential Methods in Statistics.* Methuen's Monographs, London.

Wilcoxon, F. (1945) Individual comparisons by ranking methods. *Biometrics* **1**, p. 80.

Wilcoxon, F. (1947) Probability tables for individual comparisons by ranking methods. *Biometrics* **3**, p. 119.

Wilks, S. S. (1942) Statistical prediction with special reference to the problem of tolerance limits. *Ann. Maths. Statist.* **13**, p. 400.

Youden, W. J. (1937) Use of incomplete block replication in estimating tobacco-mosaic virus. *Contr. Boyce Thompson Inst.* **9**, p. 41.

TABLES SECTION

Interpolation in Statistical Tables

It will be noticed that the degrees of freedom included in statistical tables include all values from 1 to 24, 30 or sometimes 40, followed by a few other values at unequal intervals. These values are chosen so that when they are divided into the value 120 the resulting values are equally spaced. Interpolation in the degrees of freedom scale is performed in this resulting linear scale, and the degrees of freedom 24, 30, 40, 60, 120 and ∞ correspond to 5, 4, 3, 2, 1 and 0. Interpolation for 48 degrees of freedom, for example, is carried out linearly for the value 2·5 between values 2 and 3. Similarly interpolation for 150 degrees of freedom is carried out linearly for the value 0·8 between the values 0 and 1.

ACKNOWLEDGEMENTS

Some of the tables in the following section appeared originally in other publications. The two principal sources were: Biometrika Tables for Statisticians, vol. 1 (ed. E. S. Pearson, 3rd ed. 1966), published by Cambridge University Press on behalf of the Biometrika Trust, and Handbook of Statistical Tables by D. B. Owen, Addison-Wesley (1962). Acknowledgements are made to the authors and publishers who gave their kind permission for the reproduction of their tables.

TABLE 1. THE NORMAL DISTRIBUTION

$$\text{Ordinate } Z(X) = \frac{1}{\sqrt{2\pi}} e^{-\frac{1}{2}X^2} \qquad \text{Area } P(X) = \int_{-\infty}^{X} Z(u)\, du$$

X	P(X)	Z(X)	X	P(X)	Z(X)	X	P(X)	Z(X)
0·00	0·500000	0·398942	0·45	0·673645	0·360527	0·90	0·815940	0·266085
0·01	0·503989	0·398922	0·46	0·677242	0·358890	0·91	0·818589	0·263688
0·02	0·507978	0·398862	0·47	0·680822	0·357225	0·92	0·821214	0·261286
0·03	0·511966	0·398763	0·48	0·684386	0·355533	0·93	0·823814	0·258881
0·04	0·515953	0·398623	0·49	0·687933	0·353812	0·94	0·826391	0·256471
0·05	0·519939	0·398444	0·50	0·691462	0·352065	0·95	0·828944	0·254059
0·06	0·523922	0·398225	0·51	0·694974	0·350292	0·96	0·831472	0·251644
0·07	0·527903	0·397966	0·52	0·698468	0·348493	0·97	0·833977	0·249228
0·08	0·531881	0·397668	0·53	0·701944	0·346668	0·98	0·836457	0·246809
0·09	0·535856	0·397330	0·54	0·705401	0·344818	0·99	0·838913	0·244390
0·10	0·539828	0·396953	0·55	0·708840	0·342944	1·00	0·841345	0·241971
0·11	0·543795	0·396536	0·56	0·712260	0·341046	1·01	0·843752	0·239551
0·12	0·547758	0·396080	0·57	0·715661	0·339124	1·02	0·846136	0·237132
0·13	0·551717	0·395585	0·58	0·719043	0·337180	1·03	0·848495	0·234714
0·14	0·555670	0·395052	0·59	0·722405	0·335213	1·04	0·850830	0·232297
0·15	0·559618	0·394479	0·60	0·725747	0·333225	1·05	0·853141	0·229882
0·16	0·563559	0·393868	0·61	0·729069	0·331215	1·06	0·855428	0·227470
0·17	0·567495	0·393219	0·62	0·732371	0·329184	1·07	0·857690	0·225060
0·18	0·571424	0·392531	0·63	0·735653	0·327133	1·08	0·859929	0·222653
0·19	0·575345	0·391806	0·64	0·738914	0·325062	1·09	0·862143	0·220251
0·20	0·579260	0·391043	0·65	0·742154	0·322972	1·10	0·864334	0·217852
0·21	0·583166	0·390242	0·66	0·745373	0·320864	1·11	0·866500	0·215458
0·22	0·587064	0·389404	0·67	0·748571	0·318737	1·12	0·868643	0·213069
0·23	0·590954	0·388529	0·68	0·751748	0·316593	1·13	0·870762	0·210686
0·24	0·594835	0·387617	0·69	0·754903	0·314432	1·14	0·872857	0·208308
0·25	0·598706	0·386668	0·70	0·758036	0·312254	1·15	0·874928	0·205936
0·26	0·602568	0·385683	0·71	0·761148	0·310060	1·16	0·876976	0·203571
0·27	0·606420	0·384663	0·72	0·764238	0·307851	1·17	0·879000	0·201214
0·28	0·610261	0·383606	0·73	0·767305	0·305627	1·18	0·881000	0·198863
0·29	0·614092	0·382515	0·74	0·770350	0·303389	1·19	0·882977	0·196520
0·30	0·617911	0·381388	0·75	0·773373	0·301137	1·20	0·884930	0·194186
0·31	0·621720	0·380226	0·76	0·776373	0·298872	1·21	0·886861	0·191860
0·32	0·625516	0·379031	0·77	0·779350	0·296595	1·22	0·888768	0·198543
0·33	0·629300	0·377801	0·78	0·782305	0·294305	1·23	0·890651	0·187235
0·34	0·633072	0·376537	0·79	0·785236	0·292004	1·24	0·892512	0·184937
0·35	0·636831	0·375240	0·80	0·788145	0·289692	1·25	0·894350	0·182649
0·36	0·640576	0·373911	0·81	0·791030	0·287369	1·26	0·896165	0·180371
0·37	0·644309	0·372548	0·82	0·793892	0·285036	1·27	0·897958	0·178104
0·38	0·648027	0·371154	0·83	0·796731	0·282694	1·28	0·899727	0·175847
0·39	0·651732	0·369728	0·84	0·799546	0·280344	1·29	0·901475	0·173602
0·40	0·655422	0·368270	0·85	0·802337	0·277985	1·30	0·903200	0·171369
0·41	0·659097	0·366782	0·86	0·805105	0·275618	1·31	0·904902	0·169147
0·42	0·662757	0·365263	0·87	0·807850	0·273244	1·32	0·906582	0·166937
0·43	0·666402	0·363714	0·88	0·810570	0·270864	1·33	0·908241	0·164740
0·44	0·670031	0·362135	0·89	0·813267	0·268477	1·34	0·909877	0·162555

TABLE 1 (*cont.*)

X	P(X)	Z(X)	X	P(X)	Z(X)	X	P(X)	Z(X)
1·35	0·911492	0·160383	1·80	0·964070	0·078950	2·25	0·987776	0·031740
1·36	0·913085	0·158225	1·81	0·964852	0·077538	2·26	0·988089	0·031032
1·37	0·914657	0·156080	1·82	0·965620	0·076143	2·27	0·988396	0·030337
1·38	0·916207	0·153948	1·83	0·966375	0·074766	2·28	0·988696	0·029655
1·39	0·917736	0·151831	1·84	0·967116	0·073407	2·29	0·988989	0·028985
1·40	0·919243	0·149727	1·85	0·967843	0·072065	2·30	0·989276	0·028327
1·41	0·920730	0·147639	1·86	0·968557	0·070740	2·31	0·989556	0·027682
1·42	0·922196	0·145564	1·87	0·969258	0·069433	2·32	0·989830	0·027048
1·43	0·923641	0·143505	1·88	0·969946	0·068144	2·33	0·990097	0·026426
1·44	0·925066	0·141460	1·89	0·970621	0·066871	2·34	0·990358	0·025817
1·45	0·926471	0·139431	1·90	0·971283	0·065616	2·35	0·990613	0·025218
1·46	0·927855	0·137417	1·91	0·971933	0·064378	2·36	0·990863	0·024631
1·47	0·929219	0·135418	1·92	0·972571	0·063157	2·37	0·991106	0·024056
1·48	0·930563	0·133435	1·93	0·973197	0·061952	2·38	0·991344	0·023491
1·49	0·931888	0·131468	1·94	0·973810	0·060765	2·39	0·991576	0·022937
1·50	0·933193	0·129518	1·95	0·974412	0·059595	2·40	0·991802	0·022395
1·51	0·934478	0·127583	1·96	0·975002	0·058441	2·41	0·992024	0·021862
1·52	0·935745	0·125665	1·97	0·975581	0·057304	2·42	0·992240	0·021341
1·53	0·936992	0·123763	1·98	0·976148	0·056183	2·43	0·992451	0·020829
1·54	0·938220	0·121878	1·99	0·976705	0·055079	2·44	0·992656	0·020328
1·55	0·939429	0·120009	2·00	0·977250	0·053991	2·45	0·992857	0·019837
1·56	0·940620	0·118157	2·01	0·977784	0·052919	2·46	0·993053	0·019356
1·57	0·941792	0·116323	2·02	0·978308	0·051864	2·47	0·993244	0·018885
1·58	0·942947	0·114505	2·03	0·978822	0·050824	2·48	0·993431	0·018423
1·59	0·944083	0·112604	2·04	0·979325	0·049800	2·49	0·993613	0·017971
1·60	0·945201	0·110921	2·05	0·979818	0·048792	2·50	0·993790	0·017528
1·61	0·946301	0·109155	2·06	0·980301	0·047800	2·51	0·993963	0·017095
1·62	0·947384	0·107406	2·07	0·980774	0·046823	2·52	0·994132	0·016670
1·63	0·948449	0·105675	2·08	0·981237	0·045861	2·53	0·994297	0·016254
1·64	0·949497	0·103961	2·09	0·981691	0·044915	2·54	0·994457	0·015848
1·65	0·950529	0·102265	2·10	0·982136	0·043984	2·55	0·994614	0·015449
1·66	0·951543	0·100586	2·11	0·982571	0·043067	2·56	0·994766	0·015060
1·67	0·952540	0·098925	2·12	0·982997	0·042166	2·57	0·994915	0·014678
1·68	0·953521	0·097282	2·13	0·983414	0·041280	2·58	0·995060	0·014305
1·69	0·954486	0·095657	2·14	0·983823	0·040408	2·59	0·995201	0·013940
1·70	0·955435	0·094049	2·15	0·984222	0·039550	2·60	0·995339	0·013583
1·71	0·956367	0·092459	2·16	0·984614	0·038707	2·61	0·995473	0·013234
1·72	0·957284	0·090887	2·17	0·984997	0·037878	2·62	0·995604	0·012892
1·73	0·958185	0·089333	2·18	0·985371	0·037063	2·63	0·995731	0·012558
1·74	0·959070	0·087796	2·19	0·985738	0·036262	2·64	0·995855	0·012232
1·75	0·959941	0·086277	2·20	0·986097	0·035475	2·65	0·995975	0·011912
1·76	0·960796	0·084776	2·21	0·986447	0·034701	2·66	0·996093	0·011600
1·77	0·961636	0·083293	2·22	0·986791	0·033941	2·67	0·996207	0·011295
1·78	0·962462	0·081828	2·23	0·987126	0·033194	2·68	0·996319	0·010997
1·79	0·963273	0·080380	2·24	0·987455	0·032460	2·69	0·996427	0·010706

TABLE 1 (*cont.*)

X	P(X)	Z(X)	X	P(X)	Z(X)	X	P(X)	Z(X)
2·70	0·996533	0·010421	3·15	0·999184	0·002794	3·60	0·999841	0·000612
2·71	0·996636	0·010143	3·16	0·999211	0·002707	3·61	0·999847	0·000590
2·72	0·996736	0·009871	3·17	0·999238	0·002623	3·62	0·999853	0·000569
2·73	0·996833	0·009606	3·18	0·999264	0·002541	3·63	0·999858	0·000549
2·74	0·996928	0·009347	3·19	0·999289	0·002461	3·64	0·999864	0·000529
2·75	0·997020	0·009094	3·20	0·999313	0·002384	3·65	0·999869	0·000510
2·76	0·997110	0·008846	3·21	0·999336	0·002309	3·66	0·999874	0·000492
2·77	0·997197	0·008605	3·22	0·999359	0·002236	3·67	0·999879	0·000474
2·78	0·997282	0·008370	3·23	0·999381	0·002165	3·68	0·999883	0·000457
2·79	0·997365	0·008140	3·24	0·999402	0·002096	3·69	0·999888	0·000441
2·80	0·997445	0·007915	3·25	0·999423	0·002029	3·70	0·999892	0·000425
2·81	0·997523	0·007697	3·26	0·999443	0·001964	3·71	0·999896	0·000409
2·82	0·997599	0·007483	3·27	0·999462	0·001901	3·72	0·999900	0·000394
2·83	0·997673	0·007274	3·28	0·999481	0·001840	3·73	0·999904	0·000380
2·84	0·997744	0·007071	3·29	0·999499	0·001780	3·74	0·999908	0·000366
2·85	0·997814	0·006873	3·30	0·999517	0·001723	3·75	0·999912	0·000353
2·86	0·997882	0·006679	3·31	0·999534	0·001667	3·76	0·999915	0·000340
2·87	0·997948	0·006491	3·32	0·999550	0·001612	3·77	0·999918	0·000327
2·88	0·998012	0·006307	3·33	0·999566	0·001560	3·78	0·999922	0·000315
2·89	0·998074	0·006127	3·34	0·999581	0·001508	3·79	0·999925	0·000303
2·90	0·998134	0·005953	3·35	0·999596	0·001459	3·80	0·999928	0·000292
2·91	0·998193	0·005782	3·36	0·999610	0·001411	3·81	0·999931	0·000281
2·92	0·998250	0·005616	3·37	0·999624	0·001364	3·82	0·999933	0·000271
2·93	0·998305	0·005454	3·38	0·999638	0·001319	3·83	0·999936	0·000260
2·94	0·998359	0·005296	3·39	0·999651	0·001275	3·84	0·999938	0·000251
2·95	0·998411	0·005143	3·40	0·999663	0·001232	3·85	0·999941	0·000241
2·96	0·998462	0·004993	3·41	0·999675	0·001191	3·86	0·999943	0·000232
2·97	0·998511	0·004847	3·42	0·999687	0·001151	3·87	0·999946	0·000223
2·98	0·998559	0·004705	3·43	0·999698	0·001112	3·88	0·999948	0·000215
2·99	0·998605	0·004567	3·44	0·999709	0·001075	3·89	0·999950	0·000207
3·00	0·998650	0·004432	3·45	0·999720	0·001038	3·90	0·999952	0·000199
3·01	0·998694	0·004301	3·46	0·999730	0·001003	3·91	0·999954	0·000191
3·02	0·998736	0·004173	3·47	0·999740	0·000969	3·92	0·999956	0·000184
3·03	0·998777	0·004049	3·48	0·999749	0·000936	3·93	0·999958	0·000177
3·04	0·998817	0·003928	3·49	0·999758	0·000904	3·94	0·999959	0·000170
3·05	0·998856	0·003810	3·50	0·999767	0·000873	3·95	0·999961	0·000163
3·06	0·998893	0·003695	3·51	0·999776	0·000843	3·96	0·999963	0·000157
3·07	0·998930	0·003584	3·52	0·999784	0·000814	3·97	0·999964	0·000151
3·08	0·998965	0·003475	3·53	0·999792	0·000785	3·98	0·999966	0·000145
3·09	0·998999	0·003370	3·54	0·999800	0·000758	3·99	0·999967	0·000139
3·10	0·999032	0·003267	3·55	0·999807	0·000732			
3·11	0·999065	0·003167	3·56	0·999815	0·000706			
3·12	0·999096	0·003070	3·57	0·999822	0·000681			
3·13	0·999126	0·002975	3·58	0·999828	0·000657			
3·14	0·999155	0·002884	3·59	0·999835	0·000634			

TABLE 2. SIGNIFICANCE POINTS FOR STUDENT'S t-DISTRIBUTION
Upper 0·05, 0·025, 0·01 and 0·005 significance points

α ν	0·05	0·025	0·01	0·005
1	6·3138	12·7062	31·8207	63·6574
2	2·9200	4·3017	6·9646	9·9248
3	2·3534	3·1824	4·5407	5·8409
4	2·1318	2·7764	3·7469	4·6041
5	2·0150	2·5706	3·3649	4·0322
6	1·9432	2·4469	3·1427	3·7074
7	1·8946	2·3646	2·9980	3·4995
8	1·8595	2·3060	2·8965	3·3554
9	1·8331	2·2622	2·8214	3·2498
10	1·8125	2·2281	2·7638	3·1693
11	1·7959	2·2010	2·7181	3·1058
12	1·7823	2·1788	2·6810	3·0545
13	1·7709	2·1604	2·6503	3·0123
14	1·7613	2·1448	2·6245	2·9768
15	1·7531	2·1315	2·6025	2·9467
16	1·7459	2·1199	2·5835	2·9208
17	1·7396	2·1098	2·5669	2·8982
18	1·7341	2·1009	2·5524	2·8784
19	1·7291	2·0930	2·5395	2·8609
20	1·7247	2·0860	2·5280	2·8453
21	1·7207	2·0796	2·5177	2·8314
22	1·7171	2·0739	2·5083	2·8188
23	1·7139	2·0687	2·4999	2·8073
24	1·7109	2·0639	2·4922	2·7969
25	1·7081	2·0595	2·4851	2·7874
26	1·7056	2·0555	2·4786	2·7787
27	1·7033	2·0518	2·4727	2·7707
28	1·7011	2·0484	2·4671	2·7633
29	1·6991	2·0452	2·4620	2·7564
30	1·6973	2·0423	2·4573	2·7500
31	1·6955	2·0395	2·4528	2·7440
32	1·6939	2·0369	2·4487	2·7385
33	1·6924	2·0345	2·4448	2·7333
34	1·6909	2·0322	2·4411	2·7284
35	1·6896	2·0301	2·4377	2·7238
36	1·6883	2·0281	2·4345	2·7195
37	1·6871	2·0262	2·4314	2·7154
38	1·6860	2·0244	2·4286	2·7116
39	1·6849	2·0227	2·4258	2·7079
40	1·6839	2·0211	2·4233	2·7045
60	1·6706	2·0003	2·3901	2·6613
120	1·6577	1·9799	2·3578	2·6174
∞	1·6449	1·9600	2·3263	2·5758

TABLE 3. NON-CENTRAL t-DISTRIBUTION

where $t_{0.05}$ is the upper 0.05 significance point of the central t-distribution

$$\text{Area (power)} = \int_{t_{0.05}}^{\infty} p(t|\nu,\delta)\,dt$$

ν	$t_{0.05}$	δ (non-centrality parameter)								
		0·0	0·5	1·0	1·5	2·0	2·5	3·0	3·5	4·0
1	6·314	0·050	0·087	0·135	0·189	0·247	0·305	0·361	0·416	0·469
2	2·920	0·050	0·099	0·174	0·271	0·384	0·503	0·617	0·719	0·803
3	2·353	0·050	0·106	0·196	0·318	0·461	0·606	0·735	0·837	0·908
4	2·132	0·050	0·111	0·210	0·346	0·504	0·660	0·791	0·886	0·944
5	2·015	0·050	0·113	0·219	0·364	0·531	0·692	0·822	0·910	0·960
6	1·943	0·050	0·115	0·225	0·377	0·550	0·713	0·840	0·923	0·968
7	1·895	0·050	0·117	0·230	0·386	0·563	0·727	0·853	0·932	0·973
8	1·860	0·050	0·118	0·233	0·393	0·573	0·738	0·862	0·938	0·977
9	1·833	0·050	0·119	0·236	0·398	0·580	0·746	0·868	0·942	0·979
10	1·812	0·050	0·119	0·238	0·403	0·586	0·752	0·874	0·946	0·981
11	1·796	0·050	0·120	0·240	0·406	0·591	0·757	0·878	0·948	0·982
12	1·782	0·050	0·120	0·242	0·409	0·595	0·761	0·881	0·950	0·983
13	1·771	0·050	0·121	0·243	0·412	0·599	0·765	0·884	0·952	0·984
14	1·761	0·050	0·121	0·244	0·414	0·601	0·768	0·886	0·953	0·984
15	1·753	0·050	0·121	0·245	0·416	0·604	0·770	0·888	0·955	0·985
16	1·746	0·050	0·122	0·246	0·417	0·606	0·773	0·890	0·956	0·985
17	1·740	0·050	0·122	0·247	0·419	0·608	0·774	0·891	0·956	0·986
18	1·734	0·050	0·122	0·248	0·420	0·610	0·776	0·892	0·957	0·986
19	1·729	0·050	0·123	0·248	0·421	0·611	0·778	0·894	0·958	0·986
20	1·725	0·050	0·123	0·249	0·422	0·613	0·779	0·895	0·958	0·987
21	1·721	0·050	0·123	0·249	0·423	0·614	0·780	0·896	0·959	0·987
22	1·717	0·050	0·123	0·250	0·424	0·615	0·781	0·896	0·959	0·987
23	1·714	0·050	0·123	0·250	0·425	0·616	0·782	0·897	0·960	0·987
24	1·711	0·050	0·123	0·250	0·426	0·617	0·783	0·898	0·960	0·987
25	1·708	0·050	0·123	0·251	0·426	0·618	0·784	0·898	0·961	0·988
26	1·706	0·050	0·123	0·251	0·427	0·619	0·785	0·899	0·961	0·988
27	1·703	0·050	0·123	0·251	0·428	0·620	0·786	0·899	0·961	0·988
28	1·701	0·050	0·124	0·252	0·428	0·620	0·786	0·900	0·962	0·988
29	1·699	0·050	0·124	0·252	0·429	0·621	0·787	0·900	0·962	0·988
30	1·697	0·050	0·124	0·252	0·429	0·622	0·788	0·901	0·962	0·988
31	1·696	0·050	0·124	0·253	0·429	0·622	0·788	0·901	0·962	0·988
32	1·694	0·050	0·124	0·253	0·430	0·623	0·789	0·902	0·962	0·988
33	1·692	0·050	0·124	0·253	0·430	0·623	0·789	0·902	0·963	0·988
34	1·691	0·050	0·124	0·243	0·431	0·624	0·790	0·902	0·963	0·988
35	1·690	0·050	0·124	0·253	0·431	0·624	0·790	0·903	0·963	0·989
36	1·688	0·050	0·124	0·253	0·432	0·625	0·790	0·903	0·963	0·989
37	1·687	0·050	0·124	0·254	0·432	0·625	0·791	0·903	0·963	0·989
38	1·686	0·050	0·124	0·254	0·432	0·625	0·791	0·903	0·963	0·989
39	1·685	0·050	0·124	0·254	0·432	0·626	0·791	0·903	0·964	0·989
40	1·684	0·050	0·124	0·254	0·432	0·626	0·792	0·904	0·964	0·989
60	1·671	0·050	0·125	0·256	0·436	0·630	0·796	0·907	0·965	0·989
120	1·658	0·050	0·126	0·258	0·439	0·635	0·800	0·910	0·967	0·990
∞	1·646	0·050	0·126	0·259	0·442	0·638	0·803	0·912	0·968	0·991

$$\text{Area (power)} = \int_{t_{0 \cdot 01}}^{\infty} p(t|\nu, \delta)\, dt \quad \text{where } t_{0 \cdot 01} \text{ is the upper } 0 \cdot 01 \text{ significance point of the central } t\text{-distribution}$$

ν	$t_{0 \cdot 01}$	δ (non-centrality parameter)								
		0·0	0·5	1·0	1·5	2·0	2·5	3·0	3·5	4·0
1	31·821	0·010	0·017	0·027	0·038	0·050	0·063	0·075	0·088	0·100
2	6·965	0·010	0·021	0·038	0·062	0·095	0·134	0·180	0·231	0·286
3	4·541	0·010	0·023	0·047	0·085	0·139	0·210	0·295	0·389	0·487
4	3·747	0·010	0·025	0·054	0·103	0·176	0·272	0·386	0·508	0·627
5	3·365	0·010	0·026	0·059	0·117	0·204	0·319	0·453	0·590	0·715
6	3·143	0·010	0·027	0·063	0·128	0·226	0·355	0·502	0·647	0·771
7	2·998	0·010	0·028	0·066	0·136	0·243	0·383	0·538	0·686	0·809
8	2·896	0·010	0·029	0·069	0·143	0·257	0·404	0·566	0·716	0·834
9	2·821	0·010	0·029	0·071	0·149	0·268	0·422	0·587	0·738	0·853
10	2·764	0·010	0·030	0·073	0·154	0·278	0·436	0·605	0·755	0·867
11	2·718	0·010	0·030	0·075	0·158	0·285	0·448	0·619	0·768	0·878
12	2·681	0·010	0·030	0·076	0·161	0·292	0·457	0·630	0·779	0·886
13	2·650	0·010	0·031	0·077	0·164	0·298	0·466	0·640	0·789	0·893
14	2·624	0·010	0·031	0·078	0·167	0·303	0·473	0·648	0·796	0·899
15	2·602	0·010	0·031	0·079	0·169	0·307	0·480	0·656	0·803	0·903
16	2·583	0·010	0·031	0·080	0·171	0·311	0·485	0·662	0·808	0·907
17	2·567	0·010	0·031	0·080	0·173	0·314	0·490	0·667	0·813	0·911
18	2·552	0·010	0·032	0·081	0·174	0·317	0·494	0·672	0·817	0·914
19	2·539	0·010	0·032	0·081	0·176	0·320	0·498	0·677	0·821	0·916
20	2·528	0·010	0·032	0·082	0·177	0·323	0·502	0·680	0·825	0·919
21	2·518	0·010	0·032	0·082	0·178	0·325	0·505	0·684	0·827	0·921
22	2·508	0·010	0·032	0·083	0·180	0·327	0·508	0·687	0·830	0·922
23	2·500	0·010	0·032	0·083	0·181	0·329	0·511	0·690	0·833	0·924
24	2·492	0·010	0·032	0·084	0·181	0·331	0·513	0·693	0·835	0·925
25	2·485	0·010	0·032	0·084	0·182	0·332	0·515	0·695	0·837	0·927
26	2·479	0·010	0·032	0·084	0·183	0·334	0·517	0·697	0·839	0·928
27	2·473	0·010	0·032	0·084	0·184	0·335	0·519	0·699	0·840	0·929
28	2·467	0·010	0·032	0·085	0·185	0·336	0·521	0·701	0·842	0·930
29	2·462	0·010	0·032	0·085	0·185	0·338	0·523	0·703	0·843	0·931
30	2·457	0·010	0·032	0·085	0·186	0·339	0·524	0·705	0·845	0·932
31	2·453	0·010	0·032	0·086	0·186	0·340	0·526	0·706	0·846	0·933
32	2·449	0·010	0·032	0·086	0·187	0·341	0·527	0·707	0·847	0·933
33	2·445	0·010	0·032	0·086	0·187	0·342	0·528	0·709	0·848	0·934
34	2·441	0·010	0·032	0·086	0·188	0·343	0·530	0·710	0·849	0·935
35	2·438	0·010	0·032	0·086	0·188	0·343	0·531	0·711	0·850	0·935
36	2·434	0·010	0·033	0·087	0·189	0·344	0·532	0·712	0·851	0·936
37	2·431	0·010	0·033	0·087	0·189	0·345	0·533	0·713	0·852	0·936
38	2·429	0·010	0·033	0·087	0·190	0·346	0·534	0·714	0·853	0·937
39	2·426	0·010	0·033	0·087	0·190	0·346	0·535	0·715	0·853	0·937
40	2·423	0·010	0·033	0·089	0·190	0·347	0·535	0·716	0·854	0·938
60	2·390	0·010	0·033	0·091	0·195	0·355	0·547	0·728	0·863	0·943
120	2·358	0·010	0·033	0·091	0·200	0·364	0·558	0·739	0·872	0·948
∞	2·330	0·010	0·034	0·092	0·204	0·371	0·568	0·748	0·879	0·952

TABLE 4. SIGNIFICANCE POINTS FOR THE CHI-SQUARED DISTRIBUTION
Upper 0·05, 0·025, 0·01 and 0·005 significance points

α ν	0·05	0·025	0·01	0·005
1	3·841	5·024	6·635	7·879
2	5·991	7·378	9·210	10·597
3	7·815	9·348	11·345	12·838
4	9·488	11·143	13·277	14·860
5	11·071	12·833	15·086	16·750
6	12·592	14·449	16·812	18·548
7	14·067	16·013	18·475	20·278
8	15·507	17·535	20·090	21·955
9	16·919	19·023	21·666	23·589
10	18·307	20·483	23·209	25·188
11	19·675	21·920	24·725	26·757
12	21·026	23·337	26·217	28·299
13	22·362	24·736	27·688	29·819
14	23·685	26·119	29·141	31·319
15	24·996	27·488	30·578	32·801
16	26·296	28·845	32·000	34·267
17	27·587	30·191	33·409	35·718
18	28·869	31·526	34·805	37·156
19	30·144	32·852	36·191	38·582
20	31·410	34·170	37·566	39·997
21	32·671	35·479	38·932	41·401
22	33·924	36·781	30·289	42·796
23	35·172	38·076	41·638	44·181
24	36·415	39·364	42·980	45·559
25	37·652	40·646	44·314	46·928
26	38·885	41·923	45·642	48·290
27	40·113	43·194	46·963	49·645
28	41·337	44·461	48·278	50·993
29	42·557	45·722	49·588	52·336
30	43·773	46·979	50·892	53·672
31	44·985	48·232	52·191	55·003
32	46·194	49·480	53·486	56·328
33	47·400	50·725	54·776	57·648
34	48·602	51·966	56·061	58·964
35	49·802	53·203	57·342	60·275
36	50·998	54·437	58·619	61·581
37	52·192	55·668	59·892	62·883
38	53·384	56·896	61·162	64·181
39	54·572	58·120	62·428	65·476
40	55·758	59·342	63·691	66·766
60	79·082	83·298	88·379	91·952
120	146·567	152·211	158·950	163·648

Approximations: $\chi_\alpha^2 = \nu \left\{ 1 - \dfrac{2}{9\nu} + X_\alpha \sqrt{\dfrac{2}{9\nu}} \right\}^3$ or $\chi_\alpha^2 = \frac{1}{2} \left\{ X_\alpha + \sqrt{(2\nu - 1)} \right\}^2$

where X_α is the corresponding standardized Normal upper α significance level. First approximation is closer than the second.

TABLES SECTION

TABLE 5. SIGNIFICANCE POINTS FOR WELCH'S TEST
Upper 0·05 significance point

ν_1	ν_2	C										
		0·0	0·1	0·2	0·3	0·4	0·5	0·6	0·7	0·8	0·9	1·0
6	6	1·94	1·90	1·85	1·80	1·76	1·74	1·76	1·80	1·85	1·90	1·94
	8	1·94	1·90	1·85	1·80	1·76	1·73	1·74	1·76	1·79	1·82	1·86
	10	1·94	1·90	1·85	1·80	1·76	1·73	1·73	1·74	1·76	1·78	1·81
	15	1·94	1·90	1·85	1·80	1·76	1·73	1·71	1·71	1·72	1·73	1·75
	20	1·94	1·90	1·85	1·80	1·76	1·73	1·71	1·70	1·70	1·71	1·72
	∞	1·94	1·90	1·85	1·80	1·76	1·72	1·69	1·67	1·66	1·65	1·64
8	6	1·86	1·82	1·79	1·76	1·74	1·73	1·76	1·80	1·85	1·90	1·94
	8	1·86	1·82	1·79	1·76	1·73	1·73	1·73	1·76	1·79	1·82	1·86
	10	1·86	1·82	1·79	1·76	1·73	1·72	1·72	1·74	1·76	1·78	1·81
	15	1·86	1·82	1·79	1·76	1·73	1·71	1·71	1·71	1·72	1·73	1·75
	20	1·86	1·82	1·79	1·76	1·73	1·71	1·70	1·70	1·70	1·71	1·72
	∞	1·86	1·82	1·79	1·75	1·72	1·70	1·68	1·66	1·65	1·65	1·64
10	6	1·81	1·78	1·76	1·74	1·73	1·73	1·76	1·80	1·85	1·90	1·94
	8	1·81	1·78	1·76	1·74	1·72	1·72	1·73	1·76	1·79	1·82	1·86
	10	1·81	1·78	1·76	1·73	1·72	1·71	1·72	1·73	1·76	1·78	1·81
	15	1·81	1·78	1·76	1·73	1·72	1·70	1·70	1·71	1·72	1·73	1·75
	20	1·81	1·78	1·76	1·73	1·71	1·70	1·69	1·69	1·70	1·71	1·72
	∞	1·81	1·78	1·76	1·73	1·71	1·69	1·67	1·66	1·65	1·65	1·64
15	6	1·75	1·73	1·72	1·71	1·71	1·73	1·76	1·80	1·85	1·90	1·94
	8	1·75	1·73	1·72	1·71	1·71	1·71	1·73	1·76	1·79	1·82	1·86
	10	1·75	1·73	1·72	1·71	1·70	1·70	1·72	1·73	1·76	1·78	1·81
	15	1·75	1·73	1·72	1·70	1·70	1·69	1·70	1·70	1·72	1·73	1·75
	20	1·75	1·73	1·72	1·70	1·69	1·69	1·69	1·69	1·70	1·71	1·72
	∞	1·75	1·73	1·72	1·70	1·68	1·67	1·66	1·65	1·65	1·65	1·64
20	6	1·72	1·71	1·70	1·70	1·71	1·73	1·76	1·80	1·85	1·90	1·94
	8	1·72	1·71	1·70	1·70	1·70	1·71	1·73	1·76	1·79	1·82	1·86
	10	1·72	1·71	1·70	1·69	1·69	1·70	1·71	1·73	1·76	1·78	1·81
	15	1·72	1·71	1·70	1·69	1·69	1·69	1·69	1·70	1·72	1·73	1·75
	20	1·72	1·71	1·70	1·69	1·68	1·68	1·68	1·69	1·70	1·71	1·72
	∞	1·72	1·71	1·70	1·68	1·67	1·66	1·66	1·65	1·65	1·65	1·64
∞	6	1·64	1·65	1·66	1·67	1·69	1·72	1·76	1·80	1·85	1·90	1·94
	8	1·64	1·65	1·65	1·66	1·68	1·70	1·72	1·75	1·79	1·82	1·86
	10	1·64	1·65	1·65	1·66	1·67	1·69	1·71	1·73	1·76	1·78	1·81
	15	1·64	1·65	1·65	1·65	1·66	1·67	1·68	1·70	1·72	1·73	1·75
	20	1·64	1·65	1·65	1·65	1·66	1·66	1·67	1·68	1·70	1·71	1·72
	∞	1·64	1·64	1·64	1·64	1·64	1·64	1·64	1·64	1·64	1·64	1·64

TABLE 5 (cont.)

Upper 0·01 significance point

v_1	v_2	C										
		0·0	0·1	0·2	0·3	0·4	0·5	0·6	0·7	0·8	0·9	1·0
10	10	2·76	2·70	2·63	2·56	2·51	2·50	2·51	2·56	2·63	2·70	2·76
	12	2·76	2·70	2·63	2·56	2·51	2·49	2·49	2·52	2·57	2·62	2·68
	15	2·76	2·70	2·63	2·56	2·51	2·48	2·47	2·48	2·52	2·56	2·60
	20	2·76	2·70	2·63	2·56	2·51	2·47	2·45	2·45	2·47	2·49	2·53
	30	2·76	2·70	2·63	2·56	2·50	2·46	2·43	2·42	2·42	2·44	2·46
	∞	2·76	2·70	2·63	2·56	2·50	2·44	2·40	2·36	2·34	2·33	2·33
12	10	2·68	2·62	2·57	2·52	2·49	2·49	2·51	2·56	2·63	2·70	2·76
	12	2·68	2·62	2·57	2·52	2·48	2·47	2·48	2·52	2·57	2·62	2·68
	15	2·68	2·62	2·57	2·52	2·48	2·46	2·46	2·48	2·52	2·56	2·60
	20	2·68	2·62	2·57	2·52	2·48	2·45	2·44	2·45	2·47	2·49	2·53
	30	2·68	2·62	2·57	2·52	2·47	2·44	2·42	2·41	2·42	2·44	2·46
	∞	2·68	2·62	2·57	2·51	2·46	2·42	2·38	2·36	2·34	2·33	2·33
15	10	2·60	2·56	2·52	2·48	2·47	2·48	2·51	2·56	2·63	2·70	2·76
	12	2·60	2·56	2·52	2·48	2·46	2·46	2·48	2·52	2·57	2·62	2·68
	15	2·60	2·56	2·51	2·48	2·45	2·45	2·45	2·48	2·51	2·56	2·60
	20	2·60	2·56	2·51	2·48	2·45	2·43	2·43	2·44	2·46	2·49	2·53
	30	2·60	2·56	2·51	2·47	2·44	2·42	2·41	2·41	2·42	2·44	2·46
	∞	2·60	2·56	2·51	2·47	2·43	2·40	2·37	2·35	2·34	2·33	2·33
20	10	2·53	2·49	2·47	2·45	2·45	2·47	2·51	2·56	2·63	2·70	2·76
	12	2·53	2·49	2·47	2·45	2·44	2·45	2·48	2·52	2·57	2·62	2·68
	15	2·53	2·49	2·46	2·44	2·43	2·43	2·45	2·48	2·51	2·56	2·60
	20	2·53	2·49	2·46	2·44	2·42	2·42	2·42	2·44	2·46	2·49	2·53
	30	2·53	2·49	2·46	2·44	2·42	2·40	2·40	2·40	2·42	2·43	2·46
	∞	2·53	2·49	2·46	2·43	2·40	2·38	2·36	2·34	2·33	2·33	2·33
30	10	2·46	2·44	2·42	2·42	2·43	2·46	2·50	2·56	2·63	2·70	2·76
	12	2·46	2·44	2·42	2·41	2·42	2·44	2·47	2·52	2·57	2·62	2·68
	15	2·46	2·44	2·42	2·41	2·41	2·42	2·44	2·47	2·51	2·56	2·60
	20	2·46	2·43	2·42	2·40	2·40	2·40	2·42	2·44	2·46	2·49	2·53
	30	2·46	2·43	2·42	2·40	2·39	2·39	2·40	2·40	2·42	2·43	2·46
	∞	2·46	2·43	2·41	2·39	2·37	2·36	2·35	2·34	2·33	2·33	2·33
∞	10	2·33	2·33	2·34	2·36	2·40	2·44	2·50	2·56	2·63	2·70	2·76
	12	2·33	2·33	2·34	2·36	2·38	2·42	2·46	2·51	2·57	2·62	2·68
	15	2·33	2·33	2·34	2·35	2·37	2·40	2·43	2·47	2·51	2·56	2·60
	20	2·33	2·33	2·33	2·34	2·36	2·38	2·40	2·43	2·46	2·49	2·53
	30	2·33	2·33	2·33	2·34	2·35	2·36	2·37	2·39	2·41	2·43	2·46
	∞	2·33	2·33	2·33	2·33	2·33	2·33	2·33	2·33	2·33	2·33	2·33

TABLE 7. SIGNIFICANCE POINTS FOR THE SAMPLE CORRELATION COEFFICIENT WHEN $\varrho = 0$

Upper significance points of r

n \ α	0·25	0·10	0·05	0·025	0·01	0·005
3	0·7071	0·9511	0·9877	0·9969	0·9995	0·9999
4	0·5000	0·8000	0·9000	0·9500	0·9800	0·9900
5	0·4040	0·6870	0·8054	0·8783	0·9343	0·9587
6	0·3473	0·6084	0·7293	0·8114	0·8822	0·9172
7	0·3091	0·5509	0·6694	0·7545	0·8329	0·8745
8	0·2811	0·5067	0·6215	0·7067	0·7887	0·8343
9	0·2596	0·4716	0·5822	0·6664	0·7498	0·7977
10	0·2423	0·4428	0·5493	0·6319	0·7155	0·7646
11	0·2281	0·4187	0·5214	0·6021	0·6851	0·7348
12	0·2161	0·3981	0·4973	0·5760	0·6581	0·7079
13	0·2058	0·3802	0·4762	0·5529	0·6339	0·6835
14	0·1968	0·3646	0·4575	0·5324	0·6120	0·6614
15	0·1890	0·3507	0·4409	0·5140	0·5923	0·6411
16	0·1820	0·3383	0·4259	0·4973	0·5742	0·6226
17	0·1757	0·3271	0·4124	0·4822	0·5577	0·6055
18	0·1700	0·3170	0·4000	0·4683	0·5426	0·5897
19	0·1649	0·3077	0·3887	0·4555	0·5285	0·5751
20	0·1602	0·2992	0·3783	0·4438	0·5155	0·5614
21	0·1558	0·2914	0·3687	0·4329	0·5034	0·5487
22	0·1518	0·2841	0·3598	0·4227	0·4921	0·5368
23	0·1481	0·2774	0·3515	0·4132	0·4815	0·5256
24	0·1447	0·2711	0·3438	0·4044	0·4716	0·5151
25	0·1415	0·2653	0·3365	0·3961	0·4622	0·5052
30	0·1281	0·2407	0·3061	0·3610	0·4226	0·4629
35	0·1179	0·2220	0·2826	0·3338	0·3916	0·4296
40	0·1098	0·2070	0·2638	0·3120	0·3665	0·4026
45	0·1032	0·1947	0·2483	0·2940	0·3457	0·3801
50	0·0976	0·1843	0·2353	0·2787	0·3281	0·3610
60	0·0888	0·1678	0·2144	0·2542	0·2997	0·3301
70	0·0820	0·1550	0·1982	0·2352	0·2776	0·3060
80	0·0765	0·1448	0·1852	0·2199	0·2597	0·2864
90	0·0720	0·1364	0·1745	0·2072	0·2449	0·2702
100	0·0682	0·1292	0·1654	0·1966	0·2324	0·2565

TABLE 6 (*cont.*) Upper 0·01 significance points

ν_2 \ ν_1	1	2	3	4	5	6	7	8	9	10	12	15	20	24	30	40	60	120	∞	ν_2
1	4052	4999·5	5403	5625	5764	5859	5928	5982	6022	6056	6106	6157	6209	6235	6261	6287	6313	6339	6366	1
2	98·50	99·00	99·17	99·25	99·30	99·33	99·36	99·37	99·39	99·40	99·42	99·43	99·45	99·46	99·47	99·47	99·48	99·49	99·50	2
3	34·12	30·82	29·46	28·71	28·24	27·91	27·67	27·49	27·35	27·23	27·05	26·87	26·69	26·60	26·50	26·41	26·32	26·22	26·13	3
4	21·20	18·00	16·69	15·98	15·52	15·21	14·98	14·80	14·66	14·55	14·37	14·20	14·02	13·93	13·84	13·75	13·65	13·56	13·46	4
5	16·26	13·27	12·06	11·39	10·97	10·67	10·46	10·29	10·16	10·05	9·89	9·72	9·55	9·47	9·38	9·29	9·20	9·11	9·02	5
6	13·75	10·92	9·78	9·15	8·75	8·47	8·26	8·10	7·98	7·87	7·72	7·56	7·40	7·31	7·23	7·14	7·06	6·97	6·88	6
7	12·25	9·55	8·45	7·85	7·46	7·19	6·99	6·84	6·72	6·62	6·47	6·31	6·16	6·07	5·99	5·91	5·82	5·74	5·65	7
8	11·26	8·65	7·59	7·01	6·63	6·37	6·18	6·03	5·91	5·81	5·67	5·52	5·36	5·28	5·20	5·12	5·03	4·95	4·86	8
9	10·56	8·02	6·99	6·42	6·06	5·80	5·61	5·47	5·35	5·26	5·11	4·96	4·81	4·73	4·65	4·57	4·48	4·40	4·31	9
10	10·04	7·56	6·55	5·99	5·64	5·39	5·20	5·06	4·94	4·85	4·71	4·56	4·41	4·33	4·25	4·17	4·08	4·00	3·91	10
11	9·65	7·21	6·22	5·67	5·32	5·07	4·89	4·74	4·63	4·54	4·40	4·25	4·10	4·02	3·94	3·86	3·78	3·69	3·60	11
12	9·33	6·93	5·95	5·41	5·06	4·82	4·64	4·50	4·39	4·30	4·16	4·01	3·86	3·78	3·70	3·62	3·54	3·45	3·36	12
13	9·07	6·70	5·74	5·21	4·86	4·62	4·44	4·30	4·19	4·10	3·96	3·82	3·66	3·59	3·51	3·43	3·34	3·25	3·17	13
14	8·86	6·51	5·56	5·04	4·69	4·46	4·28	4·14	4·03	3·94	3·80	3·66	3·51	3·43	3·35	3·27	3·18	3·09	3·00	14
15	8·68	6·36	5·42	4·89	4·56	4·32	4·14	4·00	3·89	3·80	3·67	3·52	3·37	3·29	3·21	3·13	3·05	2·96	2·87	15
16	8·53	6·23	5·29	4·77	4·44	4·20	4·03	3·89	3·78	3·69	3·55	3·41	3·26	3·18	3·10	3·02	2·93	2·84	2·75	16
17	8·40	6·11	5·18	4·67	4·34	4·10	3·93	3·79	3·68	3·59	3·46	3·31	3·16	3·08	3·00	2·92	2·83	2·75	2·65	17
18	8·29	6·01	5·09	4·58	4·25	4·01	3·84	3·71	3·60	3·51	3·37	3·23	3·08	3·00	2·92	2·84	2·75	2·66	2·57	18
19	8·18	5·93	5·01	4·50	4·17	3·94	3·77	3·63	3·52	3·43	3·30	3·15	3·00	2·92	2·84	2·76	2·67	2·58	2·49	19
20	8·10	5·85	4·94	4·43	4·10	3·87	3·70	3·56	3·46	3·37	3·23	3·09	2·94	2·86	2·78	2·69	2·61	2·52	2·42	20
21	8·02	5·78	4·87	4·37	4·04	3·81	3·64	3·51	3·40	3·31	3·17	3·03	2·88	2·80	2·72	2·64	2·55	2·46	2·36	21
22	7·95	5·72	4·82	4·31	3·99	3·76	3·59	3·45	3·35	3·26	3·12	2·98	2·83	2·75	2·67	2·58	2·50	2·40	2·31	22
23	7·88	5·66	4·76	4·26	3·94	3·71	3·54	3·41	3·30	3·21	3·07	2·93	2·78	2·70	2·62	2·54	2·45	2·35	2·26	23
24	7·82	5·61	4·72	4·22	3·90	3·67	3·50	3·36	3·26	3·17	3·03	2·89	2·74	2·66	2·58	2·49	2·40	2·31	2·21	24
25	7·77	5·57	4·68	4·18	3·85	3·63	3·46	3·32	3·22	3·13	2·99	2·85	2·70	2·62	2·54	2·45	2·36	2·27	2·17	25
26	7·72	5·53	4·64	4·14	3·82	3·59	3·42	3·29	3·18	3·09	2·96	2·81	2·66	2·58	2·50	2·42	2·33	2·23	2·13	26
27	7·68	5·49	4·60	4·11	3·78	3·56	3·39	3·26	3·15	3·06	2·93	2·78	2·63	2·55	2·47	2·38	2·29	2·20	2·10	27
28	7·64	5·45	4·57	4·07	3·75	3·53	3·36	3·23	3·12	3·03	2·90	2·75	2·60	2·52	2·44	2·35	2·26	2·17	2·06	28
29	7·60	5·42	4·54	4·04	3·73	3·50	3·33	3·20	3·09	3·00	2·87	2·73	2·57	2·49	2·41	2·33	2·23	2·14	2·03	29
30	7·56	5·39	4·51	4·02	3·70	3·47	3·30	3·17	3·07	2·98	2·84	2·70	2·55	2·47	2·39	2·30	2·21	2·11	2·01	30
40	7·31	5·18	4·31	3·83	3·51	3·29	3·12	2·99	2·89	2·80	2·66	2·52	2·37	2·29	2·20	2·11	2·02	1·92	1·80	40
60	7·08	4·98	4·13	3·65	3·34	3·12	2·95	2·82	2·72	2·63	2·50	2·35	2·20	2·12	2·03	1·94	1·84	1·73	1·60	60
120	6·85	4·79	3·95	3·48	3·17	2·96	2·79	2·66	2·56	2·47	2·34	2·19	2·03	1·95	1·86	1·76	1·66	1·53	1·38	120
∞	6·63	4·61	3·78	3·32	3·02	2·80	2·64	2·51	2·41	2·32	2·18	2·04	1·88	1·79	1·70	1·59	1·47	1·32	1·00	∞

$F = \dfrac{s_1^2}{s_2^2} = \dfrac{S_1}{\nu_1} \bigg/ \dfrac{S_2}{\nu_2}$, where $s_1^2 = S_1/\nu_1$ and $s_2^2 = S_2/\nu_2$ are independent mean squares estimating a common variance σ^2 and based on ν_1 and ν_2 degrees of freedom, respectively.

TABLE 6. SIGNIFICANCE POINTS FOR THE F-DISTRIBUTION (VARIANCE RATIO)

Upper 0.05 significance points

$\nu_2 \backslash \nu_1$	1	2	3	4	5	6	7	8	9	10	12	15	20	24	30	40	60	120	∞
1	161·4	199·5	215·7	224·6	230·2	234·0	236·8	238·9	240·5	241·9	243·9	245·9	248·0	249·1	250·1	251·1	252·2	253·3	254·3
2	18·51	19·00	19·16	19·25	19·30	19·33	19·35	19·37	19·38	19·40	19·41	19·43	19·45	19·45	19·46	19·47	19·48	19·49	19·50
3	10·13	9·55	9·28	9·12	9·01	8·94	8·89	8·85	8·81	8·79	8·74	8·70	8·66	8·64	8·62	8·59	8·57	8·55	8·53
4	7·71	6·94	6·59	6·39	6·26	6·16	6·09	6·04	6·00	5·96	5·91	5·86	5·80	5·77	5·75	5·72	5·69	5·66	5·63
5	6·61	5·79	5·41	5·19	5·05	4·95	4·88	4·82	4·77	4·74	4·68	4·62	4·56	4·53	4·50	4·46	4·43	4·40	4·36
6	5·99	5·14	4·76	4·53	4·39	4·28	4·21	4·15	4·10	4·06	4·00	3·94	3·87	3·84	3·81	3·77	3·74	3·70	3·67
7	5·59	4·74	4·35	4·12	3·97	3·87	3·79	3·73	3·68	3·64	3·57	3·51	3·44	3·41	3·38	3·34	3·30	3·27	3·23
8	5·32	4·46	4·07	3·84	3·69	3·58	3·50	3·44	3·39	3·35	3·28	3·22	3·15	3·12	3·08	3·04	3·01	2·97	2·93
9	5·12	4·26	3·86	3·63	3·48	3·37	3·29	3·23	3·18	3·14	3·07	3·01	2·94	2·90	2·86	2·83	2·79	2·75	2·71
10	4·96	4·10	3·71	3·48	3·33	3·22	3·14	3·07	3·02	2·98	2·91	2·85	2·77	2·74	2·70	2·66	2·62	2·58	2·54
11	4·84	3·98	3·59	3·36	3·20	3·09	3·01	2·95	2·90	2·85	2·79	2·72	2·65	2·61	2·57	2·53	2·49	2·45	2·40
12	4·75	3·89	3·49	3·26	3·11	3·00	2·91	2·85	2·80	2·75	2·69	2·62	2·54	2·51	2·47	2·43	2·38	2·34	2·30
13	4·67	3·81	3·41	3·18	3·03	2·92	2·83	2·77	2·71	2·67	2·60	2·53	2·46	2·42	2·38	2·34	2·30	2·25	2·21
14	4·60	3·74	3·34	3·11	2·96	2·85	2·76	2·70	2·65	2·60	2·53	2·46	2·39	2·35	2·31	2·27	2·22	2·18	2·13
15	4·54	3·68	3·29	3·06	2·90	2·79	2·71	2·64	2·59	2·54	2·48	2·40	2·33	2·29	2·25	2·20	2·16	2·11	2·07
16	4·49	3·63	3·24	3·01	2·85	2·74	2·66	2·59	2·54	2·49	2·42	2·35	2·28	2·24	2·19	2·15	2·11	2·06	2·01
17	4·45	3·59	3·20	2·96	2·81	2·70	2·61	2·55	2·49	2·45	2·38	2·31	2·23	2·19	2·15	2·10	2·06	2·01	1·96
18	4·41	3·55	3·16	2·93	2·77	2·66	2·58	2·51	2·46	2·41	2·34	2·27	2·19	2·15	2·11	2·06	2·02	1·97	1·92
19	4·38	3·52	3·13	2·90	2·74	2·63	2·54	2·48	2·42	2·38	2·31	2·23	2·16	2·11	2·07	2·03	1·98	1·93	1·88
20	4·35	3·49	3·10	2·87	2·71	2·60	2·51	2·45	2·39	2·35	2·28	2·20	2·12	2·08	2·04	1·99	1·95	1·90	1·84
21	4·32	3·47	3·07	2·84	2·68	2·57	2·49	2·42	2·37	2·32	2·25	2·18	2·10	2·05	2·01	1·96	1·92	1·87	1·81
22	4·30	3·44	3·05	2·82	2·66	2·55	2·46	2·40	2·34	2·30	2·23	2·15	2·07	2·03	1·98	1·94	1·89	1·84	1·78
23	4·28	3·42	3·03	2·80	2·64	2·53	2·44	2·37	2·32	2·27	2·20	2·13	2·05	2·01	1·96	1·91	1·86	1·81	1·76
24	4·26	3·40	3·01	2·78	2·62	2·51	2·42	2·36	2·30	2·25	2·18	2·11	2·03	1·98	1·94	1·89	1·84	1·79	1·73
25	4·24	3·39	2·99	2·76	2·60	2·49	2·40	2·34	2·28	2·24	2·16	2·09	2·01	1·96	1·92	1·87	1·82	1·77	1·71
26	4·23	3·37	2·98	2·74	2·59	2·47	2·39	2·32	2·27	2·22	2·15	2·07	1·99	1·95	1·90	1·85	1·80	1·75	1·69
27	4·21	3·35	2·96	2·73	2·57	2·46	2·37	2·31	2·25	2·20	2·13	2·06	1·97	1·93	1·88	1·84	1·79	1·73	1·67
28	4·20	3·34	2·95	2·71	2·56	2·45	2·36	2·29	2·24	2·19	2·12	2·04	1·96	1·91	1·87	1·82	1·77	1·71	1·65
29	4·18	3·33	2·93	2·70	2·55	2·43	2·35	2·28	2·22	2·18	2·10	2·03	1·94	1·90	1·85	1·81	1·75	1·70	1·64
30	4·17	3·32	2·92	2·69	2·53	2·42	2·33	2·27	2·21	2·16	2·09	2·01	1·93	1·89	1·84	1·79	1·74	1·68	1·62
40	4·08	3·23	2·84	2·61	2·45	2·34	2·25	2·18	2·12	2·08	2·00	1·92	1·84	1·79	1·74	1·69	1·64	1·58	1·51
60	4·00	3·15	2·76	2·53	2·37	2·25	2·17	2·10	2·04	1·99	1·92	1·84	1·75	1·70	1·65	1·59	1·53	1·47	1·39
120	3·92	3·07	2·68	2·45	2·29	2·17	2·09	2·02	1·96	1·91	1·83	1·75	1·66	1·61	1·55	1·50	1·43	1·35	1·25
∞	3·84	3·00	2·60	2·37	2·21	2·10	2·01	1·94	1·88	1·83	1·75	1·67	1·57	1·52	1·46	1·39	1·32	1·22	1·00

$$F = \frac{s_1^2}{s_2^2} = \frac{S_1}{\nu_1}\Big/\frac{S_2}{\nu_2},$$ where $s_1^2 = S_1/\nu_1$ and $s_2^2 = S_2/\nu_2$ are independent mean squares estimating a common variance σ^2 and based on ν_1 and ν_2 degrees of freedom, respectively.

Table 6 (cont.) Upper 0·01 significance points

ν_2 \ ν_1	1	2	3	4	5	6	7	8	9	10	12	15	20	24	30	40	60	120	∞	ν_2
1	4052	4999·5	5403	5625	5764	5859	5928	5982	6022	6056	6106	6157	6209	6235	6261	6287	6313	6339	6366	1
2	98·50	99·00	99·17	99·25	99·30	99·33	99·36	99·37	99·39	99·40	99·42	99·43	99·45	99·46	99·47	99·47	99·48	99·49	99·50	2
3	34·12	30·82	29·46	28·71	28·24	27·91	27·67	27·49	27·35	27·23	27·05	26·87	26·69	26·60	26·50	26·41	26·32	26·22	26·13	3
4	21·20	18·00	16·69	15·98	15·52	15·21	14·98	14·80	14·66	14·55	14·37	14·20	14·02	13·93	13·84	13·75	13·65	13·56	13·46	4
5	16·26	13·27	12·06	11·39	10·97	10·67	10·46	10·29	10·16	10·05	9·89	9·72	9·55	9·47	9·38	9·29	9·20	9·11	9·02	5
6	13·75	10·92	9·78	9·15	8·75	8·47	8·26	8·10	7·98	7·87	7·72	7·56	7·40	7·31	7·23	7·14	7·06	6·97	6·88	6
7	12·25	9·55	8·45	7·85	7·46	7·19	6·99	6·84	6·72	6·62	6·47	6·31	6·16	6·07	5·99	5·91	5·82	5·74	5·65	7
8	11·26	8·65	7·59	7·01	6·63	6·37	6·18	6·03	5·91	5·81	5·67	5·52	5·36	5·28	5·20	5·12	5·03	4·95	4·86	8
9	10·56	8·02	6·99	6·42	6·06	5·80	5·61	5·47	5·35	5·26	5·11	4·96	4·81	4·73	4·65	4·57	4·48	4·40	4·31	9
10	10·04	7·56	6·55	5·99	5·64	5·39	5·20	5·06	4·94	4·85	4·71	4·56	4·41	4·33	4·25	4·17	4·08	4·00	3·91	10
11	9·65	7·21	6·22	5·67	5·32	5·07	4·89	4·74	4·63	4·54	4·40	4·25	4·10	4·02	3·94	3·86	3·78	3·69	3·60	11
12	9·33	6·93	5·95	5·41	5·06	4·82	4·64	4·50	4·39	4·30	4·16	4·01	3·86	3·78	3·70	3·62	3·54	3·45	3·36	12
13	9·07	6·70	5·74	5·21	4·86	4·62	4·44	4·30	4·19	4·10	3·96	3·82	3·66	3·59	3·51	3·43	3·34	3·25	3·17	13
14	8·86	6·51	5·56	5·04	4·69	4·46	4·28	4·14	4·03	3·94	3·80	3·66	3·51	3·43	3·35	3·27	3·18	3·09	3·00	14
15	8·68	6·36	5·42	4·89	4·56	4·32	4·14	4·00	3·89	3·80	3·67	3·52	3·37	3·29	3·21	3·13	3·05	2·96	2·87	15
16	8·53	6·23	5·29	4·77	4·44	4·20	4·03	3·89	3·78	3·69	3·55	3·41	3·26	3·18	3·10	3·02	2·93	2·84	2·75	16
17	8·40	6·11	5·18	4·67	4·34	4·10	3·93	3·79	3·68	3·59	3·46	3·31	3·16	3·08	3·00	2·92	2·83	2·75	2·65	17
18	8·29	6·01	5·09	4·58	4·25	4·01	3·84	3·71	3·60	3·51	3·37	3·23	3·08	3·00	2·92	2·84	2·75	2·66	2·57	18
19	8·18	5·93	5·01	4·50	4·17	3·94	3·77	3·63	3·52	3·43	3·30	3·15	3·00	2·92	2·84	2·76	2·67	2·58	2·49	19
20	8·10	5·85	4·94	4·43	4·10	3·87	3·70	3·56	3·46	3·37	3·23	3·09	2·94	2·86	2·78	2·69	2·61	2·52	2·42	20
21	8·02	5·78	4·87	4·37	4·04	3·81	3·64	3·51	3·40	3·31	3·17	3·03	2·88	2·80	2·72	2·64	2·55	2·46	2·36	21
22	7·95	5·72	4·82	4·31	3·99	3·76	3·59	3·45	3·35	3·26	3·12	2·98	2·83	2·75	2·67	2·58	2·50	2·40	2·31	22
23	7·88	5·66	4·76	4·26	3·94	3·71	3·54	3·41	3·30	3·21	3·07	2·93	2·78	2·70	2·62	2·54	2·45	2·35	2·26	23
24	7·82	5·61	4·72	4·22	3·90	3·67	3·50	3·36	3·26	3·17	3·03	2·89	2·74	2·66	2·58	2·49	2·40	2·31	2·21	24
25	7·77	5·57	4·68	4·18	3·85	3·63	3·46	3·32	3·22	3·13	2·99	2·85	2·70	2·62	2·54	2·45	2·36	2·27	2·17	25
26	7·72	5·53	4·64	4·14	3·82	3·59	3·42	3·29	3·18	3·09	2·96	2·81	2·66	2·58	2·50	2·42	2·33	2·23	2·13	26
27	7·68	5·49	4·60	4·11	3·78	3·56	3·39	3·26	3·15	3·06	2·93	2·78	2·63	2·55	2·47	2·38	2·29	2·20	2·10	27
28	7·64	5·45	4·57	4·07	3·75	3·53	3·36	3·23	3·12	3·03	2·90	2·75	2·60	2·52	2·44	2·35	2·26	2·17	2·06	28
29	7·60	5·42	4·54	4·04	3·73	3·50	3·33	3·20	3·09	3·00	2·87	2·73	2·57	2·49	2·41	2·33	2·23	2·14	2·03	29
30	7·56	5·39	4·51	4·02	3·70	3·47	3·30	3·17	3·07	2·98	2·84	2·70	2·55	2·47	2·39	2·30	2·21	2·11	2·01	30
40	7·31	5·18	4·31	3·83	3·51	3·29	3·12	2·99	2·89	2·80	2·66	2·52	2·37	2·29	2·20	2·11	2·02	1·92	1·80	40
60	7·08	4·98	4·13	3·65	3·34	3·12	2·95	2·82	2·72	2·63	2·50	2·35	2·20	2·12	2·03	1·94	1·84	1·73	1·60	60
120	6·85	4·79	3·95	3·48	3·17	2·96	2·79	2·66	2·56	2·47	2·34	2·19	2·03	1·95	1·86	1·76	1·66	1·53	1·38	120
∞	6·63	4·61	3·78	3·32	3·02	2·80	2·64	2·51	2·41	2·32	2·18	2·04	1·88	1·79	1·70	1·59	1·47	1·32	1·00	∞

$F = \dfrac{s_1^2}{s_2^2} = \dfrac{S_1}{\nu_1} \bigg/ \dfrac{S_2}{\nu_2}$, where $s_1^2 = S_1/\nu_1$ and $s_2^2 = S_2/\nu_2$ are independent mean squares estimating a common variance σ^2 and based on ν_1 and ν_2 degrees of freedom, respectively.

TABLE 8. SIGNIFICANCE POINTS FOR THE MAXIMUM F-RATIO
Upper 0·05 significance point (first line)
Upper 0·01 significance point (second line)

k \ ν	2	3	4	5	6	7	8	9
2	39·0	15·4	9·60	7·15	5·82	4·99	4·43	4·03
	199	47·5	23·2	14·9	11·1	8·89	7·50	6·54
3	87·5	27·8	15·5	10·8	8·38	6·94	6·00	5·34
	448	85	37	22	15·5	12·1	9·9	8·5
4	142	39·2	20·6	13·7	10·4	8·44	7·18	6·31
	729	120	49	28	19·1	14·5	11·7	9·9
5	202	50·7	25·2	16·3	12·1	9·70	8·12	7·11
	1036	151	59	33	22	16·5	13·2	11·1
6	266	62·0	29·5	18·7	13·7	10·8	9·03	7·80
	1362	184	69	38	25	18·4	14·5	12·1
7	333	72·9	33·6	20·8	15·0	11·8	9·78	8·41
	1705	216	79	42	27	20	15·8	13·1
8	403	83·5	37·5	22·8	16·3	12·7	10·5	8·95
	2063	249	89	46	30	22	16·9	13·9
9	475	93·9	41·1	24·7	17·5	13·5	11·1	9·45
	2432	281	97	50	32	23	17·9	14·7
10	550	104	44·6	26·5	18·6	14·3	11·7	9·91
	2813	310	106	54	34	24	18·9	15·3
11	626	114	48·0	28·2	19·7	15·1	12·2	10·3
	3204	337	113	47	36	26	19·8	16·0
12	704	124	51·4	29·9	20·7	15·8	12·7	10·7
	3605	361	120	60	37	27	21	16·6

k \ ν	10	12	15	20	30	60	∞	
2	3·72	3·28	2·86	2·46	2·07	1·67	1·00	
	5·85	4·91	4·07	3·32	2·63	1·96	1·00	
3	4·85	4·16	3·54	2·95	2·40	1·85	1·00	
	7·4	6·1	4·9	3·8	3·0	2·2	1·0	
4	5·67	4·79	4·01	3·29	2·61	1·96	1·00	
	8·6	6·9	5·5	4·3	3·3	2·3	1·0	
5	6·34	5·30	4·37	3·54	2·78	2·04	1·00	
	9·6	7·6	6·0	4·6	3·4	2·4	1·0	
6	6·92	5·72	4·68	3·76	2·91	2·11	1·00	
	10·4	8·2	6·4	4·9	3·6	2·4	1·0	
7	7·42	6·09	4·95	3·94	3·02	2·17	1·00	
	11·1	8·7	6·7	5·1	3·7	2·5	1·0	
8	7·87	6·42	5·19	4·10	3·12	2·22	1·00	
	11·8	9·1	7·1	5·3	3·8	2·5	1·0	
9	8·28	6·72	5·40	4·24	3·21	2·26	1·00	
	12·4	9·5	7·3	5·5	3·9	2·6	1·0	
10	8·66	7·00	5·59	4·37	3·29	2·30	1·00	
	12·9	9·9	7·5	5·6	4·0	2·6	1·0	
11	9·01	7·25	5·77	4·49	3·36	2·33	1·00	
	13·4	10·2	7·8	5·8	4·1	2·7	1·0	
12	9·34	7·48	5·93	4·59	3·39	2·36	1·00	
	13·9	10·6	8·0	5·9	4·2	2·7	1·0	

TABLE 9. SIGNIFICANCE POINTS FOR WILCOXON'S TEST IN MANN–WHITNEY FORM

n	m	Lower significance points				Upper significance points			
		0·01	0·025	0·05	0·10	0·10	0·05	0·025	0·01
3	2	—	—	—	0	6	—	—	—
	3	—	—	0	1	8	9	—	—
4	2	—	—	—	0	8	—	—	—
	3	—	—	0	1	11	12	—	—
	4	—	0	1	3	13	15	16	—
5	2	—	—	0	1	9	10	—	—
	3	—	0	1	2	13	14	15	—
	4	0	1	2	4	16	18	19	20
	5	1	2	4	5	20	21	23	24
6	2	—	—	0	1	11	12	—	—
6	3	—	1	2	3	15	16	17	—
	4	1	2	3	5	19	21	22	23
	5	2	3	5	7	23	25	27	28
	6	3	5	7	9	27	29	31	33
7	2	—	—	0	1	13	14	—	—
7	3	0	1	2	4	17	19	20	21
	4	1	3	4	6	22	24	25	27
	5	3	5	6	8	27	29	30	32
	6	4	6	8	11	31	34	36	38
	7	6	8	11	13	36	38	41	43
8	2	—	0	1	2	14	15	16	—
	3	0	2	3	5	19	21	22	24
	4	2	4	5	7	25	27	28	30
	5	4	6	8	10	30	32	34	36
	6	6	8	10	13	35	38	40	42
8	7	7	10	13	16	40	43	46	49
8	8	9	13	15	19	45	49	51	55
9	1	—	—	—	0	9	—	—	—
	2	—	0	1	2	16	17	18	—
9	3	1	2	4	5	22	23	25	26
	4	3	4	6	9	27	30	32	33
	5	5	7	9	12	33	36	38	40
	6	7	10	12	15	39	42	44	47
	7	9	12	15	18	45	48	51	54
	8	11	15	18	22	50	54	57	61
9	9	14	17	21	25	56	60	64	67
10	1	—	—	—	0	10	—	—	—
	2	—	0	1	3	17	19	20	—
	3	1	3	4	6	24	26	27	29
	4	3	5	7	10	30	33	35	37
10	5	6	8	11	13	37	39	42	44
	6	8	11	14	17	43	46	49	52
	7	11	14	17	21	49	53	56	59
	8	13	17	20	24	56	60	63	67
	9	16	20	24	28	62	66	70	74
10	10	19	23	27	32	68	73	77	81
11	1	—	—	—	0	11	—	—	—
	2	—	0	1	3	19	21	22	—
	3	1	3	5	7	26	28	30	32
	4	4	6	8	11	33	36	38	40
11	5	7	9	12	15	40	43	46	48
	6	9	13	16	19	47	50	53	57
	7	12	16	19	23	54	58	61	65
	8	15	19	23	27	61	65	69	73
	9	18	23	27	31	68	72	76	81
11	10	22	26	31	36	74	79	84	88
	11	25	30	34	40	81	87	91	96
12	1	—	—	—	0	12	—	—	—
	2	—	1	2	4	20	22	23	—
	3	2	4	5	8	28	31	32	34
12	4	5	7	9	12	36	39	41	43
	5	8	11	13	17	43	47	49	52
	6	11	14	17	21	51	55	58	61
	7	14	18	21	26	58	63	66	70
	8	17	22	26	30	66	70	74	79
12	9	21	26	30	35	73	78	82	87
	10	24	29	34	39	81	86	91	96
	11	28	33	38	44	88	94	99	104
	12	31	37	42	49	95	102	107	113

TABLE 9 (cont.)

n	m	Lower significance points				Upper significance points			
		0·01	0·025	0·05	0·10	0·10	0·05	0·025	0·01
13	1	—	—	—	0	13	—	—	—
13	2	0	1	2	4	22	24	25	26
	3	2	4	6	9	30	33	35	37
	4	5	8	10	13	39	42	44	47
	5	9	12	15	18	47	50	53	56
	6	12	16	19	23	55	59	62	66
13	7	16	20	24	28	63	67	71	75
	8	20	24	28	33	71	76	80	84
	9	23	28	33	38	79	84	89	94
	10	27	33	37	43	87	93	97	103
	11	31	37	42	48	95	101	106	112
13	12	35	41	47	53	103	109	115	121
	13	39	45	51	58	111	118	124	130
14	1	—	—	—	0	14	—	—	—
	2	0	1	3	4	24	25	27	28
	3	2	5	7	10	32	35	37	40
14	4	6	9	11	15	41	45	47	50
	5	10	13	16	20	50	54	57	60
	6	13	17	21	25	59	63	67	71
	7	17	22	26	31	67	72	76	81
	8	22	26	31	36	76	81	86	90
14	9	26	31	36	41	85	90	95	100
	10	30	36	41	47	93	99	104	110
	11	34	40	46	52	102	108	114	120
	12	38	45	51	58	110	117	123	130
	13	43	50	56	63	119	126	132	139
14	14	47	55	61	69	127	135	141	149
15	1	—	—	—	0	15	—	—	—
	2	0	1	3	5	25	27	29	30
	3	3	5	7	10	35	38	40	42
	4	7	10	12	16	44	48	50	53
15	5	11	14	18	22	53	57	61	64
	6	15	19	23	27	63	67	71	75
	7	19	24	28	33	72	77	81	86
	8	24	29	33	39	81	87	91	96
	9	28	34	39	45	90	96	101	107
15	10	33	30	44	51	99	106	111	117
	11	37	44	50	57	108	115	121	128
	12	42	49	55	63	117	125	131	138
	13	47	54	61	68	127	134	141	148
	14	51	59	66	74	136	144	151	159
15	15	56	64	72	80	145	153	161	169
16	1	—	—	—	0	16	—	—	—
	2	0	1	3	5	27	29	31	32
	3	3	6	8	11	37	40	42	45
	4	7	11	14	17	47	50	53	57
16	5	12	15	19	23	57	61	65	68
	6	16	21	25	29	67	71	75	80
	7	21	26	30	36	76	82	86	91
	8	26	31	36	42	86	92	97	102
	9	31	37	42	48	96	102	107	113
16	10	36	42	48	54	106	112	118	124
	11	41	47	54	61	115	122	129	135
	12	46	53	60	67	125	132	139	146
	13	51	59	65	74	134	143	149	157
	14	56	64	71	80	144	153	160	168
16	15	61	70	77	86	154	163	170	179
	16	66	75	83	93	163	173	181	190
17	1	—	—	—	0	17	—	—	—
	2	0	2	3	6	28	31	32	34
	3	4	6	9	12	39	42	45	47
17	4	8	11	15	18	50	53	57	60
	5	13	17	20	25	60	65	68	72
	6	18	22	26	31	71	76	80	84
	7	23	28	33	38	81	86	91	96
	8	28	34	39	45	91	97	102	108

TABLE 9 (*cont.*)

n	m	Lower significance points				Upper significance points			
		0·01	0·025	0·05	0·10	0·10	0·05	0·025	0·01
17	9	33	39	45	52	101	108	114	120
	10	38	45	51	58	112	119	125	132
	11	44	51	57	65	122	130	136	143
	12	49	57	64	72	132	140	147	155
	13	55	63	70	79	142	151	158	166
17	14	60	69	77	85	153	161	169	178
	15	66	75	83	92	163	172	180	189
	16	71	81	89	99	173	183	191	201
	17	77	87	96	106	183	193	202	212
18	1	—	—	—	0	18	—	—	—
	2	0	2	4	6	30	32	34	36
	3	4	7	9	13	41	45	47	50
	4	9	12	16	20	52	56	60	63
	5	14	18	22	27	63	68	72	76
	6	19	24	28	34	74	80	84	89
18	7	24	30	35	41	85	91	96	102
	8	30	36	41	48	96	103	108	114
	9	36	42	48	55	107	114	120	126
	10	41	48	55	62	118	125	132	139
	11	47	55	61	69	129	137	143	151
18	12	53	61	68	77	139	148	155	163
	13	59	67	75	84	150	159	167	175
	14	65	74	82	91	161	170	178	187
	15	70	80	88	98	172	182	190	200
	16	76	86	95	106	182	193	202	212
18	17	82	93	102	113	193	204	213	224
	18	88	99	109	120	204	215	225	236
19	1	—	—	0	1	18	19	—	—
	2	1	2	4	7	31	34	36	37
	3	4	7	10	14	43	47	50	53
19	4	9	13	17	21	55	59	63	67
	5	15	19	23	28	67	72	76	80
	6	20	25	30	36	78	84	89	94
	7	26	32	37	43	90	96	101	107
	8	32	38	44	51	101	108	114	120
19	9	38	45	51	58	113	120	126	133
	10	44	52	58	66	124	132	138	146
	11	50	58	65	73	136	144	151	159
	12	56	65	72	81	147	156	163	172
	13	63	72	80	89	158	167	175	184
19	14	69	78	87	97	169	179	188	197
	15	75	85	94	104	181	191	200	210
	16	82	92	101	112	192	203	212	222
	17	88	99	109	120	203	214	224	235
	18	94	106	116	128	214	226	236	248
19	19	101	113	123	135	226	238	248	260
20	1	—	—	0	1	19	20	—	—
	2	1	2	4	7	33	36	38	39
	3	5	8	11	15	45	49	52	55
	4	10	14	18	22	58	62	66	70
20	5	16	20	25	30	70	75	80	84
	6	22	27	32	38	82	88	93	98
	7	28	34	39	46	94	101	106	112
	8	34	41	47	54	106	113	119	126
	9	40	48	54	62	118	126	132	140
20	10	47	55	62	70	130	138	145	153
	11	53	62	69	78	142	151	158	167
	12	60	69	77	86	154	163	171	180
	13	67	76	84	94	166	176	184	193
	14	73	83	92	102	178	188	197	207
20	15	80	90	100	110	190	200	210	220
	16	87	98	107	119	201	213	222	233
	17	93	105	115	127	213	225	235	247
	18	100	112	123	135	225	237	248	260
	19	107	119	130	143	237	250	261	273
	20	114	127	138	151	249	262	273	286

INDEX

331